Ceramic Membranes for Separation and Reaction

Ceramic Membranes for Separation and Reaction

Kang Li
Department of Chemical Engineering and Chemical Technology,
Imperial College London, UK

BICENTENNIAL 1807 WILEY 2007 BICENTENNIAL

John Wiley & Sons, Ltd

Other Wiley Editorial Offices
John Wiley & Sons Inc., 111 River Street, Hoboken, NJ 07030, USA
Jossey-Bass, 989 Market Street, San Francisco, CA 94103-1741, USA
Wiley-VCH Verlag GmbH, Boschstr. 12, D-69469 Weinheim, Germany
John Wiley & Sons Australia Ltd, 42 McDougall Street, Milton, Queensland 4064, Australia
John Wiley & Sons (Asia) Pte Ltd, 2 Clementi Loop #02-01, Jin Xing Distripark, Singapore 129809
John Wiley & Sons Ltd, 6045 Freemont Blvd, Mississauga, Ontario L5R 4J3, Canada

Wiley also publishes its books in a variety of electronic formats. Some content that appears in print may not be available in electronic books.

Anniversary Logo Design: Richard J. Pacifico

Library of Congress Cataloging-in-Publication Data
Li, Kang,
Ceramic membranes for separation and reaction/ Kang Li.
p. cm.
Includes bibliographical references and index.
ISBN 978-0-470-01440-0 (cloth : alk. paper)
1. Gas separation membranes. 2. Ceramic materials. 3. Gases--Separation.
4. Membranes (Technology) I. Title.
TP159.M4L5 2007
660'.28424--dc22

2006100393

British Library Cataloguing in Publication Data
A catalogue record for this book is available from the British Library
ISBN: 978 0 470 01440 0 (cloth)

Typeset in 10/12 Times by SNP Best-set Typesetter Ltd., Hong Kong
Printed and bound in Great Britain by Antony Rowe, Chippenham, Wiltshire
This book is printed on acid-free paper responsibly manufactured from sustainable forestry in which at least two trees are planted for each one used for paper production.

Contents

Preface

Polymeric membranes are in current use for a number of process applications, including ultrafiltration, desalination, gas separation, pervaporation, biochemical reactions, etc. Ceramic membranes have been developed for similar applications, especially where high alkaline (or acid) feed solution or high temperature operation precludes the use of existing polymeric membranes. Books on fundamental science and technology of polymeric membranes are available. However, there have been only a few books which address ceramic membranes and these books are mostly edited volumes with individually contributed chapters. Also, no books, so far, are available that describe the development of ceramic hollow fibre membranes. Although structured ceramic membranes are still in an early stage of development, especially in the hollow fibre geometries, a book covering ceramic membrane preparation, characterization and application would be useful to industrial and institution researchers, scientists and engineers with an interest in ceramic membranes as separator/reactors and senior undergraduate and postgraduate students pursuing advanced separation and reaction courses.

This book is divided into eight chapters, covering different aspects of ceramic membranes. Chapter 1 provides a brief review of the current status of ceramic membrane development. Different types of morphological structures of ceramic membranes and their separation performance are given first, followed by discussion of the various ceramic membrane geometries currently being used in industry or being developed in laboratories. Various membrane processes where ceramic membranes may play a role are also outlined for readers with little knowledge of membranes and membrane processes. Chapter 2 gives an overview of various preparation techniques. The fundamental principles of preparing both membranes and membrane supports using these methods are given. In addition, a new approach to preparing ceramic hollow fibre membranes is introduced with detailed examples on the fabrication of Al_2O_3, ZrO TiO_2 and various perovskite based hollow fibre membranes. Chapter 3 is devoted to the topic of membrane characterization. Various methods, which characterize ceramic membranes with permeation related as well as structural related parameters, are introduced. The characterization of surface catalytic activity and conductivities of ion and electrons of specific gases of dense ceramic membranes (mixed conducting type) is deliberately excluded as there are already several books and detailed reports on this topic.

Chapters 4, 6 and 7 are dedicated to ceramic membranes for the separation of gases. There are two types of ceramic membranes suitable for gas separations: (1) porous, especially microporous; (2) dense membranes. Gas permeation/separation in porous ceramic membranes is discussed in Chapter 4 where detailed analyses on gas transport through various porous membranes are given. The dense ceramic membranes are made from crystalline ceramic materials such as perovskites or fluorites, which allow permeation of only oxygen or hydrogen through its crystal lattice. Therefore, they are mostly impermeable to all other gases, giving extremely high selectivity towards oxygen or hydrogen. Dense ceramic membranes for oxygen and hydrogen permeation are a rapidly growing field. Detailed coverage on these membranes for oxygen or hydrogen separation is given in Chapters 6 and 7, respectively.

Chapter 5 discusses a membrane contactor which represents a technology where porous membranes are used as "packing materials" for interphase mass transfer. Thus, all traditional gas stripping and absorption, distillation, liquid–liquid extraction as well as emulsification, crystallisation and phase transfer catalysis can be carried out in the membrane contactor. Although, discussion of the membrane contactor in Chapter 5 emphasises only gas absorption/stripping, it can be easily applied to other unit operation processes such as liquid–liquid extraction and distillation based on the results from gas absorption analysis. Finally, Chapter 8 presents a topic on ceramic membrane reactors, i.e. a device that combines a membrane separation or distribution process with a chemical reactor in one unit. Membrane reactors are capable of promoting a reaction process by: 1) selectively removing at least one of the products through the membrane from the reaction zone, making the equilibrium reaction shift to the product side; 2) supplying only a particular reactant to the reaction zone giving an optimum concentration ratio of the two reactant streams. As a result, the yield can be increased (even beyond the equilibrium value for equilibrium reactions) and/or the selectivity can be improved by suppressing other undesired side reactions or the secondary reaction of products.

I am deeply indebted to many of my colleagues and students for the completion of this book. A number of people who contributed to this book are acknowledged. They include Drs Xiaoyao Tan, Shaomin Liu, Yutie Liu and Jianfeng Kong for their assistance with various aspects of the book; Mr Xunxing Loh, Mr Chiao-chien Wei, Mr Yoong Hsiang See Toh, Mr Sirichai Koonaphapdeelert and Dr Dongfei Li for drawing figures; and Professor Ahmad Fauzi Ismail and Mr Benjamin Kingsbury for proofreading.

My special thanks also go to several staff members at John Wiley & Sons, Ltd: Ms Miriam Swift and Ms Colleen Goldring for their patience and advice, and Ms Lyn Roberts for initiating the book.

Finally, I would like to acknowledge the ongoing financial support provided by EPSRC in the United Kingdom and my first research funding in ceramic membranes provided by my previous institution, National University of Singapore.

Kang Li
Imperial College London, UK

1

Ceramic Membranes and Membrane Processes

1.1 INTRODUCTION

In general, a ceramic membrane can be described as a permselective barrier or a fine sieve. Permeability and separation factor of a ceramic membrane are the two most important performance indicators. For a porous ceramic membrane, they are typically governed by thickness, pore size and surface porosity of the membrane, while for a dense ceramic membrane, the principle for permeation and separation is more complex. In porous ceramic membranes, their applications and separation mechanisms correspond to the pore size of the ceramic membranes as shown in Table 1.1.

Ceramic membranes are usually composite ones consisting of several layers of one or more different ceramic materials. They generally have a macroporous support, one or two mesoporous intermediate layers and a microporous (or a dense) top layer. As shown in Figure 1.1, the bottom layer provides mechanical support, while the middle layers bridge the pore size differences between the support layer and the top layer where the actual separation takes place. Commonly used materials for ceramic membranes are Al_2O_3, TiO_2, ZrO_2, SiO_2 etc. or a combination of these materials. An example of the pore characteristics of a four layer alumina membrane is given in Figure 1.2. It can be seen that the pore sizes of the top separation layer, intermediate layers and bottom support layer are in the range of 6 nm (mesoporous), 0.2–0.7 μm and 10 μm, respectively. A more sophisticated multilayer membrane, consisting of a α-alumina macroporous support, two γ-alumina mesoporous intermediate layers and a microporous silica top layer was reported by Vos and Verweij [1]. As shown in Figure 1.3 (a micrograph taken by transition electron microscopy (TEM)), a very thin silica layer ~30 nm with a pore diameter of 5 Å was obtained. The TEM micrograph further indicates that the silica layer is deposited on top of the γ-Al_2O_3 layer. A clear division between silica and γ-Al_2O_3 is visible. The boundary between the first and second γ-Al_2O_3 layers at approximately 250 nm from the surface is clearly visible.

The ceramic membranes mentioned above can only be achieved through multiple steps. As illustrated in Figure 1.1, a support layer is first prepared to provide mechanical strength for the membrane, followed by coating one or more intermediate layers on the support layer before a final dense separation layer can be fabricated. Each step involves a high temperature

Ceramic Membranes for Separation and Reaction K. Li

Table 1.1 Category of ceramic membranes

Type[a]	Pore size (nm)	Mechanism	Applications
Macroporous	>50	Sieving	UF, MF
Mesoporous	2–50	Knudsen diffusion	UF, NF, Gas Separation
Microporous	<2	Micropore diffusion	Gas Separation
Dense	—	Diffusion	Gas Separation, Reaction

[a] IUPAC classification [53]

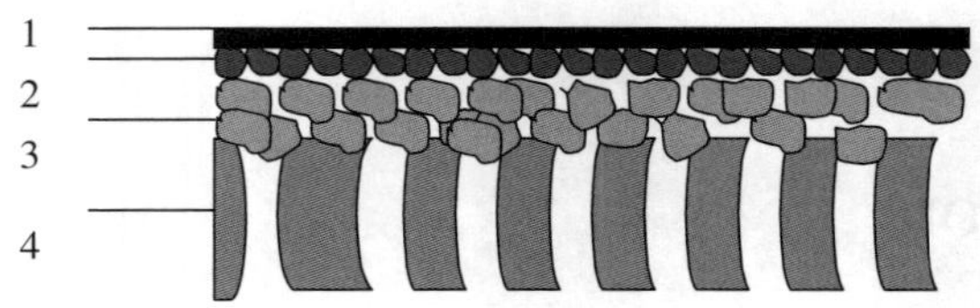

1. Modified separation layer (Dense or <2 nm)
2. Separation layer (2–50 nm)
3. Intermediate layer(s) (50–1000 nm)
4. Porous support (1–15 μm)

1 + 2 + 3+ 4	nanofiltration or gas separation membranes
2 + 3 + 4	ultrafiltration membranes
3 + 4	microfiltration membranes

Figure 1.1 Schematic representation of an asymmetric composite membrane

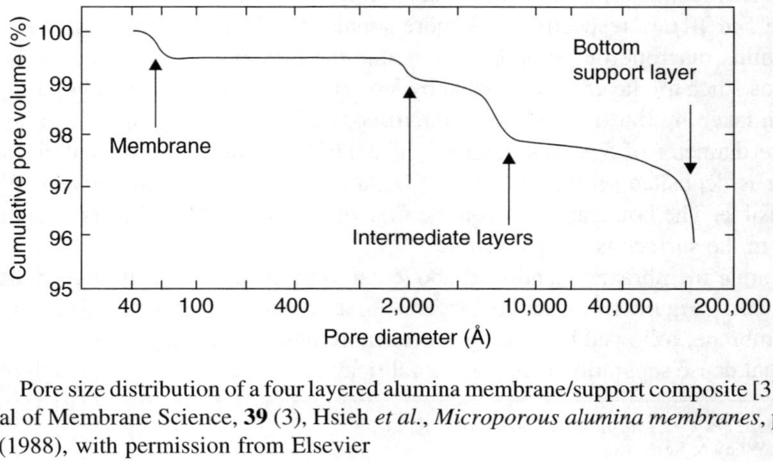

Figure 1.2 Pore size distribution of a four layered alumina membrane/support composite [3]. Reprinted from Journal of Membrane Science, **39** (3), Hsieh *et al.*, *Microporous alumina membranes*, p. 221–241. Copyright (1988), with permission from Elsevier

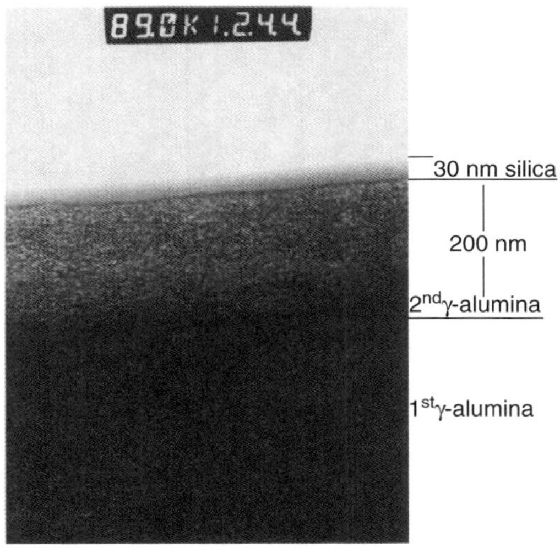

Figure 1.3 TEM micrograph of Si(400) membrane cross section showing a part of γ-Al_2O_3 layer and silica layer [1]. Reprinted from Journal of Membrane Science, **143** (1–2), Vos *et al.*, *Improved performance of silica membranes for gas separation*, p. 37–51. Copyright (1998), with permission from Elsevier

sintering treatment, making the ceramic membrane fabrication extremely expensive. Clearly, combining the multiple steps into a single one is desirable in cutting production time and costs, and hence membrane price. Li *et al.* [2] demonstrated that the above multiple step fabrication process could be combined into a single step using a phase inversion process. Figure 1.4 shows a scanning electron microscopy (SEM) micrograph of an asymmetric dense ceramic membrane prepared by Li *et al.* using such a technique. As can be seen, a dense and thin skin layer is integrated on the porous support of the same ceramic material, confirming that such a layered ceramic membrane can be prepared in one step.

Most commercial ceramic membranes are in disc, plate or tubular configuration. They are usually assembled as a plate and frame module using disc or sheet membranes or as a tubular module using membrane tubes. In order to increase the surface area to volume ratio, which gives more separation area per unit volume of membrane element, alumina multichannel monolithic elements have been developed as shown in Figure 1.5 [3]. These monolithic elements can be combined into modules. Hsieh [4] reported that the surface area to volume ratios are round 30–250 $m^2 m^{-3}$ for tubes, 130–400 $m^2 m^{-3}$ for multichannel monolithics and up to 800 $m^2 m^{-3}$ for honeycomb multichannel monolithics. Similar modules have also been developed by CeraMem Corporation as shown in Figure 1.6.

Similarly, a plate and frame ceramic module can be assembled by stacking many membrane cells (made from ceramic sheets) together. In this way, a high packing density can also be obtained for the disc or sheet membranes. The principle is shown in Figure 1.7. It can be seen that the feed flows into a porous spacer sandwiched by two membrane sheets. The fluid permeates through the membranes and the permeate flows out of the system through cell spacers which provide space for permeate flow between the cells. Detailed description of the plate and frame system can be found elsewhere [5].

To further increase the packing density, hollow fibre membrane modules, as shown in Figure 1.8, can be employed, as they offer substantially high packing density, i.e. around

Figure 1.4 SEM micrograph of a layered ceramic membrane [2]. Copyright (2006) Elsevier. Journal of Membrane Science, **272** (1–2), Li, K., Tan, X. and Liu, Y., *Single step fabrication of ceramic hollow fibres for oxygen permeation*, 1–5

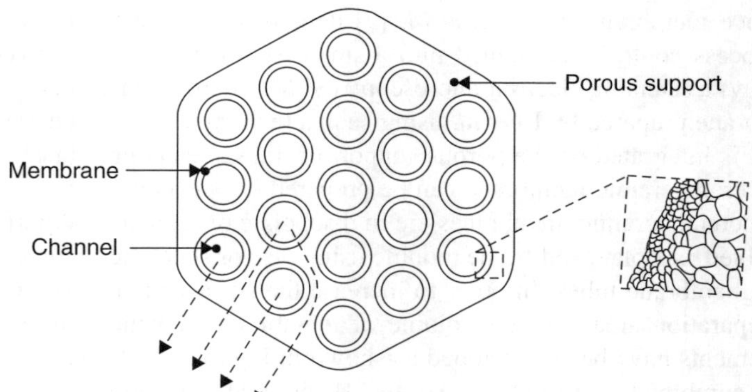

Figure 1.5 Cross section of a monolithic multichannel membrane element [3]. Reprinted from Journal of Membrane Science, **39** (3), Hsieh *et al.*, *Microporous alumina membranes*, p. 221–241. Copyright (1988), with permission from Elsevier

9000 $m^2 m^{-3}$ as compared to packing density of around 30–500 $m^2 m^{-3}$ offered by the plate and frame or tubular membrane format. The greatest challenge in the preparation of a ceramic hollow fibre membrane lies in overcoming the physical brittleness often associated with ceramic materials, especially for high temperature industrial applications. Ceramic hollow fibre membranes have been produced in porous or dense form depending on their application

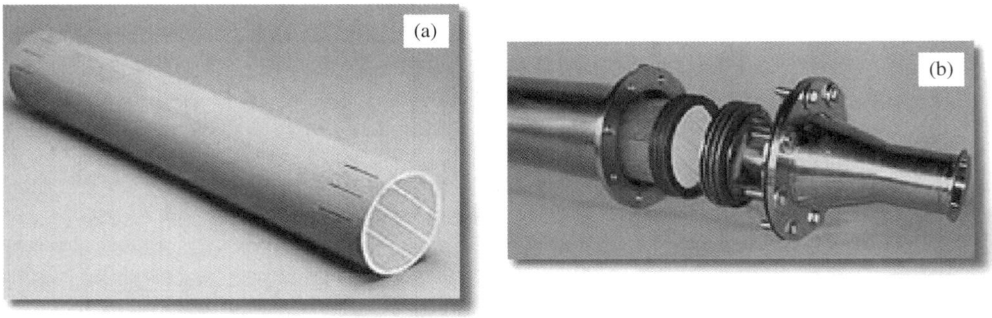

Figure 1.6 Multi-channel ceramic module: (a) membrane element 142 mm diameter × 864 mm length 10.7 m^2 membrane area; (b) membrane element in stainless steel housing with simple compressive seal design. Reprinted with permission from CeraMem Corporation

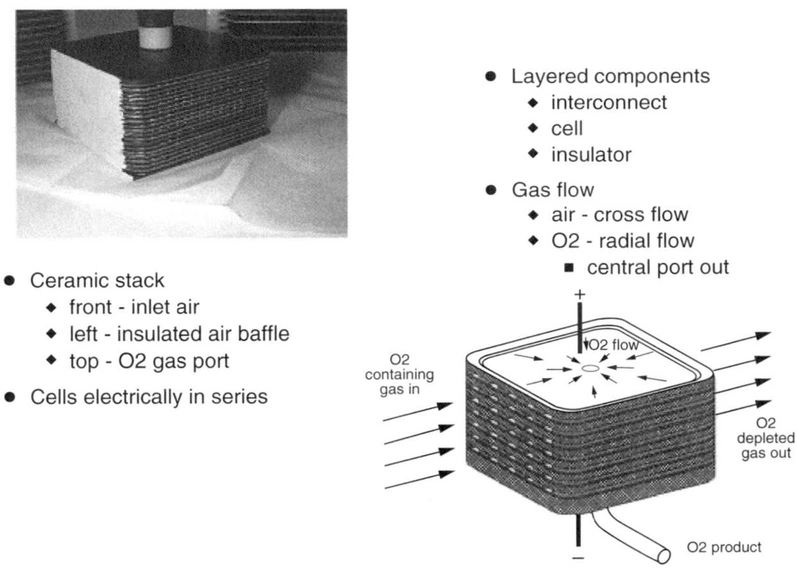

Figure 1.7 SEOS™ Oxygen Generator multiple planar cell stack [5]. Reprinted from Solid State Ionics, **134**, Dyer *et al.*, *Ion transport membrane* . . . , p. 21–23. Copyright (2000), with permission from Elsevier

requirements. They have been investigated in areas such as gas separation, membrane reactor, solvent recovery, etc. [6–9].

1.2 MEMBRANE PROCESSES

Membrane processes have become an accepted unit operation for a variety of separations in industries. The processes are driven by pressure, concentration or electric field across the

Figure 1.8 Ceramic hollow fibre modules. Reprinted with permission from Hyflux CEPA*ration* BV

membrane and can be differentiated according to type of driving force, molecular size or type of operations. Some common membrane processes are briefly introduced below.

1.2.1 Gas Separation

Gas mixtures can be separated by either dense or porous ceramic membranes. The dense ceramic membranes are made from crystalline ceramic materials such as perovskites or fluorites, which allow permeation of only oxygen or hydrogen through its crystal lattice. Therefore, they are mostly impermeable to all other gases, giving extremely high selectivity towards oxygen or hydrogen.

Oxygen permeation through a dense ceramic membrane is due to a large number of oxygen vacancies that are generated by doping and the electron holes produced by the defect reaction exist in the solid electrolyte. Under a gradient of oxygen partial pressure imposed on the membrane at a high temperature, the oxide ions are transported along with holes from the high partial pressure side to the low partial pressure side as illustrated in Figure 1.9. In addition to the bulk diffusion, oxygen permeation through a mixed ionic–electronic conducting membrane also undergoes the surface exchange reactions at both the oxygen rich and oxygen lean sides of the membrane, which involves many sub-steps such as oxygen adsorption, dissociation, recombination and charge transfer [10, 11]. Therefore, the permeation process from the high oxygen partial pressure side to the low oxygen partial pressure side includes the following steps in series: (1) mass transfer of gaseous oxygen from the gas stream to the membrane surface (high pressure side); (2) reaction between the molecular oxygen and oxygen vacancies at the membrane surface (high pressure side); (3) oxygen vacancy bulk diffusion across the membrane; (4) reaction between lattice oxygen and electron holes at the membrane surface (low pressure side) and (5) mass transfer of oxygen from the membrane surface to the gas stream (low pressure side). However, the resistances between the gas phase and membrane (steps 1 and 5) are usually small and negligible [12] and as a result, only the membrane bulk diffusion, the surface reactions need to be taken into consideration for the oxygen permeation. Similarly, when hydrogen is exposed to a mixexd proton conducting

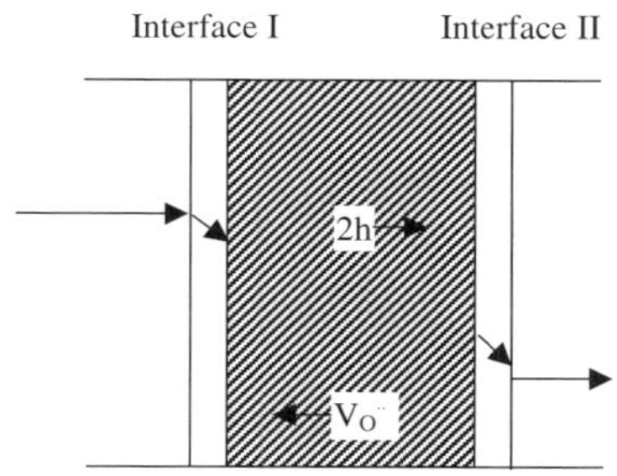

$$\tfrac{1}{2}O_2 + V_O^{\cdot\cdot} \xrightarrow{k_f} O_O^x + 2h^{\cdot} \qquad O_O^x + 2h^{\cdot} \xrightarrow{k_r} \tfrac{1}{2}O_2 + V_O^{\cdot\cdot}$$

Figure 1.9 Schematic diagram of oxygen permeation in a mixed ionic–electronic conducting membrane

membrane, it may be transferred through the membrane under a hydrogen partial pressure gradient. Again, apart from the membrane bulk diffusion, the surface reactions are also important and need to be taken into consideration for the hydrogen permeation. Dense ceramic membranes for oxygen or hydrogen separations will be discussed in detail in Chapters 6 and 7, respectively.

In microporous ceramic membranes, the gas permeation behaviour may be dominated by Knudsen diffusion, surface diffusion, multilayer diffusion, capillary condensation or molecular sieving (i.e. configurational diffusion) and is strongly dependent on the pore size and pore size distribution of the membrane, operating temperature and pressure, and the nature of the membrane and the permeating molecules [13]. The progression from Knudsen diffusion to molecular sieving is in parallel with increasing permselectivities. For pores that are large relative to the molecular size of the permeating gases, Knudsen diffusion is the likely mechanism controlling the rate of transport. In this case, the gases permeate in proportion to their molecular velocity and hence, in inverse proportion to the square root of their molecular weight. If the gas is strongly adsorbed in the membrane pores, surface diffusion will enhance the permeation rate relative to Knudsen diffusion. When the pores of the membrane are roughly the same size as the gas molecule's diameter, molecular sieving may take place. This mechanism is characterized by a strong temperature dependence and more importantly, sharp decline in permeabilities for larger gas molecules [14]. Gas separation using porous ceramic membranes is one of the important research topics and a comprehensive discussion on this will be given in Chapter 4.

1.2.2 Pervaporation

Pervaporation, as shown in Figure 1.10, is a separation process where a liquid mixture is in direct contact with one side of the membrane and where the permeate stream is removed in the vapour state from the other side of the membrane. Because of the presence of the membrane, the liquid–vapour equilibrium is perturbed as shown in Figure 1.11. Application of pervaporation includes separation of azeotropic mixtures, mixture of closed boiling point components, heat-sensitive products, etc.

Ceramic membranes for pervaporation offer some significant advantages over polymeric ones, such as much higher chemical and thermal stability than most polymeric materials.

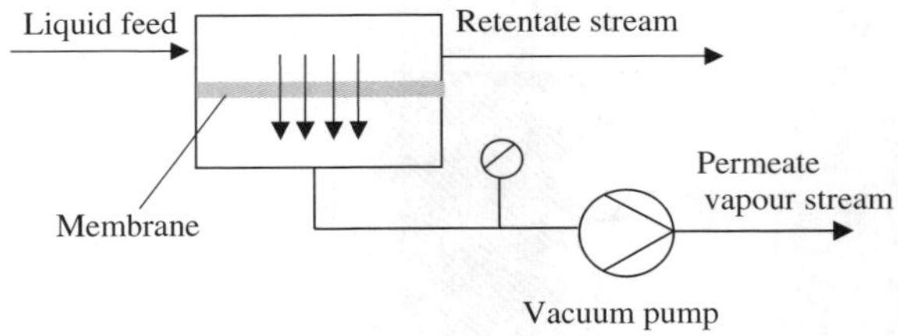

Figure 1.10 A pervaporation process

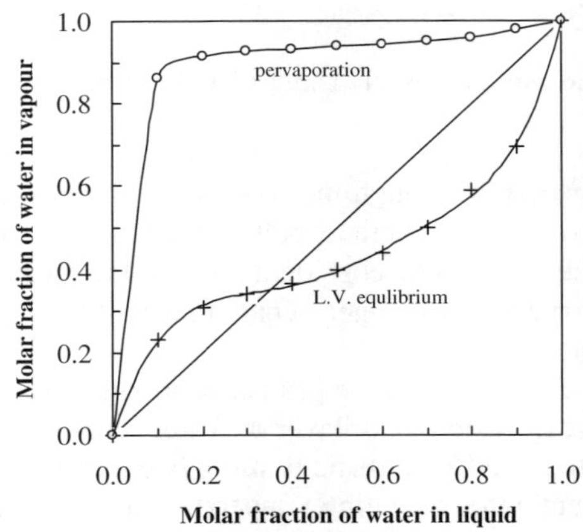

Figure 1.11 Perturbation of liquid–vapour equilibrium by membranes

Thus membranes made from ceramic materials can be operated at higher temperatures and in the presence of solvents that would cause polymeric membranes to fail. They offer much better mechanical stability and do not swell and thus achieve a more constant performance with varying feed concentration. Their ability to operate at higher temperatures with higher fluxes also reduces the required membrane area for operation much below that required for a polymeric membrane. Ceramic supported membranes are much harder than the thin polymer structure of the polymeric membranes. Ceramics offer advantages such as being chemically inert and therefore better at operating with highly reactive compounds present and in acidic or alkaline conditions.

Gallego-Lizon *et al.* [15, 16] conducted pervaporation experiments for t-butanol and iso-propanol dehydration using a range of commercially available ceramic membranes. They found that when dehydrating a 90 wt% t-butanol solution, microporous silica membranes offered the highest flux, followed by zeolites, both being better than that of polymeric membranes. The separation factors are the greatest for the zeolite membrane. Van Veen *et al.* [17] also tested ceramic membranes for pervaporation and found that the ceramic membranes show many advantages over polymeric membranes: (1) giving an extremely constant operation over several weeks, (2) allowing operation at much higher temperatures than polymeric membranes (up to 300 °C) and (3) giving much higher fluxes whilst retaining high selectivity. This, in turn, suggests that the required membrane area can be vastly reduced.

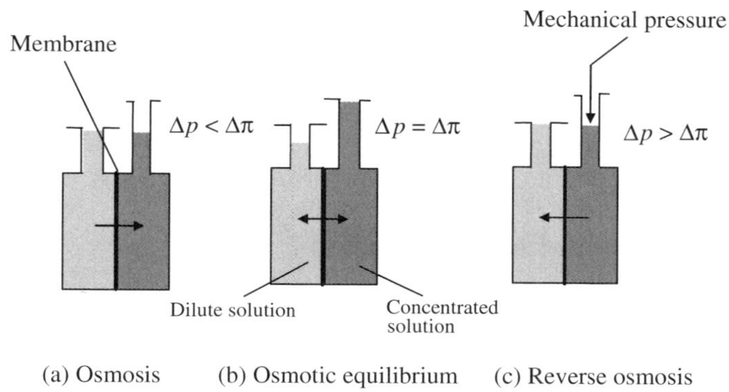

Figure 1.12 Phenomenon of osmosis

1.2.3 Reverse Osmosis and Nanofiltration

Reverse osmosis (RO) and nanofiltation (NF or loose RO) processes allow selective passage of a particular species (solvent), while other species, i.e. solutes, are retained partially or completely. Solute separation and solvent permeability are the membrane characteristics and are dependent on the membrane material and the structure of the membrane layer. The main difference between RO and nanofiltration membranes is that RO rejects all the solutes, including monovalent ions, while the nanofiltration membrane can only reject multivalent ions with no selectivity towards monovalent ions.

As illustrated in Figure 1.12, osmosis is a natural phenomenon where water passes (Figure 1.12a) through a membrane from one side with lower solute concentration to a higher solute concentration until the osmotic equilibrium is reached (Figure 1.12b). To reverse the water flow, mechanical pressure (Figure 1.12c) is applied, providing a pressure difference greater than the osmotic pressure difference; as a result, separation of water from a solution becomes possible. This phenomenon is referred to as reverse osmosis. Applications of reverse osmosis processes include: seawater desalination, waste water treatment and ultrapure water production.

Reverse osmosis (RO) is a well established membrane technology for the treatment of water in a variety of applications. Today, only polymeric RO/NF membranes are commercially available. Major problems associated with polymeric RO/NF membranes are: (1) excessive fouling due to poor feed flow hydrodynamics; (2) low resistance to chlorine and other oxidants; (3) extensive pretreatment/chemical usage and associated waste generation and (4) lack of desirable surface charge to reduce fouling potential. Ceramic membranes, in this context, display a number of performance advantages over commercially available polymeric membranes. Of particular importance in RO and NF applications are the excellent resistance to chlorine, oxidants, radiation and solvents; the high thermal and chemical stability and the long reliable life of ceramic membranes. RO/NF membranes have been available from the outset with ceramic materials. However, the high cost, low packing density and/or poor selectivity renders commercially available ceramic membrane technology economically untenable for RO/NF applications.

Research into formation of ceramic nanofiltration membranes has been carried out recently and membranes prepared from titania [18–21], zirconia [22–24], silica–zirconia [25, 26], hafnia [27, 28] and γ-alumina [19, 29] have been reported. Most of these nanofiltration membranes have been prepared for separation of nonaqueous solvent using sol-gel processes, where a mesoporous ceramic support is coated with a layer of a metal oxide which determines the final pore size. This provides a great advantage in controlling the pore diameter through the proper choice of colloidal solutions at the final coating stage. Reported MWCO of these membranes lies between 200 and 1000 $g\,mol^{-1}$.

1.2.4 Ultrafiltration and Microfiltration

Ultrafiltration is a membrane process where porous membranes are used to separate colloidal particles or large molecular weight solutes from solvent. In ultrafiltration, the mechanism for separation of the solvent from the solute/colloidal particle is similar to that of reverse osmosis/nanofiltration. Therefore, the rejection of solutes is determined based on the pore size and pore size distribution of the membranes, and surface interactions between the membrane surface and solvent/solutes. The overall solvent transfer is often dominated by mass transfer resistances in the membrane as well as at the solution boundaries. Thus, the presence of even a low concentration of retained species can have a profound effect on the solvent permeation.

Although there is no precise definition of the boundary between ultrafiltration and microfiltration as shown in Figure 1.13, microfiltration, as a membrane process, closely resembles conventional filtration processes and separates discrete particles from solution. As can be seen in Figure 1.13, there is also no clear dividing line between coarse filtration and microfiltration, but the accepted upper pore size limit is around several μm. The lower size limit is set at 0.1 μm. Microfiltration is used to separate small insoluble particles, bacteria and yeast cells from broths and aqueous streams. Conventional depth filters typically consisting of a matrix of fibres are used to perform similar tasks, but the separation is achieved by a mechanism of entrapment within the fibres and adsorption to the surface. There is no defined 'pore' size for depth filters but the voids between the fibres are greater than the size of the smallest retained particle. Thus the retention in depth filtration is a statistical function. Microfiltration membranes, on the other hand, have a well defined pore size and separation is achieved on the basis of sieving effects. Because the pore sizes in membranes are large enough, the solvent is usually transported through the pores of the membrane by convection. The rate of transport of solvent through the membrane is thus proportional to the pressure difference across the membrane and can be described by the Hagen–Poiseuille equation, if the pores of the membranes are assumed to be cylindrical.

Ultrafiltration and microfiltration membranes have been prepared from a wide range of polymers such as cellulose acetate, cellulose nitrate, polyacrylonitrile, polyamide,

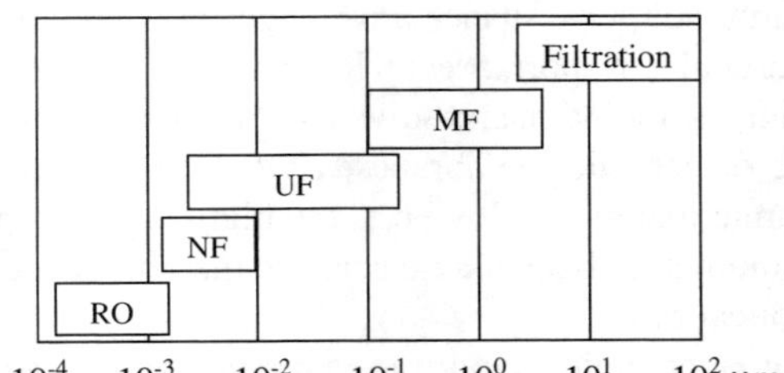

Figure 1.13 Relation between the membrane process and the membrane pore size

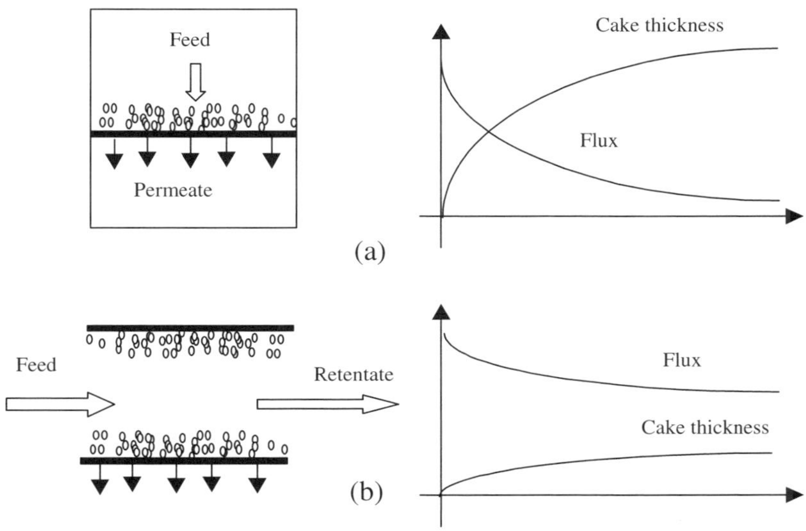

Figure 1.14 Schematic diagram of filtration processes: (a) dead-end filtration and (b) cross-flow filtration

polyethersulfone, polyimide, polysulfone, polyvinyl alcohol, polyvinylidene fluoride, etc. Ceramic membranes have also been developed for ultrafiltration and microfiltration applications. The development of ceramic membranes is mainly driven by the need to produce membranes with greater chemical and thermal tolerance, because the upper temperature limit of polymeric membranes is mostly below 200 °C. In addition, most polymers mentioned above cannot survive in solvents such as benzene and toluene. Ceramic ultrafiltration and microfiltration membranes are prepared from materials such as aluminium oxide, titanium oxide and zirconium oxide, as they can withstand high temperatures and harsh chemical environments. Typical applications of ultrafiltration and microfiltration processes using ceramic membranes can be found in the dairy, food, pharmaceutical, biological, paint, paper and water industries. A detailed review on use of ceramic membranes in these applications can be found elsewhere [4].

There are two types of operations in ultrafiltration and microfiltration: (1) dead-end filtration and (2) cross-flow filtration. As illustrated in Figure 1.14, dead-end filtration is only suitable for dealing with suspensions with a very low solid content, while cross-flow filtration can be used for much higher concentrations as the deposits on the membrane are swept away by membrane-parallel flow.

1.2.5 Dialysis

Dialysis is basically a diffusion process and it describes the separation of substances in solution by means of their unequal diffusion rate through porous membranes; therefore, dialysis is achieved by imposing a concentration gradient across the membrane. Typical application for this process is the artificial kidney as shown in Figure 1.15. The dialysis unit consists of a membrane module comprising a bundle of hollow fibres. Blood flows through the hollow fibre where the dialysis fluid flows through the shell side of the module. The dialysis fluid is

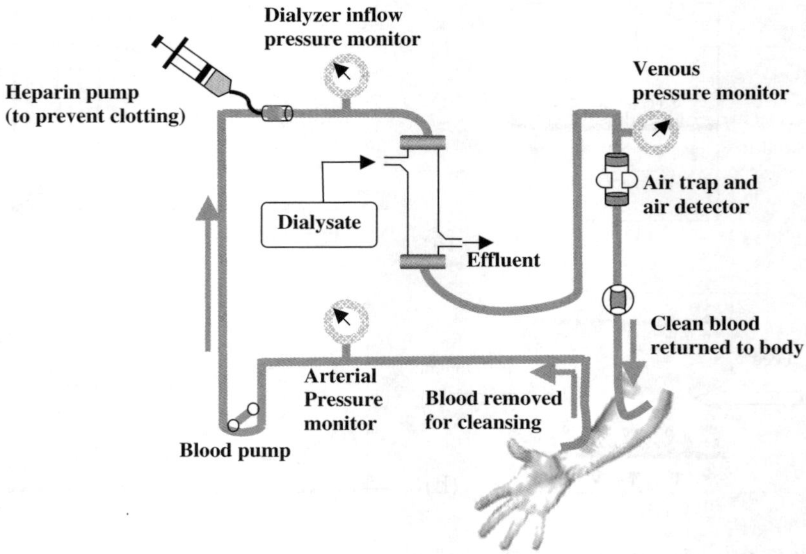

Figure 1.15 Typical dialysis

pumped at a sufficiently high rate to prevent concentration build up in the shell. The toxic species in the blood diffuse through the porous wall (i.e., membrane) and are carried away by the dialysis fluid. The membrane permeability is determined by the characteristics of the membrane and the specific solutes. However, the length, inside diameter and thickness of the membrane itself are all important design variables. Also, operating conditions such as transmembrane pressure and flow velocity in hollow fibre lumen must be optimized.

Most of the membranes used in artificial kidneys are made of polymers, mainly from cellulose based materials. More recently, synthetic membranes including polysulfone, polymethylmetherylate and polyacrilonitrile have been developed for use in dialysis as they are more biocompatible. These new membranes are all synthetic and show high fluxes.

There are so far no ceramic membranes used in dialysis. However, Bender *et al.* [30] and Czermak *et al.* [31] suggested the use of ceramic membranes for removal of endotoxins from dialysis water and dialysate for the reason that ceramic membranes are more resistant to harsh operating conditions.

1.2.6 Electrodialysis

Electrodialysis is a process where solute ions move across ion exchange membranes by application of an electrical field. Although electrodialysis was started as a modification of ordinary dialysis by adding a couple of electrodes, the two processes are distinctly different in many ways as shown in Table 1.2.

The principle of the electrodialysis has been described by Strathmann [32] with an example of desalination of brackish water. As can be seen in Figure 1.16, a typical electrodialysis process, where a series of anionic and cationic membranes are placed alternately and an electrical potential is applied from a cathode and an anode stationed at either end, is used for desalination of brackish water. When the brackish water, containing sodium chloride, is fed into the individual cells, the positively charged cations such as sodium ions are driven by the

Table 1.2 Difference between ordinary dialysis and electrodialysis

Ordinary dialysis	Electrodialysis
Based on concentration gradient	Based on external electrical field
Use of normal membranes	Use of ion exchange membranes
Flow direction:	Flow direction:
high concentration $\rightarrow$ low concentration	high concentration $\leftrightarrow$ low concentration
Concentration gradient diminish as results of mass transfer	Desired degree of separation is achievable

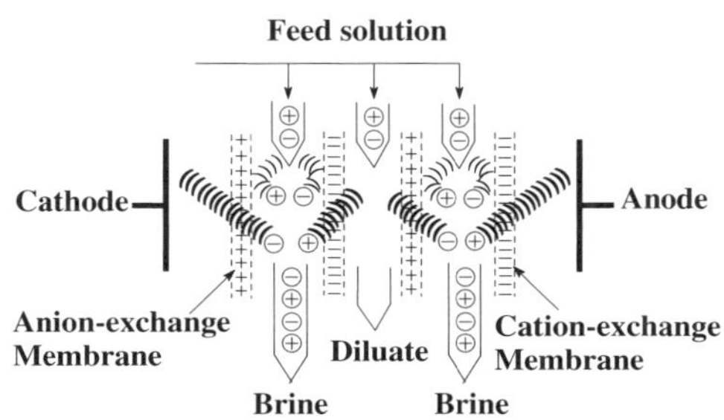

Figure 1.16 Principle of electrodialysis [32]. Reprinted from H. Strathman *Electrodialysis*, in *Synthetic Membranes: Science, Engineering and Applications*, P. M. Bungay, H. K. Lonsdale and M. N. de Pinho, Editors. (1983), D. Reidel Publishing Company: Dordrecht. p. 199. With kind permission of Springer Science and Business Media

electrical potential to the cathode. The sodium ions can permeate through the negatively charged cationic membrane, but are rejected by the positively charged anionic membrane. Similarly, the negatively charged anions such as chloride ions migrate toward the anode through the anionic membrane, but are rejected by the cationic membrane. As a result, both sodium and chloride ions are depleted in chambers referred as 'dilute' and are concentrated in the neighbouring chambers called 'brine'. Electrodialysis can thus be used either to concentrate the salt or to produce potable water from brackish water.

Ion exchange membranes have also been used in caustic soda industries for the electrolysis of sodium chloride solution to produce sodium hydroxide and chlorine. The process is schematically illustrated in Figure 1.17. As can be seen, a cationic membrane made from a perfluorocarbon polymer is placed at the centre of an electrolysis chamber. When a sodium chloride solution is introduced to the left side of the cationic membrane, the sodium ions are attracted toward the cathode and move to the right side of the membrane. On the surface of the cathode, water is decomposed into proton and hydroxyl ions. The protons are immediately reduced into hydrogen atoms by receiving one electron from the surface of the cathode. Two hydrogen atoms are combined to form a hydrogen molecule, and leave the cathode compartment. Sodium hydroxide solution is thus produced in the cathode compartment. On the other hand, chlorine ions move toward the anode. Upon reaching to the anode, the chlorine ions donate an electron to the anode and become chlorine atoms. Two chlorine atoms are then combined to form a chlorine molecule, before leaving the anode compartment.

So far, electrodialysis processes primarily employ polymeric ion selective membranes because they show desirable features, such as low electrical resistance and mechanical flexi-

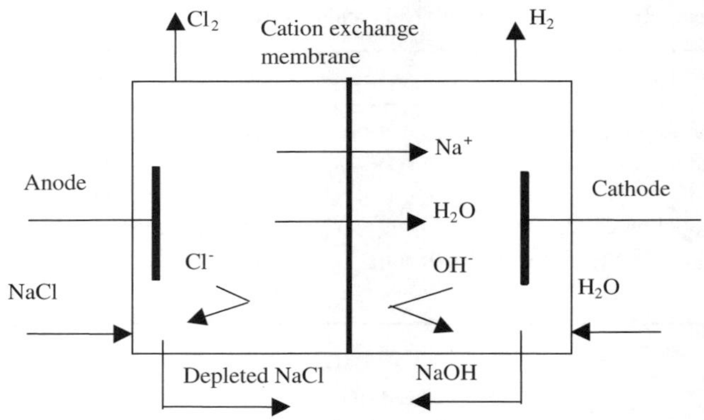

Figure 1.17 Electrodialysis for caustic soda industry

bility. The drawbacks of the polymeric ion selective membranes are their relatively low selectivity and thermal stability [33]. Sodium selective ceramic membranes, in this context, provide enhanced current efficiencies and can operate at high temperatures without being damaged [34]. However, the thickness required for self supporting ceramic membranes leads to high resistance and increases the energy consumption to unacceptable levels. Cormier *et al.* [35] proposed a composite membrane consisting of a thin film of a ceramic of NASICON composition ($Na_{1+x}Zr_2Si_xP_{3-x}O_{12}$, $0 \leq x \leq 3$) deposited on a cation selective polymeric membrane. They demonstrated that the presence of this ceramic thin film on a polymer membrane increases current efficiency and prevents fouling [36], as the membrane combines the advantages of both polymer (mechanical flexibility and low electrical resistance) and ceramic membranes (Na^+/H^+ selectivity and low multivalent ion fouling rates).

1.2.7 Membrane Contactors

Unlike conventional membrane processes, where the membrane is a selective layer towards the fluids to be separated, the membranes used in membrane contactors are nonselective. Therefore, separation achieved in membrane contactors is primarily based on the same principle as in conventional contact processes, i.e. based on phase equilibria. Figure 1.18 illustrates the principles of the process. As can be seen, the applied porous membrane separates two fluids (gas or liquid) from each other and a diffusive mass transfer takes place through the porous membrane. Depending on the membrane material, the physicochemical properties of the liquid and the operating pressures employed, the pores of the membrane can be filled with either gas or liquid, which would result in great differences in the mass transfer resistance of the membrane employed [37].

Membrane contactors represent a technology where porous membranes are used as 'packing materials' for interphase mass transfer. Therefore, all traditional gas stripping and absorption [38, 39], distillation [40, 41], liquid–liquid extraction [42, 43], as well as emulsification [44], crystallization [45, 46] and phase transfer catalysis [47] can be carried out in membrane contactors.

The performances of membrane contactors are strongly dependent on the properties of the membranes, physicochemical properties of the fluids and the operating pressures employed.

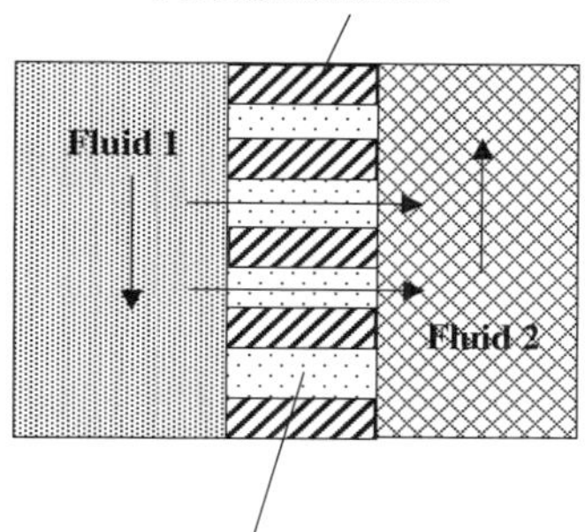

Figure 1.18 Operation principle of a membrane contactor

In general, a membrane with relatively uniform pore size and hydrophobic surface is required so that wetting and mixing between contacting phases can be prevented. Membranes with big pore size, high porosities and asymmetric structures can provide high permeation fluxes, but may cause bubble formations in gas–liquid operations. Thus, operating pressures in the liquid phase need to be properly controlled. In spite of the difficulties in membrane selection and some operational complexity, membrane contactors have been applied to almost all the unit operations due to its considerable advantages, such as larger interfacial area per unit volume, independent control of gas and liquid flow rates without any flooding, unloading, foaming, etc., known gas–liquid interfacial area and high modularity and compatibility for an easy scale up. Disadvantages are mainly related to the presence of an additional mass transport resistance and to the quite limited range of the operating pressures below the wetting pressure threshold. In addition, currently, the membrane contactors are mainly made of polymeric membrane and replacement cost is considered to be another disadvantage. However, new ceramic hollow fibre membranes [48] have been recently developed for membrane contactors. This would certainly prolong the lifetime of the membrane contactor, overcoming this disadvantage.

Zhang and Cussler [40] explored the use of hollow fibre membrane contactors as a potential route to faster distillation. Because of the foreseeable advantages in hollow fibre distillation, a system as shown in Figure 1.19, was set up. As can be seen, the system consists of a column, a reboiler and a total condenser. The only difference is that the column is not staged or filled with packing, but is built with hollow fibres. It is quite clear that such a hollow fibre column has a several advantages over a conventional distillation tower. Firstly, it offers a large interfacial area per volume. All of this area is actively involved in the mass transfer between vapour and liquid. Secondly, because the liquid always fills the hollow fibres, this contact area is maintained, even at very low flows. There is thus no constraint of column 'loading', and the turndown ratio of the column is infinite. Thirdly, no flooding is caused by any two phase flow, as the liquid and vapour flows are not in direct contact, thus, it can be routinely operated in regions which are not accessible to normal packed columns. Also, because the flows of liquid and vapour are now not around submerged objects, pressure drops in the column are more modest than in conventional equipment.

Although, as acknowledged by Zhang and Cussler [40], any improvements they obtained would not be of direct commercial value because hollow fibres used by them were made of polymers, the recent development in ceramic hollow fibre membranes may very soon put this process into commercial practice, as ceramic hollow fibre will remain intact in solvents and temperatures routinely used in distillation.

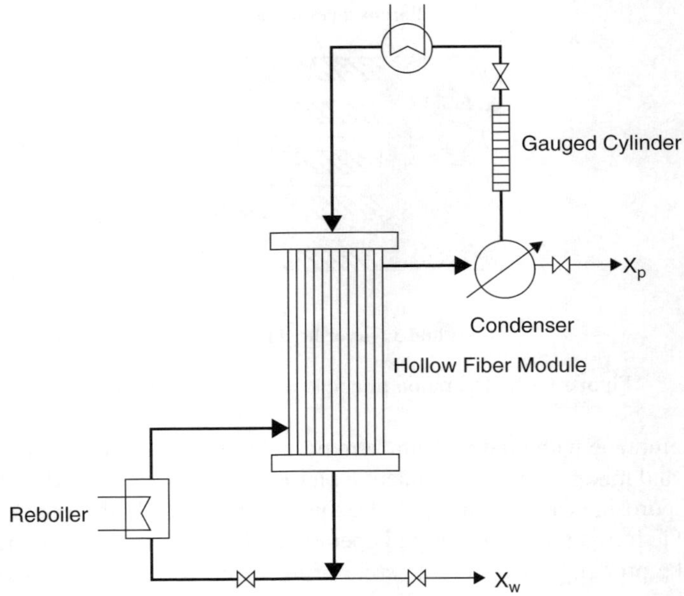

Figure 1.19 Process for hollow fibre distillation (The hollow fibre module replaces the conventional packedtower used for differential distillation) [40]. Reprinted from Zhang, G. L. and Cussler, E. L., *Distillation in hollow fibres*. American Institute of Chemical Engineers Journal, **49** (9) p. 2344–2351. Copyright (2003), with permission from John Wiley & Sons, Inc.

Detailed discussion on membrane contactors will be given in Chapter 5 with emphasis on one of the unit operations, i.e. gas absorption. Obviously, using the same approach, the results from gas absorption analysis can be easily applied to other unit operation processes such as liquid–liquid extraction and distillation.

1.2.8 Membrane Reactors

A membrane reactor is a device that combines a membrane separation or distribution process with a chemical reactor in one unit. Due to the integration of reaction and separation/distribution, chemical processes become simpler leading to a much lower processing cost. In addition, membrane reactors are capable of promoting a reaction process by: (1) selectively removing at least one of the products from the reaction zone through the membrane, making the equilibrium reaction shifting to the product side; (2) supplying only a particular reactant to the reaction zone giving an optimum concentration ratio of the two reactant streams. As a result, the yield can be increased (even beyond the equilibrium value for equilibrium reactions) and/or the selectivity can be improved by suppressing other undesired side reactions or the secondary reaction of products. Figure 1.20 illustrates the two main functions of the membrane reactors.

Use of membrane reactors to shift the equilibrium in a reversible reaction has been mainly studied for dehydrogenation reactions. For example, as the hydrogen produced in the dehydrogenation reaction is removed continuously through a permselective membrane, the reaction equilibrium shifts to the product side, resulting in a high yield even at lower temperatures, and then the deactivation of catalyst and undesirable side reactions may be avoided.

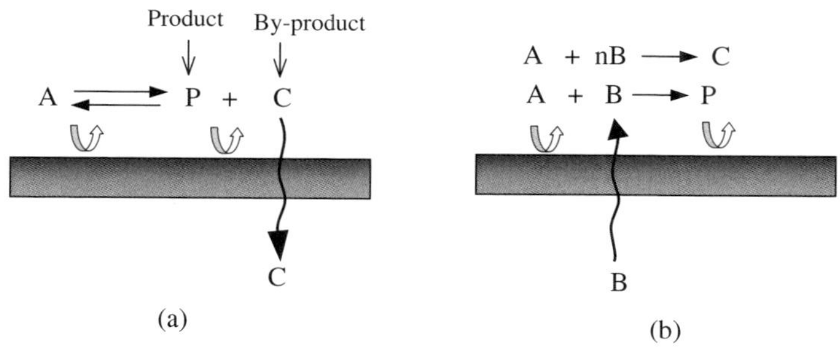

Figure 1.20 Principle of membrane reactors to promote reactions: (a) selective permeation of by-product of an equilibrium limited reaction; (b) dosing a reactant through the membrane

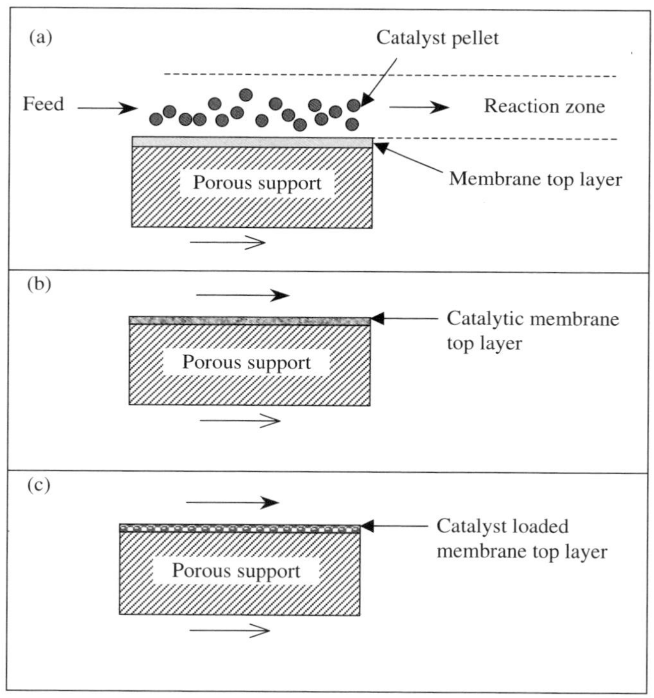

Figure 1.21 Coupling of the membrane with catalysts: (a) membrane coupled with conventional pellet catalysts; (b) membrane itself is catalytically active; (c) catalyst impregnated into the pores of micro-porous membranes

Use of membranes to control the addition and distribution of a reactant has been mainly studied for partial oxidative reactions of hydrocarbons [49]. The membrane controls the oxygen reactant supply to the hydrocarbon compartment and avoids the direct presence of gas-phase oxygen, which is often deleterious to hydrocarbon selectivity, thus, to some degree, suppressing the deep oxidation of the hydrocarbons. The use of membrane reactors also has advantages in controlling the hot spots in exothermic reactions [50, 51].

In catalytic membrane reactors, coupling of the membrane with catalysts is achieved basically in three ways [52]. As illustrated in Figure 1.21(a), the membrane is coupled with conventional pellet catalysts, the membrane forming the inner wall of the tubular reactor.

It should be noted that the membrane top layer, which facilitates the separation forms only a small part of the overall membrane thickness, with the support layer forming the major part. This configuration has been mostly applied in dehydrogenation reactions. Sometimes, the catalyst in the form of paste is coated on the membrane top layer, but it functions in a similar way to the pellet catalyst. In the second arrangement, the membrane itself is catalytically active as shown in Figure 1.21(b). The active catalyst is a thin dense membrane layer deposited on a surface of a porous support. A potential problem with this configuration is that the membrane may not have sufficient catalytic area to be totally effective. The final configuration shown in Figure 1.21(c) is for a catalyst impregnated into the pores of a microporous material either as individual particles or as a layer. This is a convenient way of introducing catalyst into the membrane and has also been used in dehydrogenation reactions.

Detailed analysis of membrane reactors as product separators and as reactant distributors will be covered in Chapter 8.

REFERENCES

1. Vos, R. M. de and Verweij, H., *Improved performance of silica membranes for gas separation.* Journal of Membrane Science, **143** (1–2): 37–51 (1998).
2. Li, K., Tan, X. and Liu, Y., *Single-step fabrication of ceramic hollow fibres for oxygen permeation.* Journal of Membrane Science, **272** (1–2): 1–5 (2006).
3. Hsieh, H. P., Bhave, R. R. and Fleming, H. L., *Microporous alumina membranes.* Journal of Membrane Science, **39** (3): 221–241 (1988).
4. Hsieh, H. P., *Inorganic Membranes for Separation and Reaction.* (1996) Amsterdam, the Netherlands: Elsevier Science B.V., 23–86.
5. Dyer, P. N., Richards, R. E., Russek, S. L. and Taylor, D. M., *Ion transport membrane technology for oxygen separation and syngas production.* Solid State Ionics, **134**: 21–33 (2000).
6. Way, J. D. and Roberts, D. L., *Hollow fibre inorganic membranes for gas separations.* Separation Science and Technology, **27** (1): 29–41 (1992).
7. Liu, S. and Gavalas, G. R., *Oxygen selective ceramic hollow fibre membranes.* Journal of Membrane Science, **246** (1): 103–108 (2005).
8. Liu, Y. T., Tan, X. Y. and Li, K., *Nonoxidative methane coupling in a SrCe0.95Yb0.05O3-alpha (SCYb) hollow fibre membrane reactor.* Industrial and Engineering Chemistry Research, **45** (11): 3782–3790 (2006).
9. Liu, S. M., Teo, W. K., Tan, X. Y. and Li, K., *Preparation of PDMSvi-Al2O3 composite hollow fibre membranes for VOC recovery from waste gas streams.* Separation and Purification Technology, **46** (1–2): 110–117 (2005).
10. Ishihara, T., Kilner, J. A., Honda, M., Sakai, N., Yokokawa, H. and Takita, Y., *Oxygen surface exchange and diffusion in $LaGaO_3$ based perovskite type oxides.* Solid State Ionics, **113–115**: 593–600 (1998).
11. Lane, J. A. and Kilner, J. A., *Oxygen surface exchange on gadolinia doped ceria.* Solid State Ionics, **136–137**: 927–932 (2000).
12. Xu, S. J. and Thomson, W. J., *Oxygen permeation rates through ion-conducting perovskite membranes.* Chemical Engineering Science, **54** (17): 3839–3850 (1999).
13. Shelekhin, A. B., Dixon, A. G. and Ma, Y. H., *Theory of gas-diffusion and permeation in inorganic molecular-sieve membranes.* American Institute of Chemical Engineers Journal, **41** (1): 58–67 (1995).
14. Pohl, P. I. and Heffelfinger, G. S., *Massively parallel molecular dynamics simulation of gas permeation across porous silica membranes.* Journal of Membrane Science, **155** (1): 1–7 (1999).

15. Gallego-Lizon, T., Edwards, E., Lobiundo, G. and Santos, L. F. dos, *Dehydration of water/t-butanol mixtures by pervaporation: comparative study of commercially available polymeric, microporous silica and zeolite membranes.* Journal of Membrane Science, **197**: 309–319 (2002).
16. Gallego-Lizon, T., Ho, Y. S. and Santos, L. F. dos, *Comparative study of commercially available polymeric and microporous silica membranes for the dehydration of IPA/water mixtures by pervaporation/vapour permeation.* Desalination, **149**: 3–8 (2002).
17. Veen, H. M. van, Delft, Y. C. van, Engelen, C. W. R. and Pex, P. P. A. C., *Dewatering of organics by pervaporation with silica membranes.* Separation and Purification Technology, **22–23**: 361–366 (2001).
18. Sekulic, J., ten Elshof, J. E. and Blank, D. H. A., *A microporous titania membrane for nanofiltration and pervaporation.* Advanced Materials, **16** (17): 1546 (2004).
19. Van Gestel, T., Van der Bruggen, B., Buekenhoudt, A., Dotremont, C., Luyten, J., Vandecasteele, C. and Maes, G., *Surface modification of γ-Al_2O_3/TiO_2 multilayer membranes for applications in non-polar organic solvents.* Journal of Membrane Science, **224** (1–2): 3–10 (2003).
20. Voigt, I., Fischer, G., Puhlfurss, P., Schleifenheimer, M. and Stahn, M., *TiO_2-NF-membranes on capillary supports.* Separation and Purification Technology, **32** (1–3): 87–91 (2003).
21. Van Gestel, T., Vandecasteele, C., Buekenhoudt, A., Dotremont, C., Luyten, J., Leysen, R., Van der Bruggen, B. and Maes, G., *Salt retention in nanofiltration with multilayer ceramic TiO_2 membranes.* Journal of Membrane Science, **209** (2): 379–389 (2002).
22. Weber, R., Chmiel, H. and Mavrov, V., *Characteristics and application of new ceramic nanofiltration membranes.* Desalination, **157** (1–3): 113–125 (2003).
23. Benfer, S., Popp, U., Richter, H., Siewert, C. and Tomandl, G., *Development and characterization of ceramic nanofiltration membranes.* Separation and Purification Technology, **22–23** (1–3): 231–237 (2001).
24. Vacassy, R., Guizard, C., Palmeri, J. and Cot, L., *Influence of the interface on the filtration performance of nanostructured zirconia ceramic.* Nanostructured Materials, **10** (1): 77–88 (1998).
25. Tsuru, T., Miyawaki, M., Yoshioka, T. and Asaeda, M., *Reverse osmosis of nonaqueous solutions through porous silica–zirconia membranes.* Aiche Journal, **52** (2): 522–531 (2006).
26. Tsuru, T., Sudoh, T., Yoshioka, T. and Asaeda, M., *Nanofiltration in non-aqueous solutions by porous silica–zirconia membranes.* Journal of Membrane Science, **185** (2): 253–261 (2001).
27. Palmeri, J., Blanc, P., Larbot, A. and David, P., *Hafnia ceramic nanofiltration membranes – part II. Modeling of pressure-driven transport of neutral solutes and ions.* Journal of Membrane Science, **179** (1–2): 243–266 (2000).
28. Blanc, P., Larbot, A., Palmeri, J., Lopez, M. and Cot, L., *Hafnia ceramic nanofiltration membranes – part I. Preparation and characterization.* Journal of Membrane Science, **149** (2): 151–161 (1998).
29. Alamiyounssi, S., Larbot, A., Persin, M., Sarrazin, J. and Cot, L., *Rejection of mineral salts on a gamma-alumina nanofiltration membrane application to environmental process.* Journal of Membrane Science, **102**: 123–129 (1995).
30. Bender, H., Pflanzel, A., Saunders, N., Czermak, P., Catapano, G. and Vienken, J., *Membranes for endotoxin removal from dialysate: considerations on feasibility of commercial ceramic membranes.* Artificial Organs, **24** (10): 826–829 (2000).
31. Czermak, P., Ebrahimi, M. and Catapano, G., *New generation ceramic membranes have the potential of removing endotoxins from dialysis water and dialysate.* International Journal of Artificial Organs, **28** (7): 694–700 (2005).
32. Strathmann, H., *Electrodialysis*, in *Synthetic Membranes: Science, Engineering and Applications*, P. M. Bungay, H. K. Lonsdale and M. N. de Pinho, Editors. (1983) D. Reidel Publishing Company: Dordrecht, p. 197.
33. Paleologou, M., Thibault, A., Wong, P. Y., Thompson, R. and Berry, R. M., *Enhancement of the current efficiency for sodium hydroxide production from sodium sulphate in a two-compartment bipolar membrane electrodialysis system.* Separation and Purification Technology, **11** (3): 159–171 (1997).

34. Sutija, D., Balagopal, S., Landro, T. and Gordon, J., *Environmental uses of sodium super-ionic conducting ceramics.* Electrochemical Society Interface, **5** (4): 26–34 (1996).
35. Cormier, L. M., Ma, F., Bah, S. T., Guetre, S., Meunier, M., Paleologou, M. and Yelon, A., *Sodium salt-splitting performance of a novel ceramic-polymer composite cation-selective membrane.* Journal of the Electrochemical Society, **149** (1): D21–D26 (2002).
36. Girard, F., Izquierdo, R., Quenneville, E., Bah, S. T., Paleologou, M., Meunier, M., Ivanov, D. and Yelon, A., *Evaluation of a ceramic-polymer composite cation-selective membrane for sodium salt splitting.* Journal of the Electrochemical Society, **146** (8): 2919–2924 (1999).
37. Karoor, S. and Sirkar, K. K., *Gas absorption studies in microporous hollow fibre membrane modules.* Industrial and Engineering Chemistry Research, **32** (4): 674–684 (1993).
38. Li, K., Tai, M. S. L. and Teo, W. K., *Design of a CO_2 scrubber for self-contained breathing systems using a microporous membrane.* Journal of Membrane Science, **86** (1–2): 119–125 (1994).
39. Tai, M. S. L., Chua, I., Li, K., Ng, W. J. and Teo, W. K., *Removal of dissolved-oxygen in ultrapure water production using microporous membrane modules.* Journal of Membrane Science, **87** (1–2): 99–105 (1994).
40. Zhang, G. L. and Cussler, E. L., *Distillation in hollow fibres.* American Institute of Chemical Engineers Journal, **49** (9): 2344–2351 (2003).
41. Chung, J. B., DeRocher, J. P. and Cussler, E. L., *Distillation with nanoporous or coated hollow fibres.* Journal of Membrane Science, **257** (1–2): 3–10 (2005).
42. Prasad, R. and Sirkar, K. K., *Hollow fibre solvent-extraction – performances and design.* Journal of Membrane Science, **50** (2): 153–175 (1990).
43. Basu, R., Prasad, R., and Sirkar, K. K., *Nondispersive membrane solvent back extraction of phenol.* American Institute of Chemical Engineers Journal, **36** (3): 450–460 (1990).
44. Gijsbertsen-Abrahamse, A. J., van der Padt, A. and Boom, R. M., *Status of cross-flow membrane emulsification and outlook for industrial application.* Journal of Membrane Science, **230** (1–2): 149–159 (2004).
45. Di Profio, G., Curcio, E., Cassetta, A., Lamba, D. and Drioli, E., *Membrane crystallization of lysozyme: kinetic aspects.* Journal of Crystal Growth, **257** (3–4): 359–369 (2003).
46. Curcio, E., Criscuoli, A. and Drioli, E., *Membrane crystallizers.* Industrial and Engineering Chemistry Research, **40** (12): 2679–2684 (2001).
47. Trusek-Holownia, A. and Noworyta, A., *Catalytic membrane preparation for enzymatic hydrolysis reactions carried out in the membrane phase contactor.* Desalination, **144** (1–3): 427–432 (2002).
48. Tan, X., Liu, S. and Li, K., *Preparation and characterization of inorganic hollow fibre membranes.* Journal of Membrane Science, **188** (1): 87–95 (2001).
49. Eng, D. and Stoukides, M., *Catalytic and electrocatalytic methane oxidation with solid oxide membranres.* Catal. Rew. Sci. Eng., **33** (3–4): 375–412 (1991).
50. Tsai, C. Y., Ma, Y. H., Moser, W. R. and Dixon, A. G., *Modeling and simulation of a nonisothermal catalytic membrane reactor.* Chemical Engineering Communications, **134**: 107–132 (1995).
51. Coronas, J., Gonzalo, A., Lafarga, D. and Menendez, M., *Effect of the membrane activity on the performance of a catalytic membrane reactor.* American Institute of Chemical Engineers Journal, **43** (11): 3095–3104 (1997).
52. Scott, K. and Hughes, R., *Industrial membrane separation technology.* 1st edn. (1996) Blackie Academic and Professional, Chapman and Hall.
53. Koros, W. J., Ma, Y. H. and Shimidzu, T., *Terminology for membranes and membrane processes.* Pure and Applied Chemistry, **68** (7): 1479–1489 (1996).

2

Preparation of Ceramic Membranes

2.1 INTRODUCTION

In general, preparation of ceramic membranes involves several steps: (1) formation of particle suspensions; (2) packing of the particles in the suspensions into a membrane precursor with a certain shape such as flat sheet, monolith or tube and (3) consolidation of the membrane precursor by a heat treatment at high temperatures. A generalized procedure for preparation of symmetric and composite ceramic membranes is illustrated in Figure 2.1 where various methods listed such as slip casting, tape casting, extrusion and pressing are the shaping techniques which turn the particle suspension (slurry or paste) into a membrane precursor. Figure 2.1 further illustrates that no matter which process is selected in preparation of the ceramic membranes, the final membrane product or membrane support can only be obtained through a firing step. Indeed, composite membranes (multi-layer membranes) can be further produced on a membrane support through a coating step using dip-coating, sol-gel, CVD or EVD process, followed by a repeated firing step as shown in Figure 2.1.

In this chapter, methods for the preparation of the ceramic membranes listed in Figure 2.1 will be first discussed with appropriate illustrations, followed by a relatively new approach for preparing ceramic hollow fibre membranes. This method, as illustrated in Figure 2.2, is a three step process comprising: (1) preparation of spinning suspension; (2) spinning of ceramic hollow fibre precursors and (3) final sintering. A detailed discussion of these three steps will be provided, which includes: (1) characteristics of the spinning suspension including particles and particle packing, dispersion of particles and rheology of suspension (step 1); (2) mechanisms of the phase inversion and its relation to the membrane asymmetry and the substrate pore structures (step 2) and (3) sintering process, i.e. presintering, thermolysis and evolution/elimination of surface pores (step 3). Examples of fabricating Al_2O_3, TiO_2 and various perovskite hollow fibre membranes will also be given.

Ceramic Membranes for Separation and Reaction K. Li

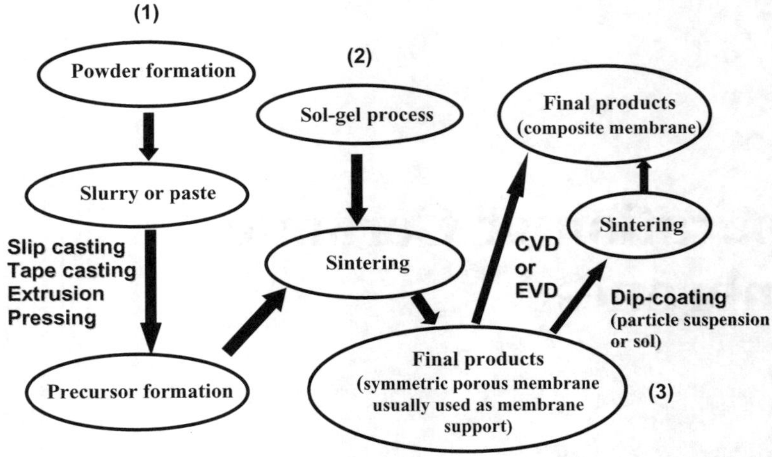

Figure 2.1 A generalized flow sheet for preparation of ceramic membranes using various conventional methods

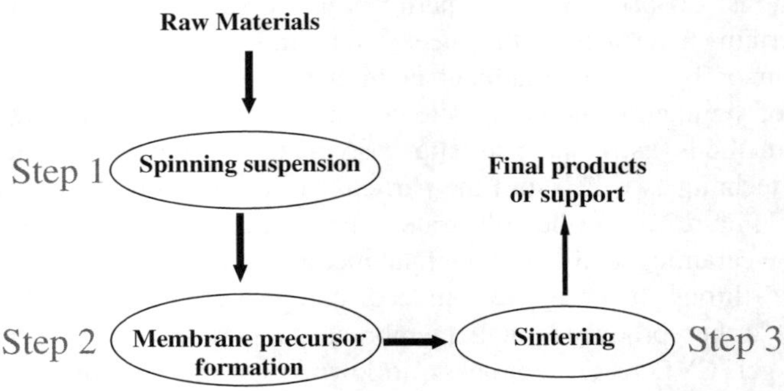

Figure 2.2 A three step procedure for preparation of hollow fibre ceramic membranes

2.2 SLIP CASTING

Slip casting is perhaps one of the most commonly used methods in membrane preparation. It is traditional, but the casting time is usually very long. Also, during the casting, the wall thickness is difficult to control and is usually thick.

As shown in Figure 2.3, when a well mixed powder suspension (slurry) is poured into a porous mould, solvent of suspension is extracted into the pores of the mould via the capillary driving force or capillary suction. The slip particles are, therefore, consolidated on the surface of the mould to form a layer of particles or a gel layer. It is important that the formation of the consolidated layer is fast so that the suspension of the particles will not penetrate into the pores of the mould. Some processing parameters such as solid concentration, particle diameter and viscosity of suspension have been studied by Leenaars and Burggraaf [1] and Uhlhorn, *et al.* [2].

A comprehensive discussion on the theory of slip casting has been given by Tiller and Tsai [3] who provided a method for selection of the mould with the optimum pore diameter. On

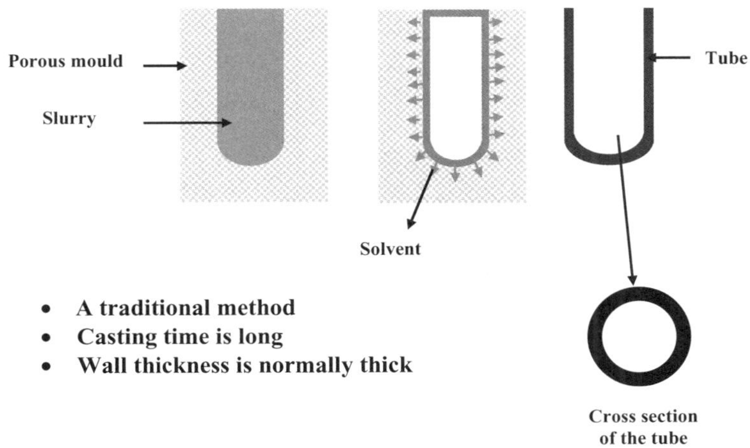

Figure 2.3 Slip casting

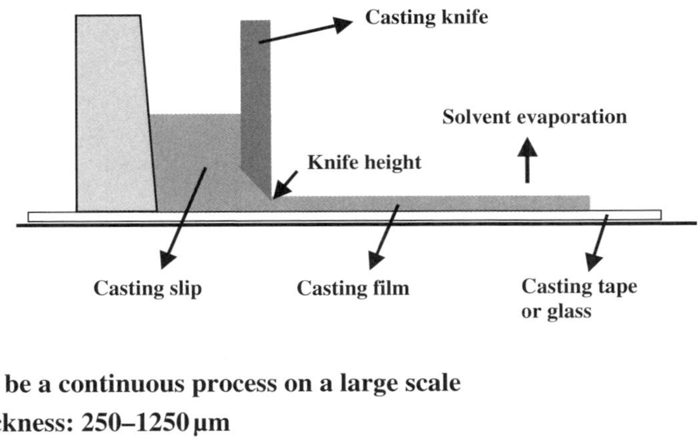

Figure 2.4 Tape casting

the other hand, the pore size of the membrane is mainly dependent on the particle size of the powders used in the slip. Nowadays, if the powders are prepared via a sol-gel process, the pore size range of the prepared membrane as small as 3–6 nm can be obtained.

2.3 TAPE CASTING

Tape casting is another well known method for producing flat sheet ceramic membranes. It is very similar to the casting method used in production of polymeric flat sheet membranes. Figure 2.4 illustrates the basic principle of the tape casting process as it currently exists in most of ceramic membrane industries. The process consists of a stationary casting knife, a reservoir for powder suspensions, a moving carrier and a drying zone.

In preparing flat sheet ceramic membranes, the powder suspension is poured into a reservoir behind the casting knife, and the carrier to be cast upon is set in motion. The casting

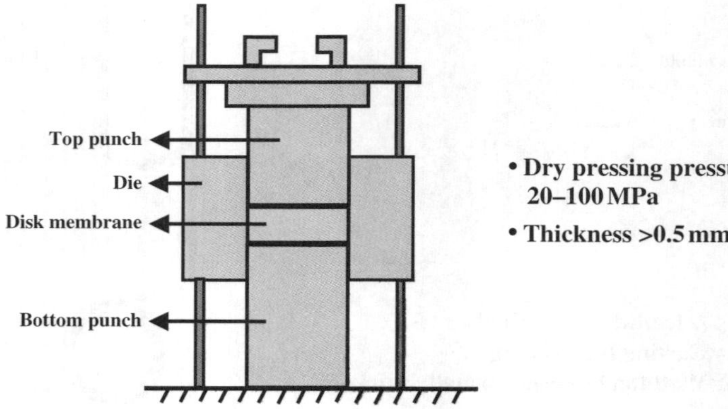

Figure 2.5 Pressing method

knife gap between the knife blade and carrier determines the thickness of the cast layer. Other variables which are important include reservoir depth, speed of carrier and viscosity of the powder suspension. The wet cast layer passes into a drying chamber, and the solvent is evaporated from surface, leaving a dry membrane precursor on the carrier surface.

The thickness of ceramic membranes prepared using the tape casting method are typically in the range of a few millimetres [4]. However, Sahibzada *et al.* [5] have achieved 5 μm thickness when preparing a $LaGaO_3$ solid electrolyte membrane.

2.4 PRESSING

Pressing is a commonly used method for preparation of disc inorganic membranes for fundamental research. As illustrated in Figure 2.5, the particle consolidation into a dense layer occurs by an applied force. This easily handled pressure press method has been frequently employed in screening new ionic and mixed conducting materials for development of oxygen or hydrogen permeable ceramic membranes. A special press machine is used to apply more than 100 MPa pressure to press powders into a compacted disc. The diameter of the disc is usually a few of cm, the thickness is often around 0.5 mm and the disc is dense after firing.

2.5 EXTRUSION

Extrusion is a very productive forming technique that is used for both large products ranging up to 1 ton and small products weighing only a few grams. Not only traditional construction materials such as brick and tile, refractories such as thermocouple protection tubes, furnace tubes, heat exchanger tubes, and kiln furniture, porcelain electrical insulators, magnets and electronic substrates, but also honeycomb cellular catalyst support and most ceramic tubular support for the manufacture of inorganic membranes are produced by extrusion. As shown in Figure 2.6, the extrusion process is similar to fibre spinning processes, but there are a few differences between extrusion and spinning. In extrusion, a stiff paste is compacted and shaped by forcing it through a nozzle. A requirement is that the precursor should exhibit

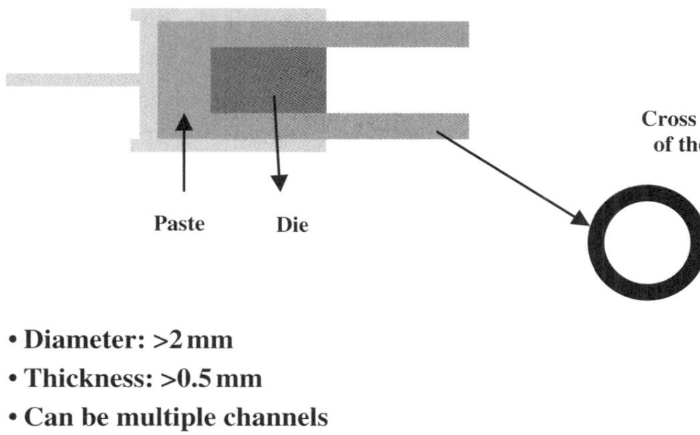

Figure 2.6 Extrusion

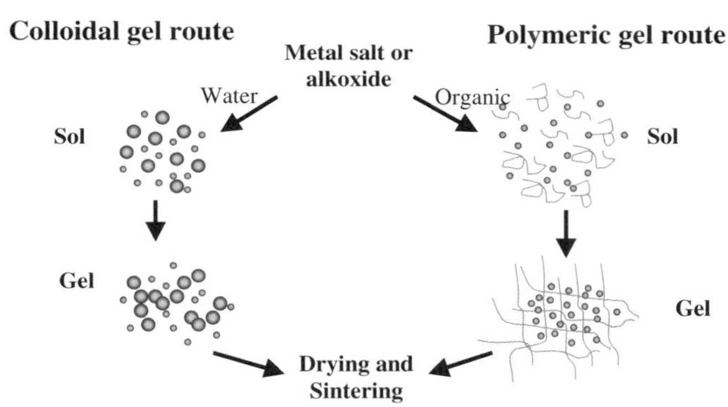

Figure 2.7 Sol-gel method

plastic behaviour, that is at lower stresses behave like a rigid solid and deform only when the stress reaches a certain value called the yield stress. In spinning, a viscous solution or suspension is transformed into a stable shape in a coagulation bath through a spinneret. In addition, the precursor made by extrusion possesses a homogeneous structure over the cross section, while it shows an asymmetric structure if prepared through the spinning process.

2.6 SOL-GEL PROCESS

The sol-gel method was first applied for development of ceramic ultrafiltration membranes by Leenaars *et al.* [6–8]. It was believed as one of the most important techniques in fabrication of ceramic membranes. More detailed description of the sol-gel technology can be found in the literature [9–13]. The advantage of the sol-gel technique is that the pore size of the membrane can be desirably controlled, especially for small pores. As shown in Figure 2.7, there are two main routes through which the sol-gel membrane is prepared:

(1) The colloidal route, in which a metal salt is mixed with water to form a sol. The sol is coated on a membrane support, where it forms a colloidal gel.

(2) The polymer route, in which metal–organic precursors are mixed with organic solvent to form a sol, which is then coated on a membrane support, where it forms a polymer gel.

Colloidal sols are the colloidal solutions of dense oxide particles such as Al_2O_3, SiO_2, TiO_2 or ZrO_2. Leenaars *et al.* developed γ-alumina membranes with pore diameters of 4–10 nm from a boehmite sol [8]. As the pore sizes of these membranes are in the ultrafiltration range, they have been used in separation of colloidal particles and large molecular weight solutes or have been used as a membrane support where smaller pore size membranes can be further developed.

For gas separation based on molecular sieving effects, ceramic membranes with pore sizes less than 1 nm must be employed. In this case, the membrane can be prepared through the polymer sol route using the γ-alumina membrane prepared from the above colloidal sol as a support. It should be noted that in the polymer sol route, the pore size of the membrane prepared is determined by the degree of branching of the inorganic polymer as shown in Figure 2.8. As can be seen, a low degree of branching would result in a narrower pore system. Polymeric sols of SiO_2 with low branched clusters have been prepared using acidic hydrolysis with the corresponding Si alkoxides [14, 15]. These sol-gel membranes show high selectivities in gas separation [15].

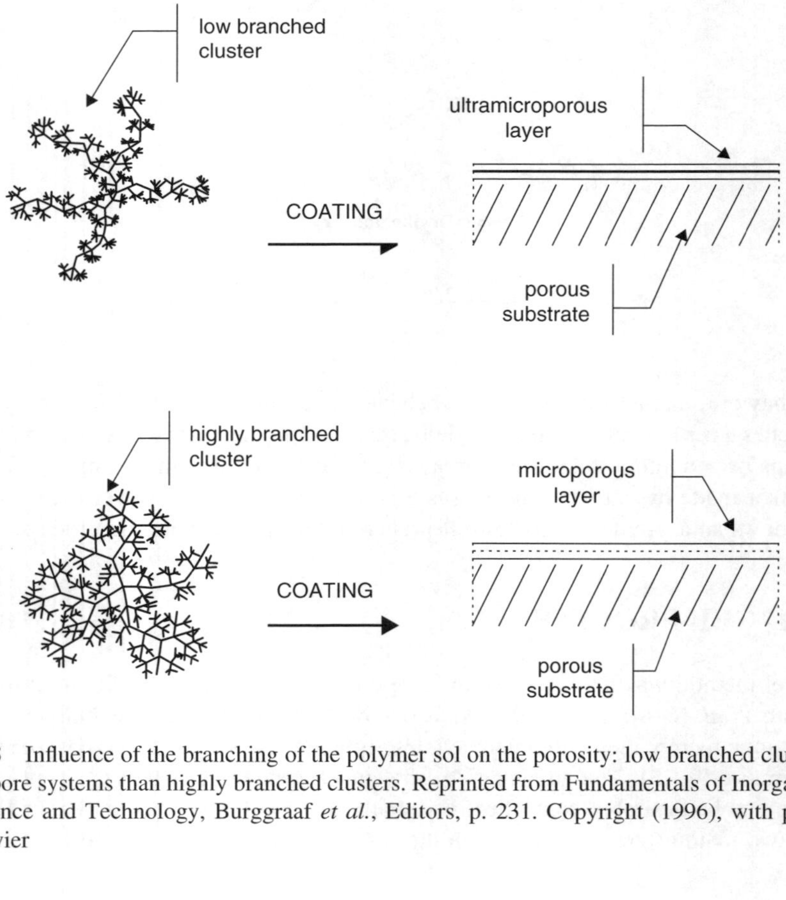

Figure 2.8 Influence of the branching of the polymer sol on the porosity: low branched clusters form narrower pore systems than highly branched clusters. Reprinted from Fundamentals of Inorganic Membrane Science and Technology, Burggraaf *et al.*, Editors, p. 231. Copyright (1996), with permission from Elsevier

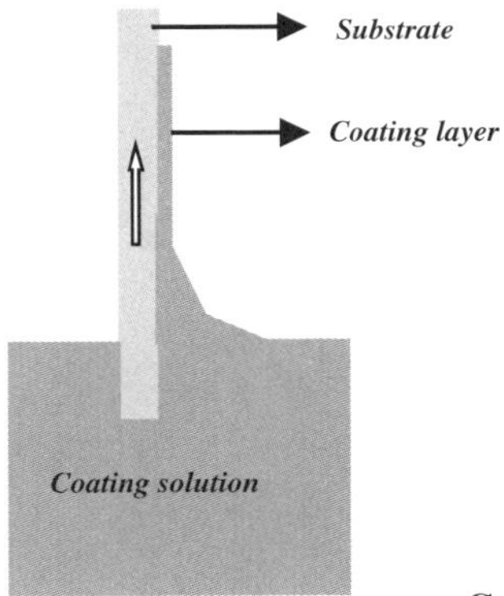

- **Coating layer <0.2 μm is possible**
- **Can be a continuous process**

Figure 2.9 Dip coating

Since sols of very small particles are prepared through hydrolysis and condensation of their corresponding alkoxides, the partial charges of the metal in the alkoxides influence the hydrolysis behaviour. Transition metals such as Ti or Zr in the alkoxides carry much higher partial charges than the Si in tetraethylorthosilicate. Transition metal alkoxides hydrolyse much faster than Si alkoxides. In addition, the properties of the type of R ligands in the alkoxide, the degree of oligomerization of the starting alkoxide, the possibility of a coordination expansion of the metal during the hydrolysis and the pH influence the hydrolysis behaviour. In the drying process of the coated sol, gelation takes place. After thermal treatment, the gel particles become sintered and the membranes are produced through the sol-gel technique.

2.7 DIP COATING

Figure 2.9 shows a simple illustration of a dip coating technique. The critical factors in dip coating are the viscosity of the particle suspension and the coating speed or time. The drying process starts simultaneously with the dip coating, when the substrate is in contact with a atmosphere that has a relative humidity below 100 %. In a multiple step process, after calcination of the first layer, the complete cycle of dipping, drying and calcination is repeated. For example, an alumina coating with a median pore size of typically 100 nm can be prepared from a suspension (in water) of commercially available sub-micron alumina powder with a mass based median diameter of 500 nm [16, 17].

2.8 CHEMICAL VAPOUR DEPOSITION (CVD)

Chemical vapour deposition is a technique which modifies the properties of membrane surfaces by depositing a layer of the same or a different compound through chemical reactions in a gaseous medium surrounding the component at an elevated temperature. Figure 2.10 shows a CVD system which includes a system of metering a mixture of reactive and carrier

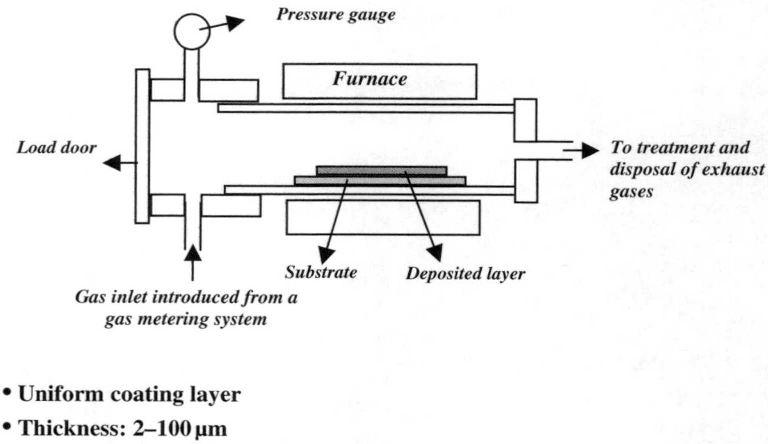

Figure 2.10 Chemical vapour deposition (CVD)

Table 2.1 Chemical reactions in CVD

Reaction	Equation
Thermal decomposition	$2Al(OC_3H_7)_2 \rightarrow Al_2O_3 + 6C_3H_6 + 3H_2O$
Oxidation	$SiH_4 + O_2 \xrightarrow{420°C} SiO_2 + 2H_2$
Hydrolysis	$2AlCl_3 + 3H_2O \rightarrow Al_2O_3 + 6HCl$
Coreduction	$TiCl_4 + 2BCl_3 + 5H_2 \rightarrow TiB_2 + 10HCl$

gases, a heated reaction chamber, and a system for the treatment and disposal of exhaust gases. The gas mixture (which typically consists of hydrogen, nitrogen or argon, and reactive gases such as metal halides and hydrocarbons) is carried into a reaction chamber that is heated to the desired temperature. Different types of CVD methods have been developed over the last few decades. These include moderate-temperature CVD, plasma-assisted CVD and laser CVD. However, all the CVD systems require a mechanism by which the products of the chemical reaction are treated. These products contain various reactive and potentially hazardous components and particles, which must be trapped and neutralized before the gases are exhausted to the atmosphere. In addition, since most CVD processes are carried out at sub-atmospheric pressures, the pumping equipment must be protected from the relatively hot, corrosive gases.

The deposition of coatings by CVD can be achieved in a number of ways such as thermal decomposition, oxidation and hydrolysis as shown in Table 2.1. It can be seen that these reactions between various constituents occur in the vapour phase over the heated substrate, and a solid film is deposited on the surface. On the other hand, a surface film can also be deposited by a reaction between the substrate surface and one or more of the constituents of the vapour phase. One example of this technique is the formation of a nickel aluminide film on the surface of nickel by a reaction of aluminum trichloride and hydrogen from the vapour with the surface of nickel at a high temperature. Such a coating may, therefore, be called a conversion coating.

2.9 PREPARATION OF HOLLOW FIBRE CERAMIC MEMBRANES

Polymeric hollow fibre membranes are in current use for a number of process applications, including filtration, desalination, gas separation, pervaporation and biochemical reactions. However, this membrane configuration is seldom found in ceramic membranes, possibly due to the lack of methods available for producing the ceramic membranes in hollow fibre geometry. It would be interesting to develop a method for preparation of hollow fibre ceramic membranes, as they would unfold their applications to areas where high alkaline (or acid) feed solution or high temperature operation precludes the use of existing polymeric hollow fibre membranes. A typical example is distillation, where hollow fibres would give better, more productive separations than distillation in random or structured packing [18]. Methods for preparing the ceramic membranes mentioned above are not suitable for production of the membrane in hollow fibre goemetry. Although some attempts have been made in fabricating a silica glass hollow fibre membrane for gas separation [19], the membrane was too fragile to be used on large industrial scales. Recently, Tan *et al.* [20] and Liu *et al.* [21–24] successfully produced hollow fibre ceramic membranes using materials such as Al_2O_3 and perovskites by employing a phase inversion method, which is a three step process including: (1) preparation of a spinning suspension; (2) spinning of ceramic hollow fibre precursors and (3) final sintering.

Preparation of ceramic membranes using the phase inversion method is a quite complex process. Many factors may affect the final membrane performance. Therefore, a detailed discussion on this method is provided according to the sequence of the three step process.

2.9.1 Preparation of Spinning Suspensions

Preparation of a spinning suspension may consist of several steps in sequence: (1) mixing a dispersant and a solvent of choice; (2) adding preconditioned particles for deflocculation and dispersion; (3) adding polymer binders and plasticizers and (4) degassing the suspension prior to the spinning of the membrane precursor. The deflocculation is a very important step, as soft agglomerates of the particles in the spinning suspension would have an effect on pore size and permeability of the final membrane product [25]. A more detailed preparation procedure will be given in the end of this section.

The main components in the spinning suspension for the fabrication of hollow fibre ceramic membranes are the ceramic powders, additives and solvents. In selection of the ceramic powders, one of the important factors is the particle sizes and their distribution as well as the shape of the particles. They have an effect on the porosity, pore size and pore size distribution of the final membrane product [25].

Additives such as deflocculants/dispersants, binders, plasticizers, anti-foaming agents, pore formers, antistatic, chelating, bactericide agents, etc. are used to give the spinning suspension the properties required. In preparing the ceramic hollow fibre membrane using the phase inversion technique, the requirement of the organic binders is that they should not only be invertible, but also must be burned out without leaving ash and tar during the calcinations. The quantity of organic binders should be as low as possible, but without affecting the inversion capability.

The solvent(s) must dissolve the additives and binders used and must show a higher exchange rate with nonsolvent (coagulants). The rates of solvent outflow and coagulant inflow have an effect on the cross sectional structures of the membrane precursor and hence the structures of the final membrane products [20].

As preparation of the spinning suspension is still in the development stage, the following few general rules are, so far, considered to be useful in preparation of a spinning suspension:

(1) The amount of dispersant must be maintained to ensure the stability of the suspension.

(2) The amount of solvent must be fixed at a minimum to maintain a homogeneous suspension.

(3) The ratio between organic components and ceramic powders must be as low as possible.

(4) The plasticizer to binder ratio must be adjusted to make the membrane precursor flexible, resistant and easy to release.

2.9.1.1 Particles and Packing of Particles

In ceramic membrane fabrication processes, the most important component in the spinning suspension is the ceramic powders. After binder removal and final sintering, the ceramic powders are the only component left and determine the properties of the ceramic membrane produced. The other components in the spinning suspension such as solvents, plasticizers, binders, dispersant, etc. are used simply to predesign the membrane morphology (i.e. pore size, pore size distribution, porosity, cross sectional structures, etc.) and to facilitate the spinning and phase inversion which is, in general, a process for obtaining and holding the powder particles in the hollow fibre configuration.

During the phase inversion (consolidation process), the particles in the spinning suspension undergo a process of particle packing. It is, therefore, possible to adjust the particles to attain high packing densities or to attain specific compact properties. The most widely used parameters of the particle packing are the packing density (or bulk density) and the coordination number. The bulk density (defined as weight of the particle bed/volume of the bed) is directly proportional to the density of the individual particles and a high bulk density may be obtained simply because the particles have a high density rather than because the particles are closely packed. In order to avoid this difficulty, an expression involving the ratio of the volume of particles composing the bed (V_p) to the volume of the bed (V_b) may be adopted. This ratio, i.e. packing fraction (defined as bulk density/particle density) is usually referred to as the packing fraction or the fractional solids content. The fractional voidage ($\varepsilon = 1 - V_p/V_b$), expressed as a percentage, is often known as the volumetric porosity.

The coordination number is defined as the number of other particles that are in contact with a selected particle. There are three main packing types, i.e. ordered, random loose and random dense are available for monosized spheres. In an analysis of particle packing, Rahaman [26] discussed various systematic assemblages of spheres and introduced several simple assemblages to characterize packing and pore structures. Figure 2.11 shows four basic packings, corresponding to cubic, orthorhombic, rhombohedral and tetragonal geometries. Table 2.2 gives the packing density and coordination number of the four regular packing arrangements with monosized spheres. It can be seen that that the packing density increases as the coordination number is increased.

The packing density can also be improved by introducing smaller sized particles. Small particles introduced into a packing of large particles are to fill its voids without increasing

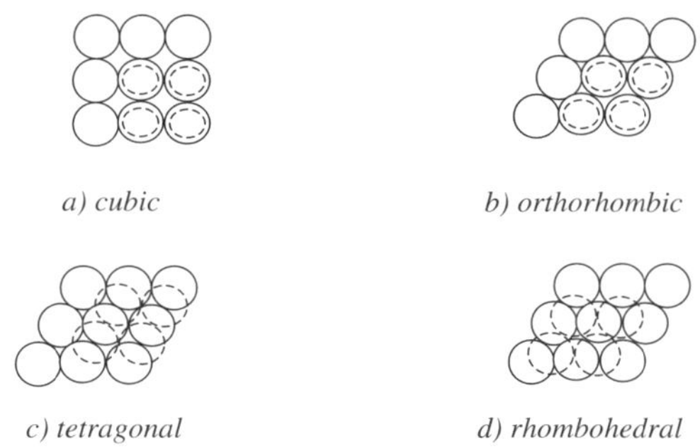

Figure 2.11 Spheres can nest together in a variety of configurations, each with a different packing density

Table 2.2 The densities and coordination numbers for ordered packings

Packing type	Coordination number	Packing density
Cubic	6	0.524
Orthorhombic	8	0.605
Tetragonal	10	0.698
Rhombohedral	12	0.740

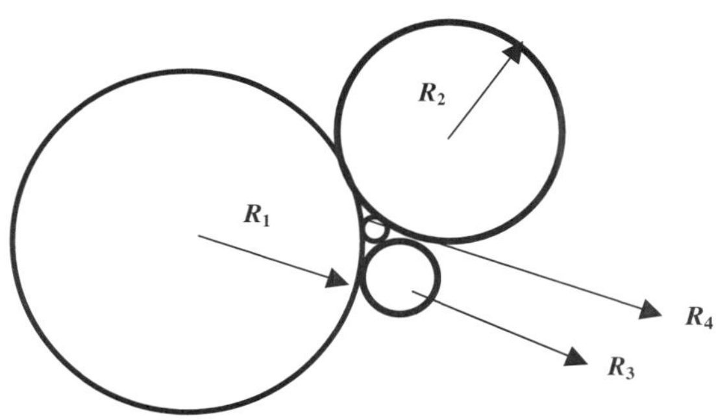

Figure 2.12 The void at a three disk junction which is just filled by a touching disk of radius R_4

the overall volume of the bed. Several studies have been conducted to view the two dimensional structure using disks and spheres [27–29]. Two dominant factors are the size ratio and composition of the mixture.

The size ratio of particles determines the packing coordination number at the highest packing density. Figure 2.12 shows the most efficient packing improvement geometry in two dimensions. The three existing disks have radii of R_1, R_2 and R_3, the radius of small disk that just touches these three disks is noted as R_4 and can be calculated as follows:

$$1/R_4 = 1/R_1 + 1/R_2 + 1/R_3 + 2[1/(R_1R_2) + 1/(R_1R_3) + 1/(R_2R_3)]^{1/2} \tag{2.1}$$

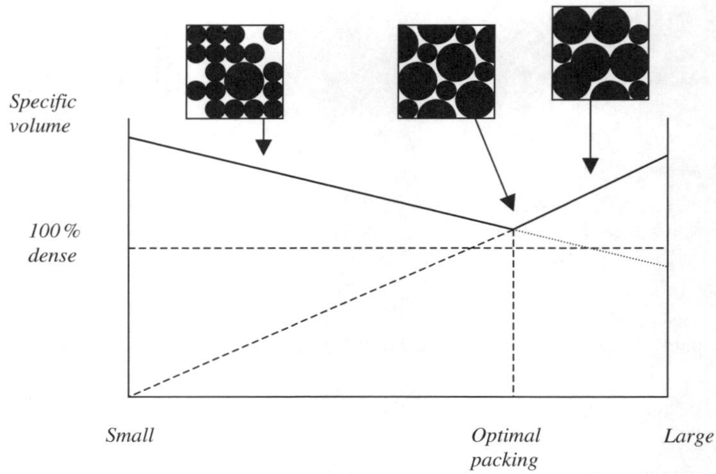

Figure 2.13 The reduction in specific volume for mixed large and small spheres, showing the condition of optimal packing where the small spheres fill all voids in the large sphere packing

For the most efficient packing improvement, defined as the gain in density per particle, the added particle should just touch the three neighbouring particles. If the three large disks are equal in size, this equation predicts R_4 will have a radius of 1/6.464 or 0.1547 of the large disk radius, and the coordination number is only three.

When two sizes of spheres are mixed, the smaller particles fill the unoccupied voids among the large particles. So as long as the proportion of the smaller spheres is sufficiently great, the proportion of large spheres could reach a stage at which they touch each other. The resulting packing is optimal. When the quantity of small particles becomes too great since all of the voids are filled, further additions force the large particles apart and no longer improve the packing density. The basic behaviour for random dense packing is sketched in Figure 2.13, which shows that the packing volume, termed the specific volume (volume to mass ratio), is a function of the composition for a mixture composed of large and small spheres. It can be seen that there is a composition of maximum packing density [30]. The relative improvement in packing density depends on the particle size ratio of the large and small particles. Detailed discussion on the packing of different particle sizes is given by Rahaman [26].

2.9.1.2 Dispersion of Particles

As mentioned above, in preparation of ceramic membranes one of the problems to be tackled is the dispersion of each individual ceramic particle in a solvent. When individual primary particles are in proximity, they often have a tendency to form loosely bound groups, often called soft agglomerates. In a fluid suspension of particles, particularly when the powder has little affinity for wetting, soft agglomerates can form to lower the free energy of the suspension by reducing the solid–liquid interface area. There are many types of surface forces acting between particles, among which electrostatic forces, van der Waals forces and steric forces are the three most important ones. Descriptions of these three forces are given in Appendix 2.1.

In order to deflocculate soft agglomerates as well as to disperse the primary particles in a solvent, addition of a dispersant is usually required. The purpose of the dispersant is to disperse the primary particles and to hold them in a homogeneous suspension. As the dispersant separates primary particles and holds them in suspension, which allows the solvent to form a separating layer between particles, the suspension usually shows a lower viscosity due to the particle mobility offered by the solvent in the interparticle layer. There are basically two different mechanisms, i.e. ionic repulsion and steric hindrance, in describing the deflocculation and dispersion. The term ionic repulsion is used to describe the charging of particle surfaces so that they magnetically repel to each other. In aqueous systems, this can be achieved by change of the pH. The surface charge of the particles can also be accomplished by introducing polyelectrolyte solutions, which coat the particle surfaces and provide the surface charge. The term steric hindrance is used to describe the separation of particles by coating a layer on one particle that would physically prohibit another particle coming in contact with it. Organic polymers are often used in both aqueous and nonaqueous systems to accomplish this particle–particle separation. A list of common dispersants/deflocculants can be found in Table 2.3. It is interesting to note that most

Table 2.3 Some reported dispersants

1. Polyisobutylene
2. Linoleic acid
3. Sodium silicate
4. Oleic acid
5. Dibutyl amine
6. Citric acid
7. Substituted imidazolines
8. Stearic acid
9. Sulfanates
10. Lanolin fatty acids
11. Aliphatic hydrocarbons
12. Salts of polyacrylic acids
13. 2-Amino-2-methyl-l-propanol
14. Salts of methacrylic acids
15. Polyethylene glycol
16. Polyvinylpyrrolidone
17. Blown menhaden fish oil
18. Polyvinyl butyral
19. Corn oil
20. Sodium sulfosuccinates
21. Safflower oil
22. Ethoxylate
23. Linseed oil
24. Phosphate ester
25. Glycerol trioleate
26. Glycerol tristearate
27. Synthetic waxy esters

of the dispersants for use in organic solvents contain blends of various fatty acids and esters.

There are several reasons for a spinning suspension to be fully dispersed. Firstly, the ceramic powders being added into the spinning suspension are usually in the form of soft agglomerates because they are normally produced through a sol-gel process and are high surface area powders. The soft agglomerates tend to trap air in the interstitial space between the primary particles. If the soft agglomerates are not broken up, the trapped air would produce bubbles in the membrane precursor, resulting in pin holes and unwanted surface porosity in the final sintered membranes. Secondly, when the polymeric binder is introduced, it will envelop the soft agglomerates instead of the individual (primary) particles. This would make the soft agglomerates a permanent group throughout the rest of the process, which gives less density to the membrane precursor, leading to high shrinkage during the final consolidation process.

The dispersion of particles in a solvent takes place in several stages. When the powder is first put in proximity to the solvent, the particle surfaces (or agglomerates) get coated with the solvent. Therefore, stage 1 of suspension is the wetting of the particle surface. After the particle surface is wetted, soft agglomerates held together by electrostatic forces or by van der Waals forces need to be separated, i.e. stage 2 of suspension is the mechanical separation of primary particles. After the completion of stage 2, the particles are evenly distributed throughout the solvent. The final stage of suspension consists of protecting and stabilizing what has been accomplished, i.e. prohibiting soft agglomerates from reforming in the spinning suspension.

Dispersion mechanisms have been fairly well studied in nonaqueous solvent systems, primarily with oxides. Although the main particle separation mechanism has been found to be steric hindrance, Lee and Rives [31] concluded that both steric and ionic mechanisms are active in the organic solvent based suspension. The ionic repulsion dispersion mechanism may come into organic solvent systems with certain additives. However, since the spinning suspension generally contains a high solid loading and the particles are in close proximity to each other, steric hindrance is still an important stabilization method [32].

One of the measurable effects of using a dispersant is a higher packed bed density. This phenomenon has been used to determine the proper dispersant level. The level or extent of dispersion can be measured by a number of techniques such as micro electrophoresis, optical microscopy and laser scattering. Targeted variables of these tests include zeta potential, iso-electric point and equivalent spherical diameter. A settling experiment is perhaps another means for the characterization of the spinning suspension, as one of the main reasons for dispersing the solution is to form a densely packed bed of particles. It follows that a screening experiment should be performed with the densest packed bed as the goal. Therefore, an experiment in which different concentrations of dispersant are used to disperse identical suspensions of powder and solvent can be carried out. After a proper screening, the extent of dispersion in powder suspensions can be observed.

2.9.1.3 General Preparation Procedure

Based on the above individual discussions, a general procedure in preparation of the spinning suspension can now be drawn. The first step is the dispersion of particles in the presence of a dispersant in the solvent(s). The usual procedure would be to dissolve a predetermined amount of dispersant in a weighed or volumetric amount of solvent by stirring or by rolling. The powder(s) are then added.

The percentage of the solids loading during dispersion is dependent on the powder density, particle size and dispersant effectiveness. It is beneficial to optimize the solids loading during this step, since any excess solvent will have to be removed during the phase inversion (consolidation step). The solid loading is often adjusted with experience, because the solvent volume also affects the final viscosity after the polymer binder and/or plasticizer(s) are added to the mixture. The viscosity of the dispersion is not a primary concern, as long as deagglomeration occurs. Viscosity of the dispersions at this stage is extremely low, but will increase when the binder is added.

The purpose of the dispersion procedure is to accomplish three goals: (1) to break apart any soft agglomerates; (2) to coat the primary particles with a dispersant; which (3) keeps the particles apart by steric effects, electrostatic effects, or both.

After the dispersion process is complete, the plasticizers (if any) are poured onto the top of the mixture, and then the weighed quantity of binder, either in powder form or dissolved in a solvent.

Once the plasticizer and binder are added, the mixture is stirred or rolled to make sure that the binder is not in large clumps. Usually the suspension becomes more viscous with the addition of a long chain polymeric binder. The only time that this does not occur is when the binder is a more powerful dispersant and replaces or overrides the dispersant of choice. This is also a good reason for using the suspension in the phase inversion process as soon as possible after the binder is mixed into the suspension. The rheological properties of the suspension can change with time. Once the plasticizer(s) and binder(s) have been mixed into the suspension, it is ready for prespinning conditioning and characterization.

The next step in that process is to degas the suspension to remove any air that may have been entrained during the mixing process. Air bubbles cause defects in the membrane precursor. Pinholes (small holes left by air bubbles) are the most common defect, and these can lead to cracking which radiates from the pinhole during the phase inversion (consolidation step). It is therefore essential that entrained air is removed before the spinning. Several techniques are used for the degassing process. The simplest technique is partial vacuum accompanied by gentle stirring or agitation. Once the spinning suspension is properly degassed, it is ready for the spinning of the hollow fibre precursors and this is discussed below.

2.9.2 Spinning of Ceramic Hollow Fibre Precursors

Spinning of ceramic hollow fibre precursors can be carried out in a spinning apparatus as shown in Figure 2.14. The hollow fibre precursors are spun through a tube-in-orifice spinneret as shown in the insert of Figure 2.14, typically with orifice diameter/inner diameter of the tube of 3.0/1.2 mm. The extrusion rate of the spinning suspension is controlled by the nitrogen pressure (or a gear pump) and an adjusting valve. The nonsolvent, usually water, is used at room temperature as the internal and external coagulants. The spinneret is arranged such that the nascent fibre precursor can be extruded vertically downwards into the coagulation bath. After coagulation (immersion induced phase inversion), the fibre precursor is guided through the wash bath, and then dried. Figure 2.15 shows a typical fibre precursor spun from a suspension containing $La_{0.6}\,Sr_{0.4}\,Co_{0.2}\,Fe_{0.8}\,O_{3-\alpha}$ (LSCF) powders. It can be seen that the precursors are very flexible because of polymeric binders. In spinning of ceramic hollow fibre precursors, there are two important parameters, i.e. spinnability and invertability of the powder suspension, which are discussed below.

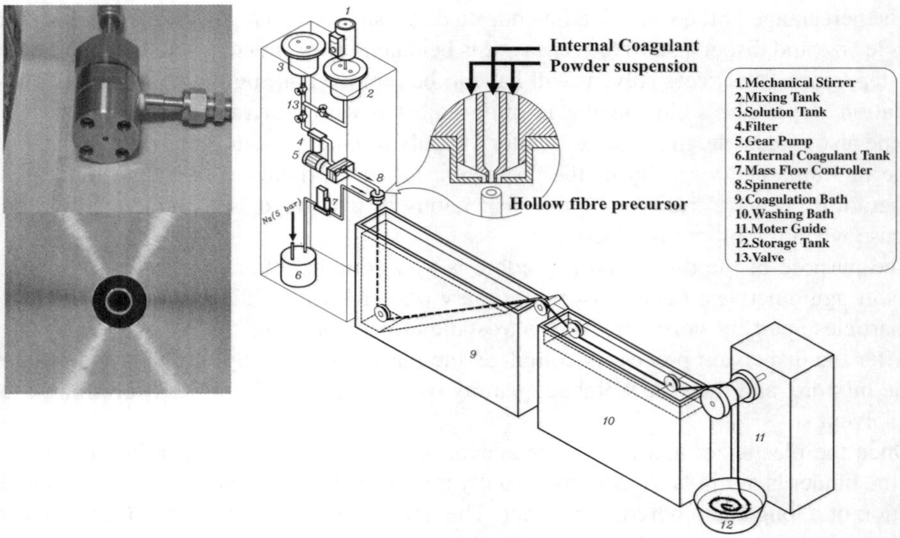

Figure 2.14 Hollow fibre ceramic membrane spinning apparatus (insert: photograph of a spinneret)

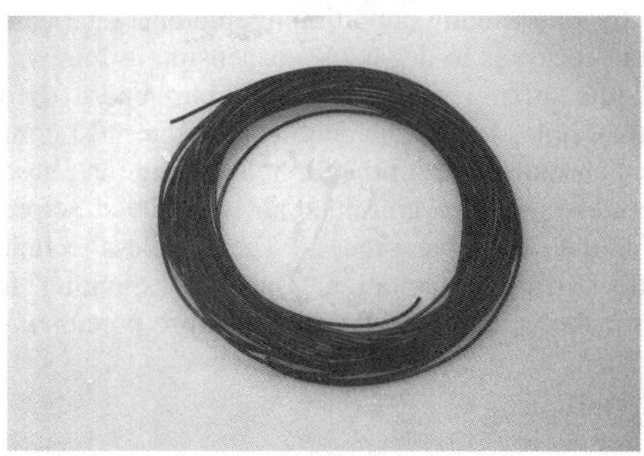

Figure 2.15 Ceramic hollow fibre precursors (prepared from LSCF)

2.9.2.1 Rheology and Spinnability

In any polymer solutions or particle suspensions, the rheological properties play a key role in controlling the shape forming behaviour [33]. A general observed phenomenon is that the viscosity increases with the increase of polymer contents in polymer solutions and increases with volume fraction of solids and finally ceases to flow at some critical maximum volume fraction of solids in particle suspensions. The rheological behaviour of a concentrated suspension of ceramic particles is affected by a number of factors including the particle size distribution, particle shape, volume fraction of solids, and the range and magnitude of the interparticle forces [34–36].

When the viscosity reaches about 10 poise for a solution, the solution becomes spinnable and fibres can be drawn through an orifice [37, 38]. For a suspension containing spherical particles, the relation between the reduced viscosity η_{sp}/c and the particle concentration c, where η_{sp} is the specific viscosity, can be examined. In principle, solutions containing spherical particles follow Einstein's formula [37]:

$$\eta_{sp}/c = k/\rho \tag{2.2}$$

where k is a constant and ρ is the density of the particles.

For a concentrated particle suspension, the best spinnability should be observed when the viscosity behaviour is highly shear thinning but not thixotropic. A very high viscosity is necessary to prevent the drawn fibre from breaking up into droplets. The high viscosity required for stable fibre formation is generally achieved by increasing the concentration of the suspension or adding some additives.

2.9.2.2 Phase Inversion

In the 1960s, Loeb and Sourirajan [39] first developed an immersion induced phase inversion method and produced an asymmetric polymeric membrane, which exhibited a high selectivity and high water flux. This membrane showed an asymmetric structure: a dense and thin skin (about 25 μm) integrated on porous substrate of the same material. The dense skin layer showed permselective properties and was suitable to separate salt from sea water, while it was also thin enough for a high water flux. The porous substrate provided the mechanical stability for the membrane. Since then, this method has been one of the most useful methods for the fabrication of polymeric membranes. Conceptually, membrane formation by phase inversion can be described in terms of a three component system: nonsolvent, solvent and polymer. After immersion of a polymeric solution into a nonsolvent bath, the solvent diffuses out of the polymer solution, while nonsolvent diffuses into the solution. This exchange of solvent and nonsolvent in general leads to a change of the polymer solution from a thermodynamically stable state to a metastable or unstable state. Then liquid–liquid demixing takes place in the solution. In the case that the solvent flux is higher than the nonsolvent flux, the polymer concentration at the interface would increase. The highly concentrated polymer layer at the surface reduces the exchange speed of the solvent and nonsolvent. Hence the polymer concentration in the sublayer is lower than the polymer concentration in the top layer. The further replacement of solvent by nonsolvent would result in solidification of the polymer rich phase and the asymmetric structure of the membrane is, thus, obtained. Figure 2.16 shows SEM photographs of a typical asymmetric structure with a finger-like support layer of a hollow fibre membrane produced by the phase inversion method.

The fact that should be noted in the phase inversion process is the concentration of polymer in the top layer, which is increased with the exchange of the solvent and nonsolvent. When the out diffusion of solvent is faster that the in diffusion of nonsolvent, the total volume of the original polymer solution is reduced. The polymer molecules are moving from the bulk to the interface. For the polymer solution system, this move would only result in an increase in the polymer concentration. But, for a suspension system containing different sized ceramic powders, the situation may be different. When the particles are well covered by a polymer binder in the suspension, the particles may be considered as polymer nodules. The rate of

Figure 2.16 The SEM photograph of a typical asymmetric structure with finger like support layer of a hollow fibre membrane produced by the phase inversion method, (a) cross section, (b) wall [56]. Copyright (2000) Elsevier, Journal of Membrane Science, **176** (2), Wang, D. L., Li, K., and Teo, W. K., *Highly permeable polyethersulfone hollow fibre gas separation membranes prepared using water as non-solvent additive*, p. 147–158

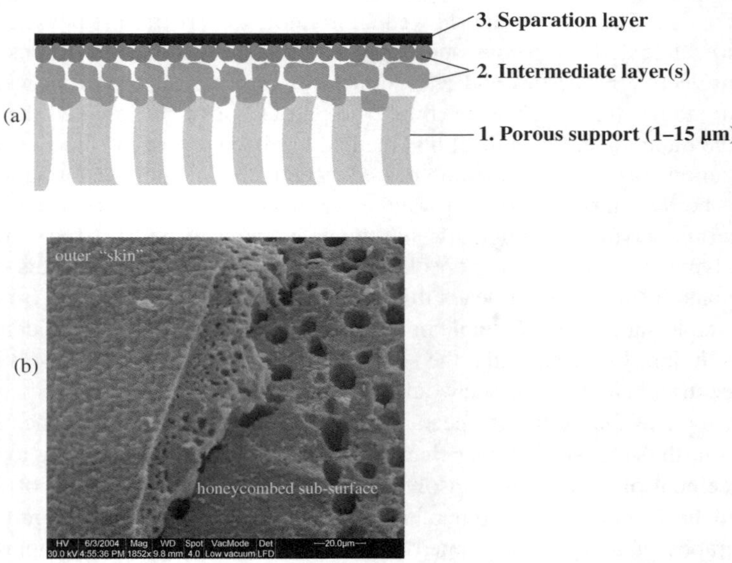

Figure 2.17 (a) A general structure of ceramic membrane membranes; (b) a multi-layered ceramic membrane prepared in single step

movement of different sized nodules may be different when exchange of solvent and nonsolvent takes place in the phase inversion process. It was postulated that the smaller particles may move faster than the larger particles [40]. Therefore, membranes with a graded structure as shown in Figure 2.17 can be prepared [40]. It thus follows that a good control of the ratio of small and large particles in the suspension and the exchange rate of the solvent and non-solvent may tailor the membrane structure, which can only be prepared with multiple steps if conventional techniques are used.

2.9.3 Sintering

Sintering processes have been studied for long time and considerable breakthroughs in understanding the fundamental aspects of the basic mechanism of sintering of ceramics have been achieved [41, 42]. In sintering the hollow fibre membrane precursors mentioned above, three main steps including presintering, thermolysis and final sintering are required and are described separately in the following sections.

2.9.3.1 Presintering

Sintering does not commonly begin until the temperature exceeds one half to two thirds of the melting temperature of the material. This temperature would be sufficient to cause a significant atomic diffusion for solid state sintering or significant diffusion and viscous flow when a liquid phase is present or produced by a chemical reaction. In the presintering process, the material changes due to heating prior to sintering mainly include vaporization of chemically combined water from the surface of particles or formed within inorganic phases containing water of crystallization. Therefore, care must be taken to prevent the precursor from cracks or fractures because of the stresses from the pressure of the vapour evolved or from differential thermal expansion of phases. The initial heating may remove any liquid remaining after forming and drying the membrane precursor, and any moisture adsorbed from the atmosphere during transporting and setting. The adsorbed moisture may persist in the membrane precursor up to a temperature exceeding 200 °C.

2.9.3.2 Thermolysis

Thermolysis is a process which burns out the organic components such as binder, dispersant, etc. It is an important step prior to densification on sintering. Incomplete binder removal and uncontrolled thermolysis may introduce membrane defects. These defects may impair the performance of the membrane. With the proper choice of binders for the membrane and controlled heating in an appropriate atmosphere, the membrane precursors should survive thermolysis without deformation, distortion, formation of cracks or expanded pores. Binder burnout is very dependent on the composition of the binder material and the composition and flow of the gas surrounding the precursor. It is also highly dependent on the microstructure of the organic, powder and porosity phases and dynamic changes in the microstructure as the binder is eliminated. A schematic diagram of a simplified model for the binder burnout process is presented in Figure 2.18. The thermochemistry of the binder and additives, binder concentration, precursor dimensions, precursor configuration, heating rate and furnace atmosphere may influence the thermolysis behaviour. Precursors containing low amounts of a binder would open pore channels sufficiently to allow the transport of vapours and gases between the reaction zone and the membrane surface. The time for thermolysis is controlled by the diffusion length for vapour phase transport rather than the characteristic dimension of the binder phase. Decomposition and vaporization of the organics causes an internal gas pressure which depends on the gas evolution rate, gas permeability and the precursor size. The gas permeation rate is much lower in dense precursors of very fine particles. In precursors containing a higher organic content, the weight loss is much higher and the time for thermolysis without bubbling is relatively long. Thermolysis of an organic binder in an inert

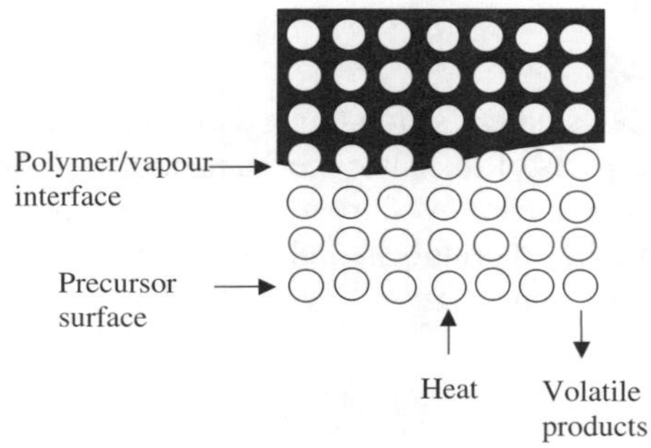

Figure 2.18 A simple model of binder burnout process

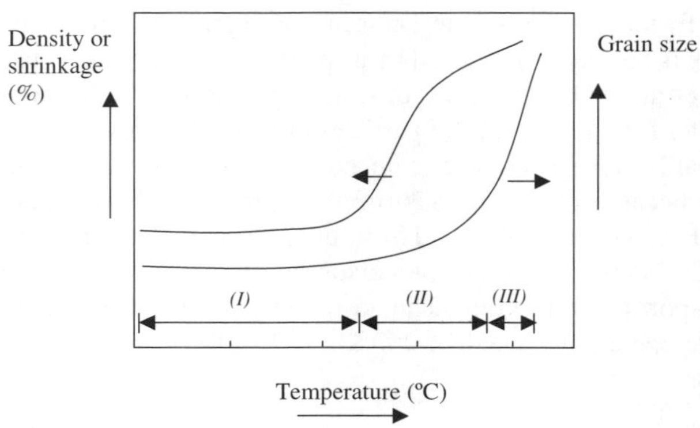

Figure 2.19 Density or shrinkage and grain size of powder compact as a function of the sintering temperature

atmosphere such as nitrogen, as is used when calcining silicon nitride and aluminium nitride ceramics, proceeds differently. Some residual carbon may be expected in the absence of oxygen. The residual carbon commonly removes adsorbed oxygen from surfaces of particles and serves as a sintering aid in carbide nitride ceramics.

2.9.3.3 Final Sintering

Sintering is normally thought to occur in three stages: (1) the initial stage; (2) the intermediate stage and (3) the final stage [43]. The movements of particles have different features at each stage and are attributed to full density, grain coarsening and pores closing. The temperature programme of the sintering process is different for different materials. Rapid densification with limited grain growth occurs at lower temperature followed by rapid grain growth with little densification at higher temperature as shown in Figure 2.19. The characteristics of each stage are summarized in Table 2.4.

There are at least six different mechanisms for mass transport during sintering as listed in Table 2.5. All of these would lead to growth of necks between particles. As shown in Figure 2.20, neck growth produces bonding between the particles, so the strength of the consolidated powders increases during sintering. Only certain mechanisms, however, lead to shrinkage

Table 2.4 Microstructural changes observed in solid state sintering

Stage	Observations
Initial: rapid interparticle neck growth by diffusion, vapour transport, plastic flow or viscous flow.	• Surface smoothing of particles • Grain boundaries form, neck growth • Rounding of interconnected, open pores • Porosity decreases <12 % and linear shrinkage of 3–5 % • Density up to 0.65 of the theoretical
Intermediate: begins when the pores have reached their equilibrium shapes as dictated by the surface and interfacial tensions. It normally covers the major part of the sintering process. Shrinkage is induced by grain growth and a change in the pore geometry. More than one mass transport mechanism may be contributing significantly to the changes in microstructure.	• Equilibrium pore shape with continuous porosity • Shrinkage of open pores intersecting grain boundaries • Mean porosity decreases significantly • Slow grain growth (differential pore shrinkage, grain growth in heterogeneous material) • Density range from 0.65–0.90 of the theoretical
Final: the pores are assumed to shrink continuously and may disappear altogether	• Equilibrium pore shape with isolated porosity • Closed pores intersect grain boundaries • Pores shrink to a limited size or disappear • Pores larger than grains shrink relatively slowly • Grains of much larger size appear rapidly • Pores within larger grains shrink relatively slowly

Table 2.5 Mass transport mechanisms in sintering

Mechanism	Densification
Surface diffusion	No
Evaporation–condensation	No
Grain boundary diffusion	Yes
Lattice diffusion	Yes
Viscous flow	Yes
Plastic flow	Yes

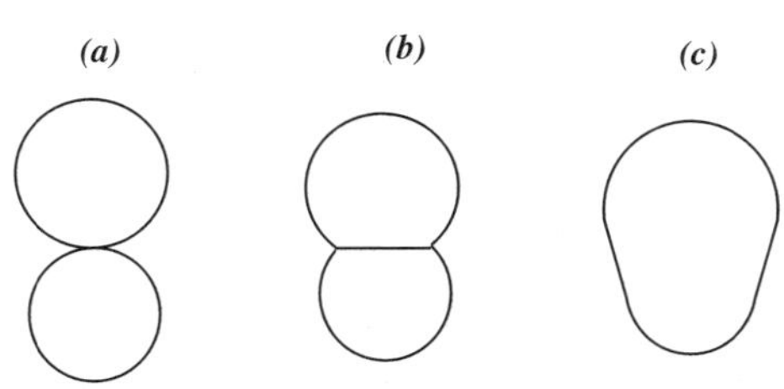

Figure 2.20 A two sphere sintering model for a qualitative mechanism for grain growth in porous powder compacts: (a) particles of slightly different size in contact; (b) neck growth by surface diffusion between particles; (c) grain growth

and densification. Surface diffusion is a general transport mechanism that can produce surface smoothing, particle joining and pore rounding, but it does not produce volume shrinkage. In materials where the vapour pressure is relatively high, sublimation and vapour transport to surfaces of lower vapour pressure also produce these effects. Diffusion along the grain boundaries and diffusion through the lattice of the grains produce both neck growth and volume shrinkage. The mechanisms of bulk viscous flow and plastic deformation may be effective when a wetting liquid is present and when a mechanical pressure is applied, respectively.

2.9.3.4 Grain Growth

The properties of ceramic membranes are strongly dependent on the microstructure. The important microstructural features are the size and shape of the grains, the porosity, the pore size and the distribution of the pores in the structure. In order to achieve the desired properties during the fabrication of ceramic membranes, the microstructural features must be controlled. For most applications, microstructural control usually means the achievement of as high a density, as small a grain size and as homogeneous a microstructure as possible. There are a few factors that should be understood, such as how the grains grow, the interaction between the pores and the moving grain boundaries, and the coarsening process.

Grain growth is the term used to describe the increase in the grain size of a single phase solid or in the matrix grain size of a solid containing second phase particles at sufficiently high temperature. For the conservation of matter, the sum of the individual grain sizes must remain constant; therefore, the increase in the average size of the grains is always accompanied by the disappearance of some grains, usually the smaller ones. In porous solids, both the grains and the pores normally increase in size while decreasing in number. Frequently the term coarsening is used to describe the progress by which grains and pores grow. The driving force for grain growth is the decrease in grain boundary energy that results from a decrease in the grain boundary area. Grain growth in ceramic membranes is generally divided into two types: normal and abnormal. In normal grain growth, the grain sizes and shapes fall within a fairly narrow range and the distribution of grain sizes at a later time is fairly similar to that at an early time. In abnormal grain growth, a few large grains develop and grow fairly rapidly at the expense of the smaller ones. The grain size distribution may change significantly.

For stable structures, the grain boundary energy of all grain boundaries should be the same. The edges must meet at an angle of 120°. Taking N as the number of edges, when $N = 6$, grain boundaries are straight. Gains with $N > 6$ have concave boundaries, and those with $N < 6$ have convex boundaries. Since the boundaries migrate toward their centre of curvature, grains with $N > 6$ tend to grow while those with $N < 6$ tend to shrink.

2.9.3.5 Pore Elimination or Evolution

Following the above description of the grain growth, pore change in the sintering process should be discussed, because it is a direct reflection of membrane performance. In the case of a grain surrounded by other grains, a pore is surrounded by N number of grains. N is normally called the pore coordination number. The pore has straight sides if $N = 6$, convex sides for $N < 6$, and concave sides for $N > 6$ as shown in Figure 2.21. Since the surface of the pore

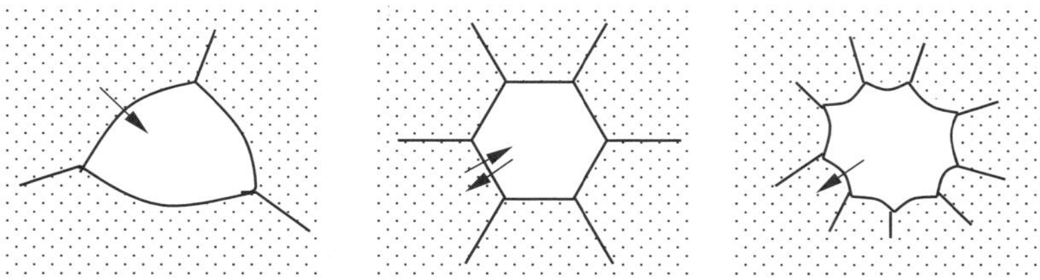

Figure 2.21 Pore stability in two dimensions for a dihedral of 120°

moves towards to its centre of curvature, a pore with $N < 6$ will shrink whereas one with $N > 6$ will grow. The pore is metastable for $N = 6$, and this number is called the critical pore coordination number, N_c. For making a dense membrane, shrinkable pores are favourable. It is helpful to obtain this pore by careful control of the particle packing and the grain growth in the sintering process.

2.9.4 Example 1: Preparation of Porous Al_2O_3 Hollow Fibre Membranes

Studies on the formation principles were first started with a spinning suspension containing only 1 μm of Al_2O_3 powders because of the complicity of the fabrication process. They were further explored using spinning suspensions containing Al_2O_3 with different sizes. By addition of different weight ratios of 1, 0.3 and 0.01 μm Al_2O_3 particles in the spinning suspension and sintering at an appropriate temperature, Al_2O_3 hollow fibre membranes with excellent properties have been prepared [44, 45].

Materials used in preparation of Al_2O_3 hollow fibre membranes were obtained commercially. Al_2O_3 powders with three different particle diameters of 0.01 μm (gamma/alpha, surface area 100 $m^2 g^{-1}$), 0.3 μm (gamma/alpha, surface area 15 $m^2 g^{-1}$) and 1 μm (alpha, surface area 10 $m^2 g^{-1}$) (Alfa AESAR, a Johnson Matthey company), polyethersulfone (PESf) (Radel A-300, Ameco Performance, USA), N-methyl-2-pyrrolidone (NMP) (Synthesis Grade, Merck) and polyvinylpyrrolidone (PVP, K90) (GAF® ISP Technologies, Inc., $M_w = 630000$) were used as membrane materials, binder, solvent and additive, respectively. Tap water was used as both the internal and external coagulants.

2.9.4.1 Morphology

A morphology study of Al_2O_3 hollow fibres has been performed using scanning electron microscopy. Figures 2.22 and 2.23 show the SEM micrographs of the Al_2O_3 hollow fibre precursors and their sintered fibres, spun from two different spinning suspensions with one containing PESf 10 %, PVP 0.5 %, NMP 39.5 %, and 1 μm Al_2O_3 50 %, and the other containing PESf 10 %, PVP 0.5 %, NMP 39.5 % and 0.3 μm Al_2O_3 50 % (weight %), respectively. The sintering process was carried out in air at a temperature of 1500 °C. It can be seen from the micrograph in Figure 2.22(AI) that the OD and ID of the fibre precursor prepared from 1 μm particles were measured to be 1287 μm and 847 μm and were shrunk to 1044 μm and 726 μm (Figure 2.22(BI)), respectively after sintering. Similarly, Figures 2.23(AI) and

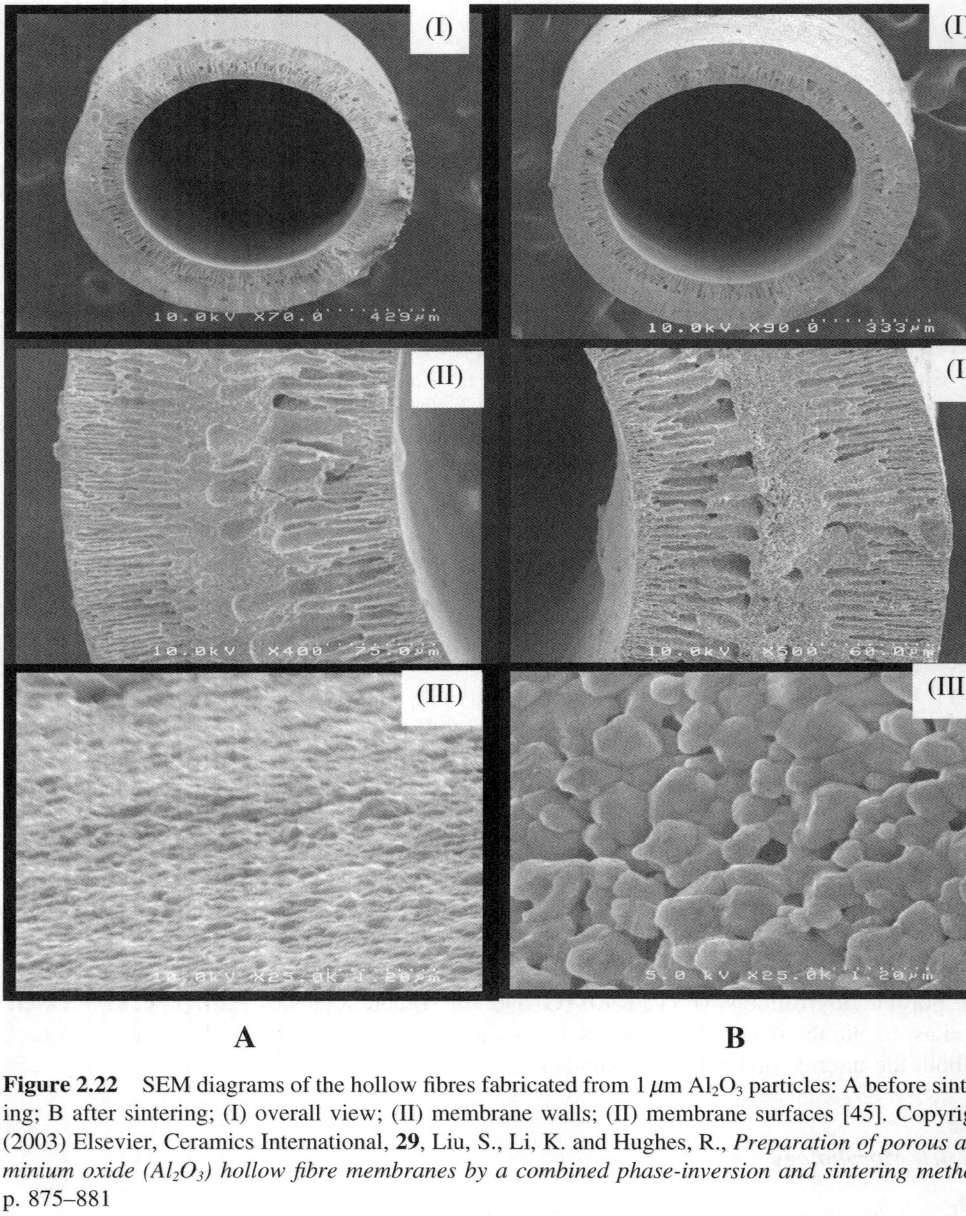

A B

Figure 2.22 SEM diagrams of the hollow fibres fabricated from 1 μm Al_2O_3 particles: A before sintering; B after sintering; (I) overall view; (II) membrane walls; (II) membrane surfaces [45]. Copyright (2003) Elsevier, Ceramics International, **29**, Liu, S., Li, K. and Hughes, R., *Preparation of porous aluminium oxide (Al_2O_3) hollow fibre membranes by a combined phase-inversion and sintering method*, p. 875–881

2.23(BI) depict that the OD and ID of the fibre precursor and the sintered fibre prepared from 0.3 μm particles were shrunk from 1705 μm and 1118 μm to 1333 μm and 911 μm, respectively. Further, comparing the fibre dimensions of the precursors and the sintered ones reveals that the fibres prepared from 0.3 μm particles have higher shrinkage during the sintering process. This is probably due to the fact that for the same weight of Al_2O_3 added into a given volume of the binder solution, the smaller particle size (0.3 μm) would leave higher voids compared to the larger particle size (1 μm). As the voids are generally occupied by the

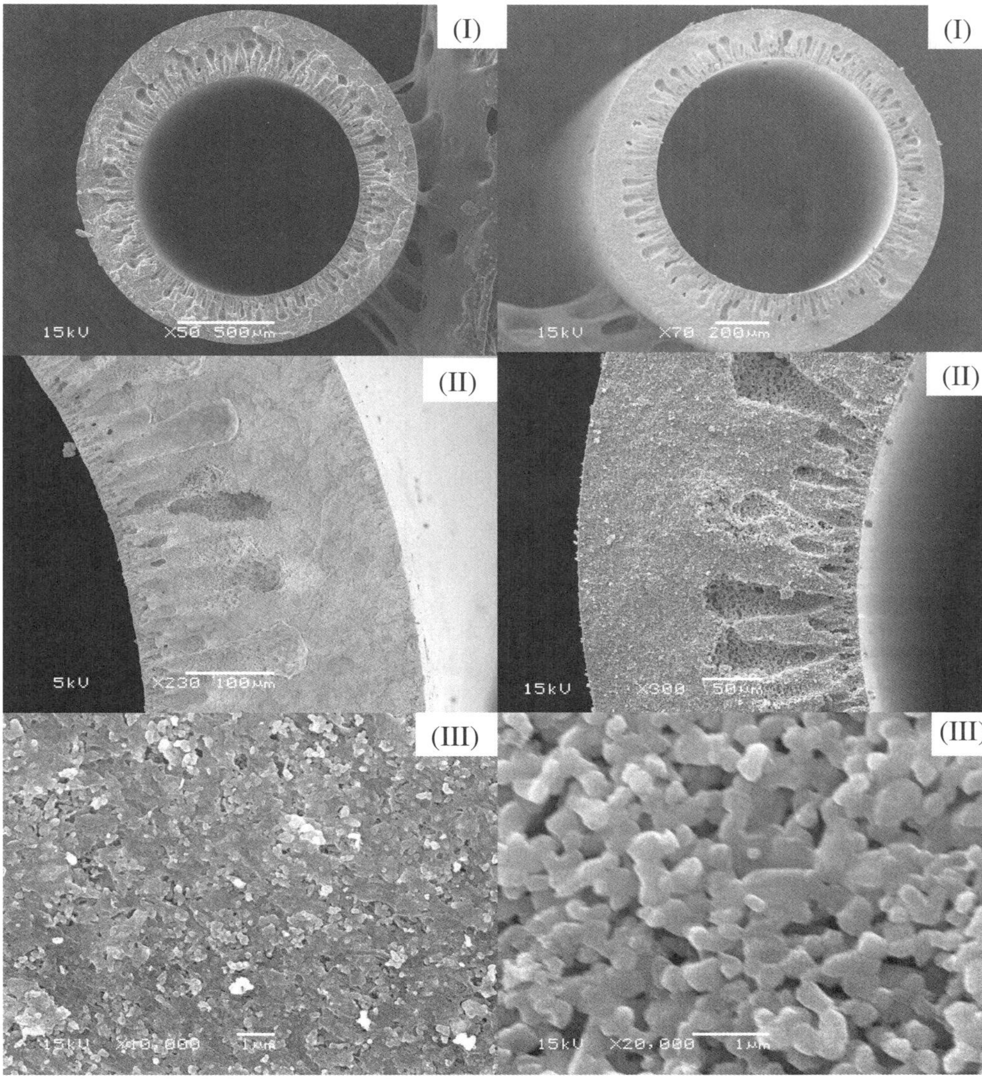

A B

Figure 2.23 SEM diagrams of the hollow fibres fabricated from 0.3 μm Al_2O_3 particles: A before sintering; B after sintering; (I) overall view; (II) membrane walls; (III) membrane surfaces [45]. Copyright (2003) Elsevier, Ceramics International, **29**, Liu, S., Li, K. and Hughes, R., *Preparation of porous aluminium oxide (Al_2O_3) hollow fibre membranes by a combined phase-inversion and sintering method*, p. 875–881

polymer binder, which is ultimately removed during the sintering process, higher shrinkage of the resulting fibres is, thus, expected for 0.3 μm particles.

Cross sectional structures of the fibre precursor and the sintered fibre prepared from 1 μm particles are shown in Figures 2.22(AII) and 2.22(BII). It can be seen from Figure 2.22(AII) that near the outer and inner walls of the fibre precursor, long finger like structures are present and that at the centre of the hollow fibre precursor, sponge like structures appear. The appearance of the fibre structures shown in the figure can be attributed to the rapid precipitation occurred at both the inner and outer fibre walls resulting in long fingers and to the slow precipitation giving the sponge like structures at the centre of the fibre. It is interesting to note that the cross sectional structure of the fibre precursor is well maintained after the sintering process as shown in Figure 2.22 (BII). Figures 2.23(AII) and 2.23(BII) depict the cross sectional structures of the fibre precursor and the sintered fibre prepared from 0.3 μm particles. Compared to the fibres prepared from 1 μm particles, it clearly shows that after the sintering, the fibre prepared from 0.3 μm particles is much denser, which probably results from higher shrinkage during the sintering process. Compared to the traditional flat and tubular membranes, Al_2O_3 hollow fibre membranes have the advantage of higher surface area/volume ratios. For example, the sintered hollow fibre membranes from Figures 2.22B and 2.23B have the surface area per membrane volume ratios of 3831 and 3000 $m^2 m^{-3}$, respectively. When these hollow fibre membranes are assembled in the modules, they are very compact and have a practical surface/volume ratio of >3000 $m^2 m^{-3}$.

Figure 2.22 (AIII) shows the surface of the hollow fibre precursor. It can be seen that the surface of the precursor is similar to conventional polymeric membranes prepared via the phase inversion technique. Further comparison of the SEM photos, especially for the surfaces of the precursors (Figures 2.22 (AIII) and 2.23 (AIII)) and the sintered fibre (Figures 2.22 (BIII) and 2.23 (BIII)) reveals that the pore quantity and pore size may have changed after the sintering process although the general structure is maintained. Such structure changes depend on the particle sizes, composition of the particle suspension and the sintering temperature, which are discussed below.

2.9.4.2 Mechanical Strength

The mechanical strength of Al_2O_3 hollow fibre membranes is one of the important issues in preparation of ceramic hollow fibre membranes. It is dependant on several preparation conditions such as Al_2O_3 particle size and its size distributions, the Al_2O_3 content in the spinning suspension and sintering environment (i.e. sintering temperature and time etc.). Experimental data on the hollow fibre mechanical strength and permeation properties is shown in Figure 2.24 where the hollow fibre membrane was prepared from 0.3 μm particles. It can be seen that an increase in sintering temperature enhances the mechanical strength. For example, at temperatures of 1300 °C and 1550 °C sintered for 10 hours, the three point (3P) values are 20.9 and 80.9 MPa, respectively. When the sintering is less than 1500 °C, the 3P value is proportional to the sintering temperature. When the sintering temperature is greater than 1550 °C, the 3P value increases sharply as the sintering temperature is further increased. It thus suggests that preparation of Al_2O_3 hollow fibre membranes with high mechanical strength is possible at a sintering temperature of 1550 °C or higher, as a slight increase in temperature dramatically increases the mechanical strength. Sintering of the hollow fibre at a temperature of 1600 °C shows, of course, an increase in the mechanical strength, however, gas permeability is decreased considerably as shown in Figure 2.24. Therefore, there is a trade-off between the mechanical strength and gas permeability. The 3P values at various sintering temperatures suggest that a sintering temperature of 1550 °C would give sufficient strength for the fibre to be fabricated into a module without breaking.

The hollow fibre precursor formed through the phase inversion technique contains the Al_2O_3 powder and the PESf binder. During the sintering process, the PESf is removed and

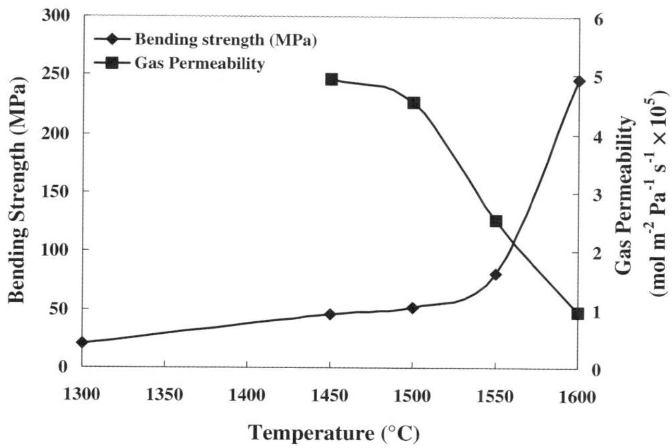

Figure 2.24 Effect of sintering temperature on the mechanical strength and gas permeability of hollow fibre membranes prepared from 0.3 μm Al_2O_3 powder at an Al_2O_3/PESf ratio of 5, sintered for 10 hours [45]. Copyright (2003) Elsevier, Ceramics International, **29**, Liu, S., Li, K. and Hughes, R., *Preparation of porous aluminium oxide (Al_2O_3) hollow fibre membranes by a combined phase-inversion and sintering method*, p. 875–881

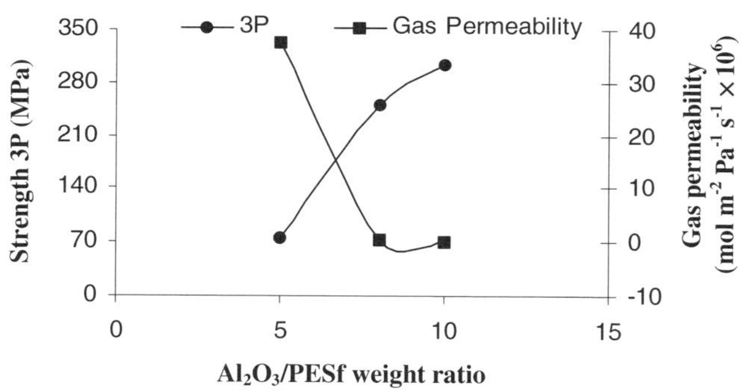

Figure 2.25 Effect of Al_2O_3 content on the mechanical strength and gas permeability of hollow fibre membranes prepared from 1.0 μm Al_2O_3 particles, sinter temperature of 1550 °C [45]. Copyright (2003) Elsevier, Ceramics International, **29**, Liu, S., Li, K. and Hughes, R., *Preparation of porous aluminium oxide (Al_2O_3) hollow fibre membranes by a combined phase-inversion and sintering method*, p. 875–881

the Al_2O_3 hollow fibre is ultimately formed. Therefore, the Al_2O_3 content in the spinning suspension plays an important role in determining its mechanical strength. Figure 2.25 illustrates the effect of Al_2O_3 content on the fibre mechanical strength and its gas permeability. It can be seen that the 3P value enhances greatly, as the Al_2O_3/PESf ratio is increased. Compared to the sintering temperature, the increase of the Al_2O_3 powder content in the spinning suspension results in a much more obvious effect on the fibre mechanical strength. It therefore follows that in order to produce an Al_2O_3 hollow fibre membrane with higher mechanical strength, a higher Al_2O_3 content in the suspension must be maintained. At an Al_2O_3/PESf ratio of 7 or grater, reduction in gas permeability is tailed, indicating that the membrane is transformed to a much denser structure.

Table 2.6 Experimental results for preparation of Al_2O_3 hollow fibre membranes

No.	Al_2O_3 powders 1/0.3/0.01 (μm)	Sintering temperature (°C)	Gas permeability mol m^{-2} Pa^{-1} s^{-1} (N_2, 1 atm)	Bending strength σ_F (MPa)
AI	100/0/0	1500	6.50E-5	48.6
BI	50/50/0	1500	3.90E-5	59.3
CI	0/100/0	1500	4.53E-5	51.3
DI	46.5/46.5/7	1500	2.78E-5	72.8
AII	100/0/0	1550	3.75E-5	76.1
BII	50/50/0	1550	3.59E-5	93.6
CII	0/100/0	1550	2.54E-5	80.9
DII	46.5/46.5/7	1550	1.60E-5	107.0
AIII	100/0/0	1600	1.15E-5	182.4
BIII	50/50/0	1600	0.98E-5	253.1
CIII	0/100/0	1600	0.95E-5	246.2
DIII	46.5/46.5/7	1600	0.22E-5	341.7

2.9.4.3 Effect of Particle Size and Size Distribution

Pores in sintered ceramic hollow fibre membranes are voids left between packed particles, having neither a regular shape, nor a regular size. The particle size and size distribution of the Al_2O_3 powders in the spinning suspension play an important role in determining the membrane wall density and gas permeability. It has been shown [20] that a higher Al_2O_3 content in the spinning suspension produces a denser membrane with a lower porosity and a smaller pore size. When the Al_2O_3 content is too high, it is difficult to form a hollow fibre precursor using the phase inversion process. An Al_2O_3/PESf ratio of 5 has, therefore, been suggested and employed for preparation of an Al_2O_3 hollow fibre membrane with a meaningful mechanical strength and appreciable permeation characteristics using a spinning suspension containing different sizes of Al_2O_3 powders. The experimental results are given in Table 2.6. It indicates that the gas permeability decreases and the mechanical strength of the hollow fibre increases as the average particle size of the Al_2O_3 powders is decreased. The results given in Table 2.6 clearly correspond with the SEM graphs presented earlier. Further comparison of the experimental results given in Figure 2.25 and Table 2.6 reveals that blending the smaller Al_2O_3 particles in the spinning suspension is a more effective way of producing an Al_2O_3 hollow fibre membrane with increased mechanical strength without considerably losing its permeation characteristics.

2.9.5 Example 2: Preparation of TiO_2/Al_2O_3 Composite Hollow Fibre Membranes

The Al_2O_3 hollow fibre membranes developed above can be used not only in microfiltration and ultrafiltration applications at high temperatures, but can also serve as porous supports for composite membrane formation. The roughness of the prepared hollow fibre substrate may decide the minimum achievable coating layer thickness. The surface roughness of the sintered membrane substrate is related to many factors, such as the Al_2O_3 content in the spinning suspension, the particle size of the primary powder, sintering temperature, sintering time and so on. In general, hollow fibres fabricated from a low Al_2O_3 content in the spinning suspension, a smaller size in the primary powder, a higher sintering temperature and a longer sintering time show a smoother surface. Although quantitative data of the membrane surface roughness for hollow fibres fabricated at different conditions is not available, hollow fibre membranes prepared from spinning suspensions containing 0.01 or 0.3 μm particles have shown a smoother outside surface compared to hollow fibres prepared using the spinning suspensions containing only 1 μm particles. Therefore, Al_2O_3 hollow fibres prepared from particle diameters of 0.3 μm have been chosen as the substrate for TiO_2/Al_2O_3 composite hollow fibre membrane preparation and the results are given below.

Direct deposition techniques in supersaturated chemical solutions to prepare metal oxide films, such as SiO_2 [46], SnO_2 [47], FeOOH [48], V_2O_5 [49] and TiO_2 [50, 51] have been recently developed. In particular, thin films of well crystallized anatase titania particles have been easily obtained from aqueous solutions of titanium tetrafluoride [51] at relatively low temperatures (i.e., 60 °C). The chemical reactions of TiF_4 hydrolysis in formation of TiO_2 are understood, and take place as in the following steps [51]:

$$TiF_4 \rightarrow Ti(OH)_{4-x}Fx \rightarrow TiO_2 \tag{2.3}$$

Using this technique, TiO_2 layers can be deposited on various substrates with complex shapes through heterogeneous nucleation. Based on the same principle, TiO_2/Al_2O_3 composite hollow fibre membranes have been prepared.

Figure 2.26 (SEMs) shows the growth of TiO_2 layers coated on the outside surface of a hollow fibre substrate with different deposition times. Comparing the microstructures of the Al_2O_3 hollow fibre membranes before and after deposition, the growth of the TiO_2 membrane is very clear, particularly for the samples with a deposition time of 96 hours (Figures 2.26e and 2.26f). Figure 2.26 also illustrates that the thickness of the TiO_2 layer increases with the deposition time. An X-ray diffraction pattern for the deposited membranes on the Al_2O_3 substrate indicated that the deposited membranes consisted of anatase particles with high crystallinity. As shown in Figure 2.27, before deposition, the crystalline phase of the fibre belongs to α alumina (Figure 2.27b) and after deposition, the small particles forming the coating layer are identified as anatase TiO_2. Due to the hydrophilic properties of the prepared Al_2O_3 hollow fibres, the formed TiO_2 membranes showed excellent adhesion to the hollow fibre substrate surface.

As suggested by Shimizu *et al.* [51], the TiO_2 growth initially occurred at the surface of the substrate and the pores between the Al_2O_3 grains in the hollow fibre surface were gradually filled up by TiO_2 particles during the deposition. Therefore, the gas permeability of the deposited hollow fibre membranes would become smaller compared to that of the original substrate. A few repetitions of the deposition and drying procedure for the same hollow fibre turned the membrane from porous to almost gas-tight as shown in Table 2.7.

2.9.6 Example 3: Preparation of Dense Perovskite Hollow Fibre Membranes

2.9.6.1 Dense $La_{0.6}Sr_{0.4}Co_{0.2}Fe_{0.8}O_{3-\alpha}$ (LSCF) Hollow Fibre Membranes

The LSCF hollow fibre membranes were prepared using the method described above. The spinning suspension consisted of 70.49 wt% LSCF powders, 5.81 wt% polyethersulfone

Figure 2.26 SEM pictures of TiO_2/Al_2O_3 composite hollow fibre membranes with different deposition times (a, b, 17 hours; c, d, 48 hours; e, f, 96 hours; a, c, e: surface view; b, d, f: cross section view; samples are dried at 60 °C for 24 hours) [23]. Copyright (2003) Elsevier, Journal of Membrane Science, **218** (1–2), Liu, S. M. and Li, K., *Preparation of* TiO_2/Al_2O_3 *composite hollow fibre membranes*, p. 269–277

(PESf), 23.26 wt% 1-methyl-2-pyrrolidinone (NMP) and 0.44 % polyvinyl pyrrolidone (PVP, K30). The membrane precursors were sintered at 1280 °C for 4 hours.

Figure 2.28 shows SEM photographs of the prepared LSCF hollow fibre membrane before and after sintering. It can be seen clearly that the asymmetric structure has been formed in the process of phase inversion, as shown in Figure 2.28(A). After sintering, the asymmetric structure is retained, but with changes in the voids due to shrinkage, as shown in Figure

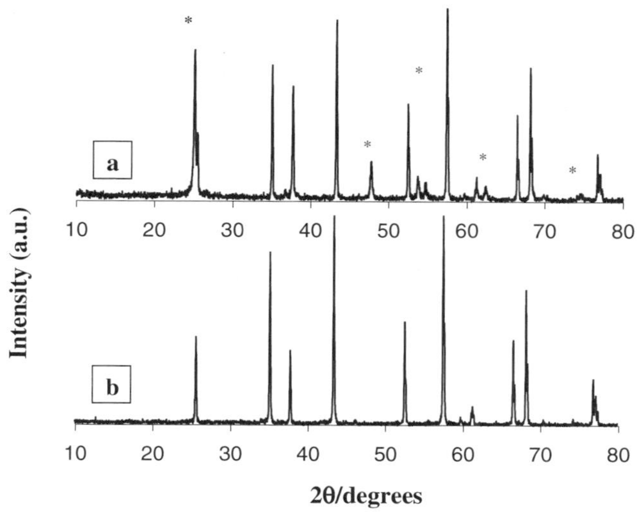

Figure 2.27 XRD patterns for (a) Al_2O_3 and anatase TiO_2 coatings (*: anatase TiO_2), and (b) Al_2O_3 [23]. Copyright (2003) Elsevier, Journal of Membrane Science, **218** (1–2), Liu, S. M. and Li, K., *Preparation of TiO_2/Al_2O_3 composite hollow fibre membranes*, p. 269–277

Table 2.7 Gas permeability changes with the number of coating times

Coating times	Deposition time (hours)	Drying temperature (°C)	Gas permeability($\times 10^7$) mol m^{-2} Pa^{-1} s^{-1} (N_2)
1st	96	100	11.7
2nd	59	100	3.7
3rd	48	100	0.1
4th	48	100	<<0.1

2.28(B). The membrane outer surface shown in Figure 2.28(C) indicates that the LSCF particles are closely connected to each other with clear boundaries.

Figure 2.29 compares the XRD patterns of the sintered membranes with the original LSCF powders. As can be seen, no changes are observed in the crystalline structure of the LSCF material due to the spinning and sintering processes after the hollow fibre membranes are formed. However, the intensities of the corresponding characteristic peaks for the hollow fibre membranes are much higher than that for the original LSCF powders. This suggests that the crystallinity in the membranes has become larger after sintering at high temperatures.

2.9.6.2 Dense $SrCe_{0.95}Yb_{0.05}O_{3-\alpha}$ (SCYb) Hollow Fibre Membranes

SCYb hollow fibre membranes were also prepared using the method described above. The spinning suspension consisted of 69 wt% SCYb powders, 7.0 wt% polyethersulfone (PESf), 23.3 wt% 1-methyl-2-pyrrolidinone (NMP) and 0.7 % polyvinyl pyrrolidone (PVP, K90). The membrane precursors were sintered at 1200 °C for 16 hours and further sintered at

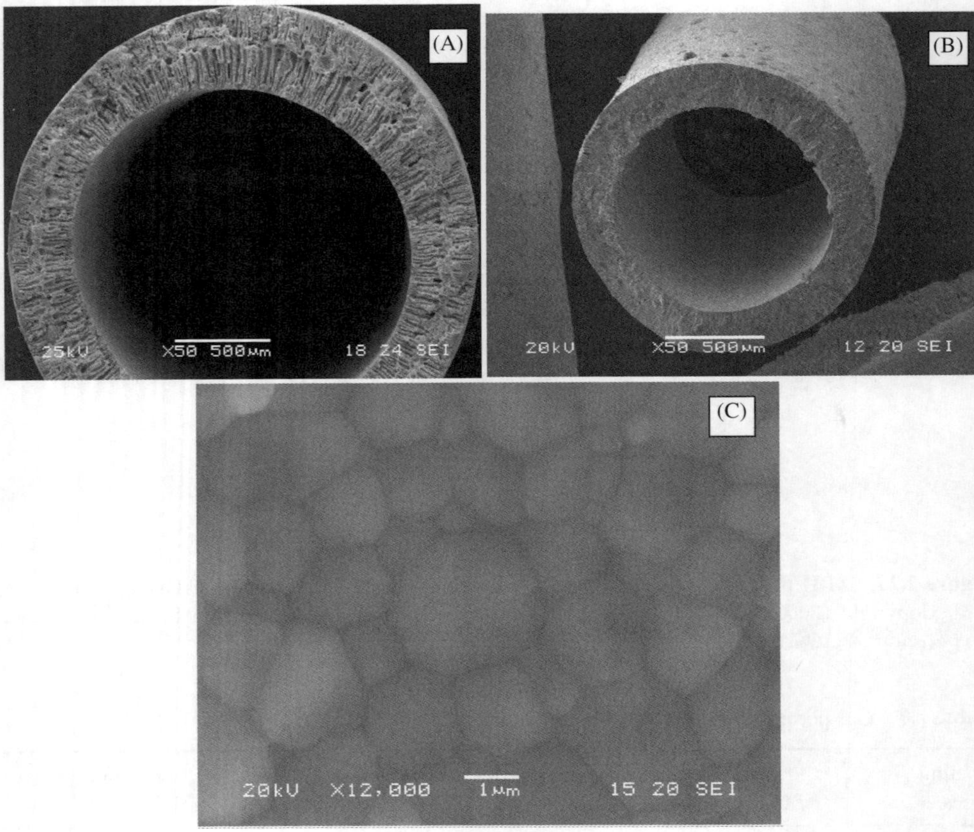

Figure 2.28 SEM micrographs of the LSCF hollow fibre: (A) before sintering; (B) after sintering; (C) membrane surface [57]. Copyright (2005) John Wiley & Sons, Ltd, American Institute of Chemical Engineers Journal, **51** (7), Tan, X., Liu, Y. and Li, K., *Mixed conducting ceramic hollow fibre membranes for air separation*, p. 1991–2000

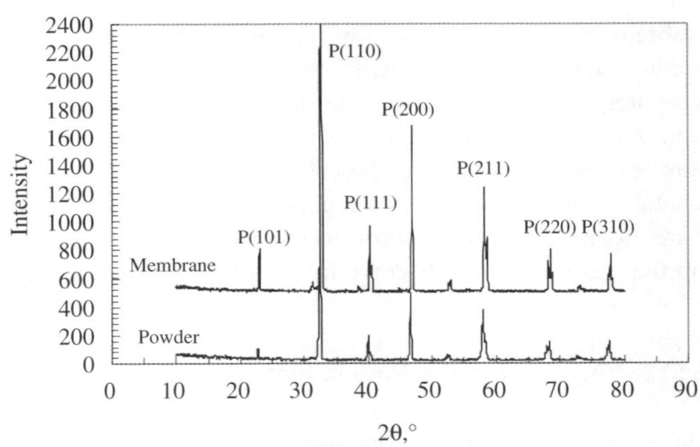

Figure 2.29 XRD patterns of the LSCF powder and of the sintered hollow fibre membrane [57]. Copyright (2005) John Wiley & Sons, Ltd, American Institute of Chemical Engineers Journal, **51** (7), Tan, X., Liu, Y. and Li, K., *Mixed conducting ceramic hollow fibre membranes for air separation*, p. 1991–2000

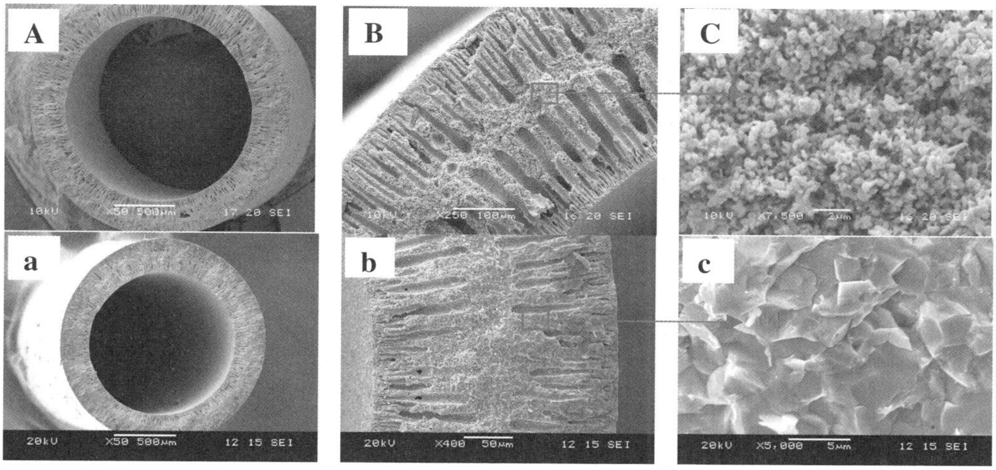

Figure 2.30 SEM photograph of the SCYb hollow fibre membranes before (A, B and C) and after (a. b and c) sintering

1350–1550 °C for another 4 hours to form the hollow fibre membrane product. The heating and cooling rates of sintering were both 5 °C min^{-1}.

Figure 2.30 shows the SEM photographs of the SCYb hollow fibre precursor (Figures 2.30A to 2.30C) and the fibre product sintered at 1450 °C (Figures 2.30a to 2.30c). As can be seen from Figures 2.30(A) and 2.30(a), the dimension shrinkage of the SCYb hollow fibre membrane was very substantial after the sintering process, e.g., the OD/ID of the hollow fibre precursor and sintered fibre were measured to be 2.08/1.50 mm and 1.65/1.20 mm, respectively. Figure 2.30(B) clearly shows the sandwich structure of the fibre precursor, i.e. a middle sponge layer is sandwiched by two integrated finger like porous substrates. Such a structure is obtained by controlling the rate of precipitation at both inner and outer walls. After sintering, the sandwich structure is retained, although the size of the fibre and pore size/numbers are changed as shown in Figures 2.30(a) and 2.30(b). Figures 2.30(C) and 2.30(c) shows the middle layers of the membrane before and after the sintering. As can be seen, the middle sponge layer is greatly densified after sintering as shown in Figure 2.30(c). The central dense layer is relatively thin and functions as the proton hole conductor at elevated temperatures, while the integrated porous support layers are thick and are not only supply the fibre's mechanical strength but also increase the effective catalytic surface areas for the both the up and the down stream surface exchange reactions.

APPENDIX 2.1: SURFACE FORCES

A2.1.1 Electrostatic Forces

Ions and polar molecules in the solution surrounding a particle will respond to the charged surface. Electrostatic forces will repel like charged ions but attract polar liquid molecules and oppositely charged ions into a region near the surface, increasing their concentration relative to that in the bulk. The relatively immobile adsorbed layer of counterions around the charged particles is called Stern layer. The model shown in Figure 2.31 is commonly called an electrical double layer model.

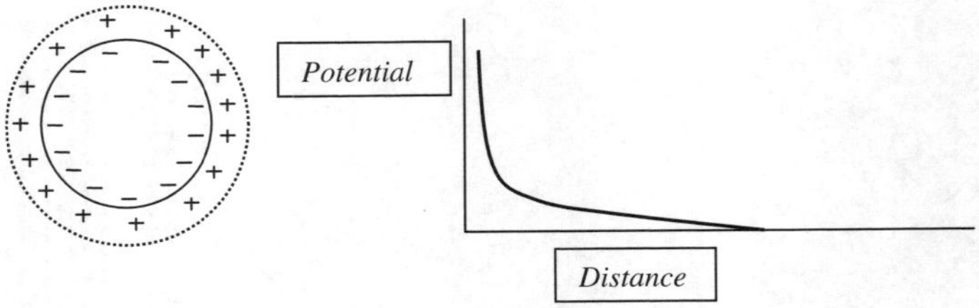

Figure 2.31 Electrical double layer model for particle charging in a polar liquid

The electrostatic forces can be modified by changing the conditions of the liquid, for example, changing the pH of the solution results in the adsorption of ion species onto the surface of the solid particles. These changes can be indicated by zeta potential.

A2.1.2 DLVO Theory and van der Waals Forces

The theory of stability of colloidal dispersion states that colloidal size particles have a tendency to aggregate because of the van der Waals attractive forces between particles, resulting in an increase in the free energy of the system when the particles are separated from each other. This theory distinguishes between lyophobic dispersions, in which the dispersed particles are not significantly solvated, and lyophilic dispersions, in which layers of solvent molecules cover the surface of the particles and help to stabilize them against aggregation. Lyophilic dispersions, which are frequently also stabilized by repulsive interactions between similarly charged electrical double layers surrounding the particles, can be aggregated only by at least partial removal of the solvent layer. This may be accomplished in aqueous media by the addition of relatively large amounts of ions that hydrate readily. The mechanisms of lyophilic dispersions are still not clear. In lyophobic dispersions, the absence of solvent layers means that stabilization against particle aggregation is due solely to repulsive interactions by similarly charged electrical double layers surrounding the particles. As a result, these systems are very sensitive to small amounts of additives, which can either increase the stability of the dispersion or cause it to aggregate, depending on whether they increase or decrease the range and/or the magnitude of the electrical double layer. In the early 1940s, Derjaguin, Landau, Verwey and Overbeek [52] developed a theory of the stability of lyophobic dispersions, therefore often called the DLVO theory. The stability of a lyophobic dispersion is dependent on the particle radius and surface potential, the ionic strength and dielectric constant of the dispersing medium, the value of the Hamaker constant [53] and the temperature. Stability is increased by an increase in the particle radius or surface potential, or in the dielectric constant of the medium and by a decrease in the effective Hamaker constant, the ionic strength of the dispersing liquid, or the temperature. The DLVO theory is very useful in predicting the effect of ionic surfactants on the electrical barriers to aggregation.

A2.1.3 Steric Hindrance

In nonaqueous media of low dielectric constant, electrical barriers to aggregation are usually ineffective and steric barriers are generally required to disperse solid particles. Another very important surface force can exist when organic molecules (surfactants and polymers) are added to a system [54, 55]. These molecules can absorb onto the surface of the particles and alter the properties of the interface. These alterations can influence the van der Waals force and electrostatic force, especially when the molecules contain polar functional groups. The adsorption of organics onto particles can prevent the close approach of particles by providing a physical barrier between the surfaces, which is termed the steric stabilization. It can be applied to most types of suspensions.

REFERENCES

1. Leenaars, A. F. M. and Burggraaf, A. J., *The preparation and characterization of alumina membranes with ultrafine pores. 2. The formation of supported membranes.* Journal of Colloid and Interface Science, **105** (1): 27–40 (1985).
2. Uhlhorn, R. J. R., Veld, M., Keizer, K. and Burggraaf, A. J., *High permselectivities of microporous silica-modified gamma-alumina membranes.* Journal of Materials Science Letters, **8** (10): 1135–1138 (1989).
3. Tiller, F. M. and Tsai, C. D., *Theory of filtration of ceramics.1. slip casting.* Journal of the American Ceramic Society, **69** (12): 882–887 (1986).
4. Larbot, A., *Ceramic processing techniques of support systems for membrane synthesis*, in *Fundamentals of Inorganic Membrane Science and Technology*, Burggraaf, A. J. and Cot, L. Editors. (1996) Amsterdam: Elsevier Science B.V., p. 130.
5. Sahibzada, M., Steele, B. C. H., Hellgardt, K., Barth, D., Effendi, A., Mantzavinos, D. and Metcalfe, I. S., *Intermediate temperature solid oxide fuel cells operated with methanol fuels.* Chemical Engineering Science, **55**: 3077–3083 (2000).
6. Leenaars, A. F. M. and Burggraaf, A. J., *The preparation and characterization of alumina membranes with ultra-fine pores.3. The permeability for pure liquids.* Journal of Membrane Science, **24** (3): 245–260 (1985).
7. Leenaars, A. F. M. and Burggraaf, A. J., *The preparation and characterization of alumina membranes with ultra-fine pores.4. ultrafiltration and hyperfiltration experiments.* Journal of Membrane Science, **24** (3): 261–270 (1985).
8. Leenaars, A. F. M., Keizer, K. and Burggraaf, A. J., *The preparation and characterization of alumina membranes with ultra-fine pores.* Journal of Materials Science, **19** (4): 1077–1088 (1984).
9. Larbot, A., Julbe, A., Guizard, C. and Cot, L., *Silica membranes by the sol-gel process.* Journal of Membrane Science, **44** (2–3): 289–303 (1989).
10. Klein, L. C. and Gallagher, D., *Pore structures of sol-gel silica membranes.* Journal of Membrane Science, **39** (3): 213–220 (1988).
11. Anderson, M. A., Gieselmann, M. J. and Xu, Q., *Titania and alumina ceramic membranes.* Journal of Membrane Science, **39** (3): 243–258 (1988).
12. Moosemiller, M. D., Hill, C. G. and Anderson, M. A., *Physicochemical properties of supported gamma-Al_2O_3 and TiO_2 ceramic membranes.* Separation Science and Technology, **24** (9–10): 641–657 (1989).
13. Gieselmann, M. J., Anderson, M. A., Moosemiller, M. D. and Hill, C. G., *Physicochemical properties of supported and unsupported gamma-Al_2O_3 and TiO_2 ceramic membranes.* Separation Science and Technology, **23** (12–13): 1695–1714 (1988).
14. Elferink, W. J., Nair, B. N., DeVos, R. M., Keizer, K. and Verweij, H., *Sol-gel synthesis and characterization of microporous silica membranes.2. Tailor-making porosity.* Journal of Colloid and Interface Science, **180** (1): 127–134 (1996).
15. DeVos, R. M. and Verweij, H., *Improved performance of silica membranes for gas separation.* Journal of Membrane Science, **143** (1–2): 37–51 (1998).

16. Bonekamp, *Fundamentals of membrane top layer synthesis and processing*, in *Fundamentals of Inorganic Membrane Science and Technology*, Burggraaf, A. J. and Cot, L. Editors. (1996) Amsterdam, the Netherlands: Elsevier Science B.V., p. 141–225.
17. Burggraaf, A. J., *Fundamentals of membrane top layer synthesis and processing*, in *Fundamentals of Inorganic Membrane Scienc and Technology*, Burggraaf, A. J. and Cot, L. Editors. (1996) Amsterdam, the Netherlands: Elsevier Science B.V., p. 259–329.
18. Zhang, G. L. and Cussler, E. L., *Distillation in hollow fibres.* American Institute of Chemical Engineers Journal, **49** (9): 2344–2351 (2003).
19. Way, J. D. and Roberts, D. L., *Hollow fibre inorganic membranes for gas separations.* Separation Science and Technology, **27** (1): 29–41 (1992).
20. Tan, X. Y., Liu, S. M. and Li, K., *Preparation and characterization of inorganic hollow fibre membranes.* Journal of Membrane Science, **188** (1): 87–95 (2001).
21. Liu, S. M., Li, K. and Hughes, R., *Preparation of SrCe0.95Yb0.05O3-alpha perovskite for use as a membrane material in hollow fibre fabrication.* Materials Research Bulletin, **39** (1): 119–133 (2004).
22. Liu, S. M., Li, K. and Hughes, R., *Preparation of porous aluminium oxide (Al_2O_3) hollow fibre membranes by a combined phase-inversion and sintering method.* Ceramics International, **29** (8): 875–881 (2003).
23. Liu, S. M. and Li, K., *Preparation of TiO_2/Al_2O_3 composite hollow fibre membranes.* Journal of Membrane Science, **218** (1–2): 269–277 (2003).
24. Liu, S. M., Tan, X. Y., Li, K. and Hughes, R., *Preparation and characterisation of SrCe0.95Yb0.05O2.975 hollow fibre membranes.* Journal of Membrane Science, **193** (2): 249–260 (2001).
25. Hsieh, H. P., *Inorganic Membranes for Separation and Reaction.* (1996) Amsterdam, the Netherlands: Elsevier Science B.V., p. 23–86.
26. Rahaman, M. N., *Ceramic Processing and Sintering.* (1995) New York: Marcel Dekker Inc., p. 265–279.
27. Bideau, D. and Troadec, J. P., *Compacity and Mean Coordination-Number of Dense Packings of Hard Disks.* Journal of Physics C, Solid State Physics, **17** (28): L731–L735 (1984).
28. Visscher, W. M. and Bolsterl, M., *Random packing of equal and unequal spheres in 2 and 3 dimensions.* Nature, **239** (5374): 504 (1972).
29. Kausch, H. H., Fesko, D. G. and Tschoegl, N. W., *Random packing of circles in a plane.* Journal of Colloid and Interface Science, **37** (3): 603 (1971).
30. Starr, T. L., *Packing density of fibre powder blends.* American Ceramic Society Bulletin, **65** (9): 1293–1296 (1986).
31. Lee, B. I. and Rives, J. P., *Dispersion of alumina powders in nonaqueous media.* Colloids and Surfaces, **56**: 25–43 (1991).
32. Mikeska, K. R. and Cannon, W. R., *Dispersants for tape casting barium titanate*, in *Advances in Ceramics*, Mangels, J. A. and Messing, G. L. Editors. (1984) Westerville: The American Ceramic Society, p. 164–183.
33. Bergstrom, L., *Rheological properties of concentrated, nonaqueous silicon nitride suspensions.* Journal of the American Ceramic Society, **79** (12): 3033–3040 (1996).
34. Chong, J. S., Christiansen, E. B. and Baer, A. D., *Rheology of concentrated suspensions.* Journal of Applied Polymer Science, **15** (8): 2007–2021 (1971).
35. Bergstrom, L., *Surface and colloid chemistry*, in *Advanced Ceramic Processing*, Pugh, P. J. and Bergstrom, L. Editors. (1994) New York: Marcel Dekker, p. 193–224.
36. Kitano, T., Kataoka, T. and Shirota, T., *An empirical-equation of the relative viscosity of polymer melts filled with various inorganic fillers.* Rheologica Acta, **20** (2): 207–209 (1981).
37. Sakka, S., *Rheology of sols in sol-gel processing*, in *Sol-Gel Science and Technology*, Aegerter, M. A. *et al.*, Editors. (1989) Singapore: World Scientific Publisher, p. 76–103.
38. Deshmukh, S. P. and Li, K., *Effect of ethanol composition in water coagulation bath on morphology of PVDF hollow fibre membranes.* Journal of Membrane Science, **150** (1): 75–85 (1998).

39. Loeb, S. and Sourirajan, S., *Sea water demineralization by means of an osmotic membrane*, in *Saline Water Conversion – II, Advances in Chemistry*. (1963) Washington DC: American Chemical Society, p. 117–132.
40. Li, K., Tan, X. and Liu, Y., *Single-step fabrication of ceramic hollow fibres for oxygen permeation*. Journal of Membrane Science, **272** (1–2): 1–5 (2006).
41. Richardson, D. W., *Morden Ceramic Engineering*. 1992, Marcel Dekker Inc: New York.
42. Binner, J. G. P., *Advanced Ceramic Processing and Technology*. (1990) New Jersey: Noyes Publications.
43. Reed, J. S., *Principles of Ceramic Processing*. 2nd edn. (1995) New York: John Wiley & Sons, Inc.
44. Tan, X., Liu, S. and Li, K., *Preparation and characterization of inorganic hollow fibre membranes*. Journal of Membrane Science, **188** (1): 87–95 (2001).
45. Liu, S., Li, K. and Hughes, R., *Preparation of porous aluminium oxide (Al_2O_3) hollow fibre membranes by a combined phase-inversion and sintering method*. Ceramics International, **29**: 875–881 (2003).
46. Nagayama, H., Honda, H. and Kawahara, H., *A new process for silica coating*. Journal of the Electrochemical Society, **135** (8): 2013–2016 (1988).
47. Tsukuma, K., Akiyama, T. and Imai, H., *Liquid phase deposition film of tin oxide*. Journal of Non-Crystalline Solids, **210** (1): 48–54 (1997).
48. Deki, S., Aoi, Y., Okibe, J., Yanagimoto, H., Kajinami, A. and Mizuhata, M., *Preparation and characterization of iron oxyhydroxide and iron oxide thin films by liquid-phase deposition*. Journal of Materials Chemistry, **7** (9): 1769–1772 (1997).
49. Deki, S., Aoi, Y. and Kajinami, A., *A novel wet process for the preparation of vanadium dioxide thin film*. Journal of Materials Science, **32** (16): 4269–4273 (1997).
50. Deki, S., Aoi, Y., Hiroi, O. and Kajinami, A., *Titanium(IV) oxide thin films prepared from aqueous solution*. Chemistry Letters, (6): 433–434 (1996).
51. Shimizu, K., Imai, H., Hirashima, H. and Tsukuma, K., *Low-temperature synthesis of anatase thin films on glass and organic substrates by direct deposition from aqueous solutions*. Thin Solid Films, **351** (1–2): 220–224 (1999).
52. Verwey, E. J. and Overbeek, J., *The Theory of the Stability of Lyophobic Colloids*. (1948) Amsterdam: Elsevier.
53. Hamaker, H. C., *The London-Van Der Waals attraction between spherical particles*. Physica, **4**: 1058–1072 (1937).
54. Casey, J. A., *Theoretical basic for control of suspension properties*, in *Dispersion and Aggregation: Fundamentals and Applications*, Moudgil, R. M. and Somasundaran, P. Editors. (1992) p. 169–182.
55. Sato, T. and Ruch, R., *Stabilization of Colloidal Dispersions by Polymer Adsorption*. (1981) Marcel Dekker.
56. Wang, D. L., Li, K. and Teo, W. K., *Highly permeable polyethersulfone hollow fibre gas separation membranes prepared using water as non-solvent additive*. Journal of Membrane Science, **176** (2): 147–158 (2000).
57. Tan, X., Liu, Y. and Li, K., *Mixed conducting ceramic hollow-fibre membranes for air separation*. American Institute of Chemical Engineers Journal, **51** (7): 1991–2000 (2005).

3

Characterization of Ceramic Membranes

3.1 INTRODUCTION

The separation performance of ceramic membranes is directly associated with the overall membrane morphology. Information about the pore size, shape, density, distribution and membrane surface properties for porous ceramic membranes and gas tightness, crystal structures and permeation characteristics of specific gases for dense ceramic membranes are therefore of importance to both membrane manufacturers and users, so as to allow a meaningful prediction of separation performances of the membranes.

For porous ceramic membranes, separation properties are directly determined by characteristics of the porous structure such as pore size, shape, connectivity, particle packing, surface area, etc. Therefore, proper characterization of the porous structures is crucial for the development of commercial ceramic membranes. In general, characterization of supported composite membranes is more demanding than that of most of the symmetric porous membranes because the separation layer is thin, often with anisotropic and microporous structure. Therefore, specialized characterization techniques are required.

For dense ceramic membranes, the gas tightness, crystal structures and microstructure as well as the catalytic activity of the membrane surfaces and conductivities of ion and electrons of specific gases and hence the permeation characteristics are important parameters.

Characterization techniques for ceramic membranes can generally be classified according to the nature of obtained parameters: (1) morphology/structure related parameters and (2) permeation related parameters. Therefore, in this chapter, characterization techniques such as scanning electron microscopy (SEM) or field emission scanning electron microscopy (FESEM), transmission electron microscopy (TEM), atomic force microscopy (AFM) and X-ray diffraction (XRD) will be discussed for morphology/structure related parameters; while for the permeation related parameters, characterization techniques such as mercury porosimetry, permeation, bubble point technique, permporometry, thermoporometry, etc. will be discussed. In addition, bending strength for determining the mechanical strength of ceramic hollow fibre membranes will be briefly introduced.

Ceramic Membranes for Separation and Reaction K. Li

3.2 MORPHOLOGY OF MEMBRANE SURFACES AND CROSS SECTIONS

Morphology of membrane surfaces and cross sections can be visually examined using techniques such as optical or electron microscopy. These techniques provide information about surface roughness and grain size (or shape) of ceramics at the membrane surface as well as layer continuity and thickness of the membrane cross section. They are considered as standard techniques in ceramic membrane characterization.

Optical microscopy is usually used to observe large defects on the membrane surface, as it can only achieve a magnification of 500–1000 times with a good resolution. If a higher magnification is needed to observe the membrane separation layer or its surface texture, electron microscopy can be employed. The most common techniques are scanning electron microscope (SEM) or field emission scanning electron microscopy (FESEM) [1], transmission electron microscopy (TEM) [1] and atomic force microscopy (AFM) [2]. The drawback of scanning and transmission electron microscopy (SEM and TEM) is that sample preparation is required, which can influence the real membrane structure.

The scanning electron microscopy (SEM) is a microscope that uses electrons rather than light to form an image. It can only give qualitative information about the membrane morphology, because the resolution is not good enough at magnifications greater than 10^5 times. Field emission scanning electron microscopy (FESEM) has a much better resolution of about 1.5 nm [3]. For SEM as well as for FESEM, membrane samples have to be dried and then coated with a charge conducting (usually gold) layer if they are nonconductive. The coated layer with a thickness of the order of a nanometer, which is also the size of the smaller pores in the membrane, may cause problems in interpreting the obtained electron micrographs.

Figure 3.1 shows an FESEM of a two layer composite membrane, consisting of an α-alumina macroporous support and a TiO_2 separation layer. The membrane has pores in the ultrafiltration range with a separation layer thickness of around 400 nm [4].

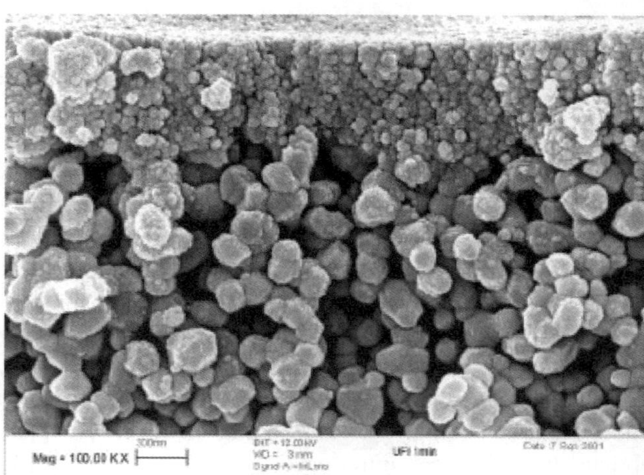

Figure 3.1 FESEM image of a TiO_2 ultrafiltration membrane on alumina support [4]. Reprinted from Advanced Engineering Materials, **6** (7), S. Benfer, P. Arki and G. Tomandl, *Ceramic membranes for filtration applications – preparation and characterization*. p. 495–500. Copyright (2004) with permission from Wiley-VCH

For transmission electron microscopy (TEM), a dry membrane has to be embedded in a resin, after which a thin slice has to be cut from the membrane. TEM has a very high resolution (about 1 nm). However, due to the limitations in sample preparation, such as small cracks in samples caused by the cutting procedure, and the unclear influence of the embedding resin on the membrane structure [1], resolution at 10 nm is often used to interpret the specimen. In TEM, the electrons which form the image must transmit through the specimen. Therefore, a very thin slice of the sample must be used.

Figures 3.2 and 3.3 illustrate a more sophisticated multilayer membrane, consisting of an α-alumina macroporous support, two γ-alumina mesoporous intermediate layers and a

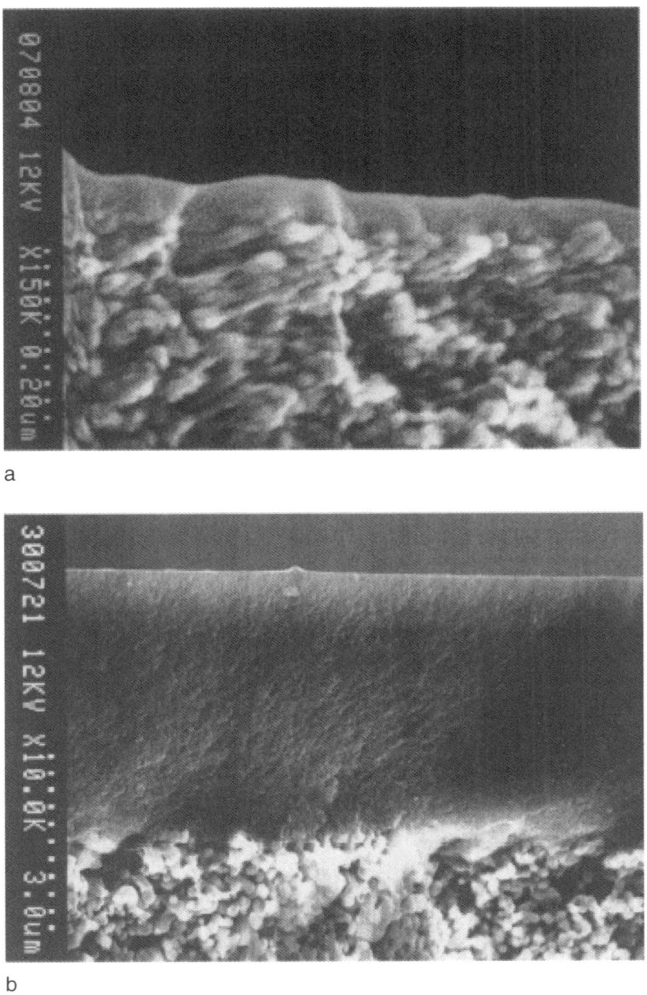

Figure 3.2 SEM micrographs of a Si(400) membrane cross section showing a part of the γ-Al_2O_3 layer and the silica layer at high magnification (a) and the α–Al_2O_3 support and γ-Al_2O_3 layers at lower magnification (b) [42]. Reprinted from Journal of Membrane Science, **143** (1–2), Vos and Verweij, *Improved performance of silica membranes for gas separation*, p. 37–51, Copyright (1998) with permission from Elsevier

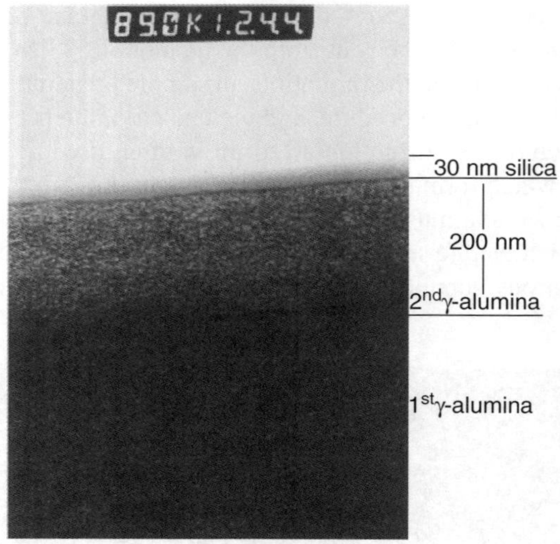

Figure 3.3 TEM micrograph of Si(400) membrane cross section showing a part of γ-Al_2O_3 layer and silica layer [42]. Reprinted from Journal of Membrane Science, **143** (1–2), Vos and Verweij, *Improved performance of silica membranes for gas separation*, p. 37–51, Copyright (1998) with permission from Elsevier

microporous silica top layer using FESEM and TEM, respectively. The separation characteristics of the composite membrane are determined by the microstructure of the top layer, which is formed during the synthesis. The prime reason of preparing a multilayer membrane is to reduce the transport resistance of the permeation species. As shown in Figures 3.2 and 3.3, a very thin silica layer of ~30 nm was obtained. The TEM micrograph indicates that the silica layer is deposited on top of the γ-Al_2O_3 layer. A clear division between silica and γ-Al_2O_3 is also visible on the FESEM. The γ-Al_2O_3 layer is about 3 μm thick. The boundary between the first and second γ-Al_2O_3 layers, at approximately 250 nm from the surface, is clearly visible. The difference between the two γ-Al_2O_3 layers, visible in the TEM micrograph, is very distinct.

Atomic force microscopy (AFM) uses a very high resolution type of scanning probe microscope. It consists of a microscale cantilever with a sharp tip at its end which is used to scan the specimen surface. The cantilever is typically silicon or silicon nitride with the sharp tip radius less than 5 nm [5]. When the tip is brought into the proximity of a sample surface, forces between the tip and the sample lead to a deflection of the cantilever according to Hooke's law [6]. Depending on the situation, forces that are measured in AFM include mechanical contact force, van der Waals forces, capillary forces, chemical bonding, electrostatic forces, solvation forces, etc. The deflection is measured using a laser spot reflected from the top of the cantilever into an array of photodiodes. Traditionally, the sample is mounted on a piezoelectric tube, that can move the sample in the z direction for maintaining a constant force, and the x and y directions for scanning the sample. The resulting map of $s(x, y)$ represents the topography of the sample.

The AFM techniques shows several advantages over the scanning electron microscope (SEM). It can produce images of materials as small as 1 nm, much smaller than that of SEM, which is around 100 nm. Unlike the electron microscope, providing only a two dimensional projection or a two dimensional image of a sample, the AFM provides a three dimensional surface image. In addition, no pretreatment is required for samples to be viewed by AFM. Also,

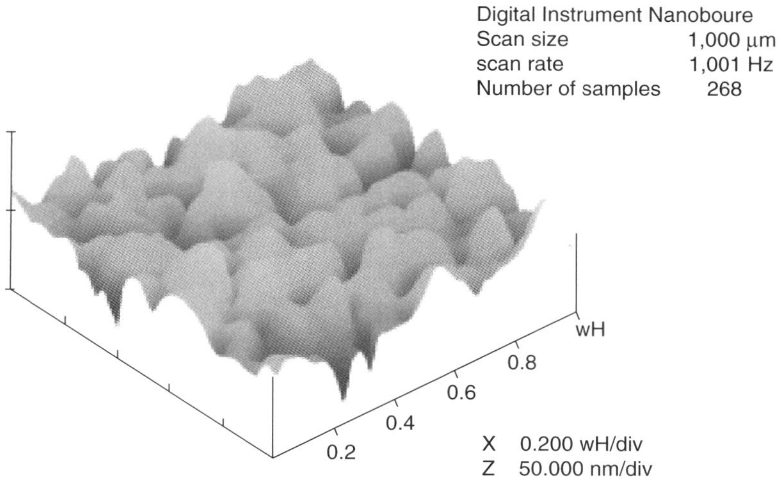

Figure 3.4 AFM images of acetic acid alumoxane derived membranes fired at 1100°C [7]. Reprinted from Journal of Membrane Science, **205** (1–2), Cortalezzi *et al.*, *Characteristics of . . .*, p. 33–43. Copyright (2002) with permission from Elsevier

an electron microscope needs an expensive vacuum environment for proper operation. However, most AFM modes can work perfectly well in ambient air or even a liquid environment.

A disadvantage of AFM compared with the scanning electron microscopy (SEM) is the image size. SEM can image an area of the order of millimetres by millimetres with a depth of field of the order of millimetres. The AFM can only image a maximum height of the order of micrometres and a maximum scanning area of around 150 by 150 μm. Another inconvenience is that at high resolution, the quality of an image is limited by the radius of curvature of the probe tip, and an incorrect choice of tip for the required resolution can lead to image artifacts. Traditionally AFM could not scan images as fast as SEM, requiring several minutes for a typical scan, while SEM is capable of scanning at near real time (although at relatively low quality) after the chamber is evacuated. Finally, high surface roughness of a sample may result in images which are difficult to be interpreted.

Figure 3.4 shows an AFM image of a surface of a prepared alumina ceramic membrane by Cortalezzi *et al.* [7]. It can be seen that the surface of the membrane is not smooth, but consists of a mass of peaks (bright regions) and valleys (dark regions). The roughness of the membranes, in terms of mean roughness, root mean square roughness or mean difference in height between the highest peaks and the lowest valleys may be obtained from the AFM images. Obviously, the membrane's roughness may not truly reflect the surface porosity of the membrane because the valleys can be through pores as well as dead pores on the membrane surface.

3.3 POROUS CERAMIC MEMBRANES

3.3.1 Gas Adsorption/desorption Isotherms

Gas adsorption/desorption is one of the most widely used techniques in characterization of porous membranes for the determination of pore size, specific surface area and pore size distribution of membranes with spherical, cylindrical or slit shaped pores. The adsorption/

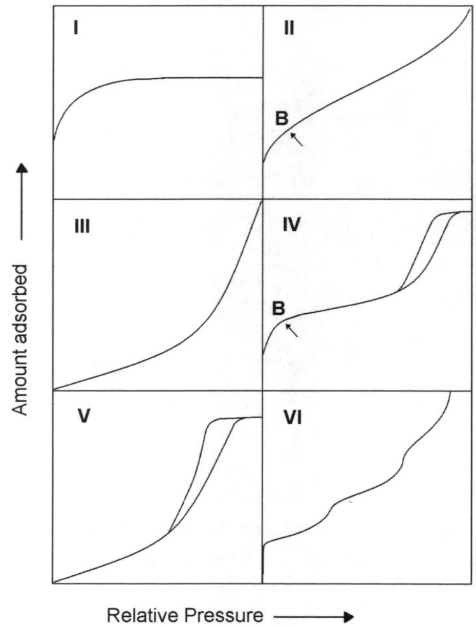

Figure 3.5 Types of physisorption isotherms [8]. Reprinted from Pure and Applied Chemistry, **57** (4), K. S. W. Sing *et al.*, *Reporting physisorption data for gas solid systems with special reference to the determination of surface area and porosity*, p. 603–619. Copyright (1985) with permission from IUPAC

desorption isotherm of a nonreactive gas (typically N_2 at 77 K) is determined as a function of the relative pressure, i.e. ratio of the applied pressure and the saturation pressure of the gas. It starts at low relative pressure. At a certain minimum pressure, the smallest pores will be filled with liquid nitrogen. As the pressure is increased further, larger pores are filled, and near the saturation pressure, all pores are filled. The total pore volume is determined by the quantity of gas absorbed near saturation pressure. Desorption occurs when the pressure is decreased from the saturation pressure downward. The majority of physisorption isotherms can be grouped into six types [8] as shown in Figure 3.5. The type I isotherm is typical for microporous solids and chemisorption isotherms. Type II is observed for finely divided non-porous solids. Types III and V are typical for vapour adsorption (e.g. water vapour on hydrophobic materials). Types IV and V feature a hysteresis loop; desorption on mesopores occurs at lower pressures than those at which adsorption in similar sized pores would take place. Finally, the rare type VI step like isotherm is seen on special types of carbons. Due to capillary condensation, membranes with mesopores exhibit a unique adsorption/desorption behaviour of types IV and V and the pore size can be determined using the Kelvin equation if the shape of the pores are assumed to be cylindrical:

$$\ln\frac{p}{p_o} = -\frac{2\sigma V}{r_K RT}\cos\theta \tag{3.1}$$

where p and p_o are the applied pressure and saturation pressure, respectively, in Pa, σ and V are the surface tension of the liquid condensate, in N m^{-1}, and the molar volume of the liquid,

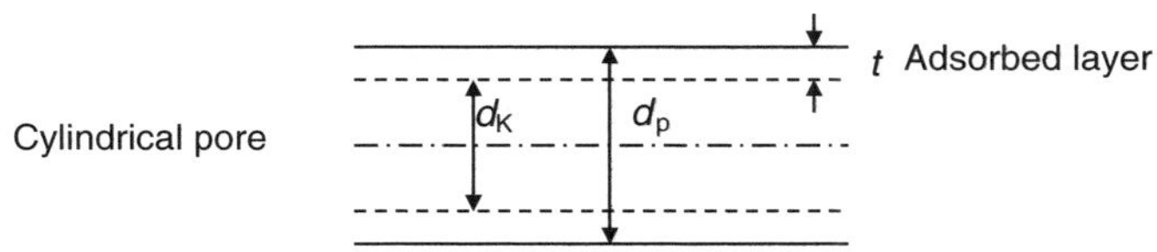

Figure 3.6 Relationship between d_K and d_p

in m^3mol^{-1}, respectively. T is the absolute temperature, in K, r_K, the Kelvin radius, in m and θ, the contact angle.

In measuring pore size of a ceramic membrane using nitrogen, the pore walls of the ceramic membrane can be easily wetted by the liquid nitrogen because of the concave meniscus of the liquid in the pores. Therefore, the contact angle, θ is 0–90° and cos θ is 0–1. With known values of $R = 8.314\,N\,m\,mol^{-1}\,K^{-1}$, $\sigma = 8.88 \times 10^{-3}\,N\,m^{-1}$, $V = 36.65 \times 10^{-6}\,m^3\,mol^{-1}$, $\theta = 0°$ and $T = 77\,K$, Equation (3.1) can be simplified as:

$$r_K = -\frac{4.15 \times 10^{-10}}{\log \frac{p}{p_o}} \tag{3.2}$$

The relationship between the pore radius, r_p and the Kelvin radius, r_K is described by the t method as illustrated in Figure 3.6. It can be seen that the pore radius can be calculated from:

$$r_p = r_K + t \tag{3.3}$$

where t is the thickness of the adsorbed layer in the pore walls, in m, and can be calculated using the Halsey equation [9, 10]:

$$t = 3.54 \times 10^{-10} \left(-\frac{5}{2.303 \log \frac{p}{p_o}} \right)^{0.333} \tag{3.4}$$

Based on Equations (3.2), (3.3) and (3.4), the relation between r_p and p/p_o can be obtained.

The above equations are useful in calculating the distributions of pore volume and surface area as a function of pore radius in the mesoporous range. The distributions of pore volume as a function of pore radius can be obtained using the BJH method, Horvath and Kawazoe method or Saito–Foley method [11–13], while the distributions of surface area can be determined by the BET method [8].

Elmarraki *et al.* [14] developed a ceramic ultrafiltration membrane using a sol-gel method. The membrane was prepared through a two step coating so that a distinctive bilayer membrane, comprising a TiO_2 layer on top of the $ZnAl_2O_4$ layer was developed. N_2 adsorption/desorption analysis indicates that the bilayer membrane shows two distribution functions as depicted in Figure 3.7. A mean pore size of 3.5 nm is found for the TiO_2 layer, while there is a mean pore size of 5.5 nm for the $ZnAl_2O_4$ layer.

In summary, gas adsorption/desorption is a simple method for obtaining pore size and pore size distribution of membranes satisfactorily if the isotherm model selected is correspondent

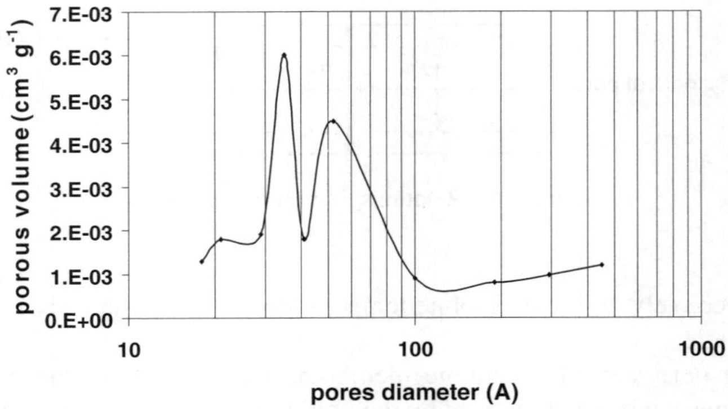

Figure 3.7 Pore size distribution of a TiO_2 and $ZnAl_2O_4$ bilayer membrane [14]. Reprinted from Materials Research Bulletin, **36** (1–2), Elmarraki *et al.*, *Elaboration and*. . . , p. 227–237. Copyright (2001), with permission from Elsevier

to the pore geometry of the membranes. The main problem of this technique is that the dead end pores, which do not participate in transport, are also measured, which may give misleading results in transport analysis.

3.3.2 Permporometry

Permporometry is a characterization technique for determining active (through) pores of porous membranes. The method was first developed by Eyraud *et al.* [15] as gas–liquid permporometry and was improved to the gas–vapour permporometry method [16]. Using this technique, pores in the mesoporous range can be determined.

Permporometry is based on controlled blocking of pores by capillary condensation with a condensable vapour combined with a simultaneous measurement of counterdiffusion of noncondensable gases through the pores that are not blocked by the condensed vapour. When a condensable vapour such as cyclohexane or ethanol is introduced at low vapour pressure, a molecular adsorption layer (i.e. *t* layer) is first formed on the inner surface of the pore walls. As the relative vapour pressure p/p_o of the condensable gas is increased from zero to one, pore blockage takes place from the smallest to the largest pores because of capillary condensation. The simultaneous counterdiffusion of noncondensable gases such as oxygen and nitrogen through the remaining open pores provides a measure for the fraction of pores with pore sizes larger than the pores that are already blocked. Upon desorption the same processes occur in reverse order. The pore size distribution is usually determined upon desorption, i.e. reducing p/p_o. Figure 3.8 illustrates the principle of the technique and experimental arrangement of the permporometry.

Capillary condensation is related to the relative vapour pressure and can be described by the Kelvin equation of Equation (3.1) if the pores are assumed to be cylindrical.

The relationship between the real pore diameter, d_p and the Kelvin radius r_K is given as:

$$d_p = 2(r_K + t) \tag{3.5}$$

where t is the thickness of the t layer formed on the inner surface of the pores, which is usually 0.3–0.5 nm.

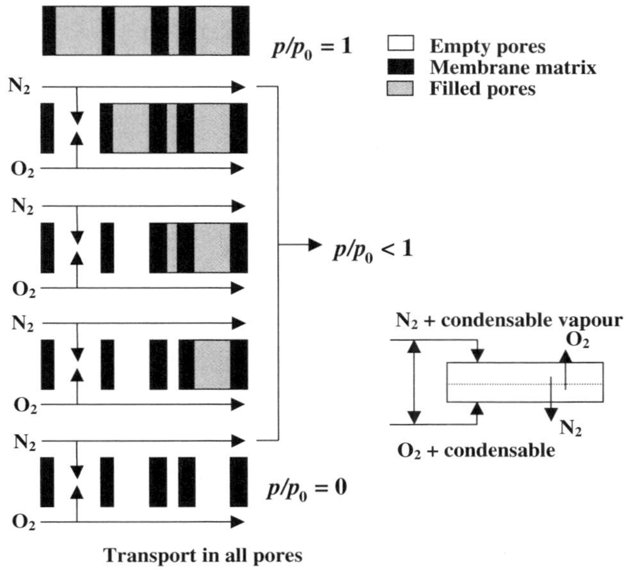

Figure 3.8 Principle of the permporometry and experimental arrangement

To determine the number of pores that are opened at a relative vapour pressure, the counterdiffusion of nitrogen and oxygen is applied. In the absence of an overall mechanical pressure difference across the membrane, diffusion with the Knudsen mechanism can be adopted to describe either O_2 or N_2 flux across the membrane:

$$J_i = \frac{2\pi n r_K^3 \Delta p_i}{3A_t RT\tau\delta}\sqrt{\frac{8RT}{\pi M}} \tag{3.6}$$

where J_i is the gas flux through the membrane, in mol m^{-2} s^{-1}, Δp_i, the gas partial pressure difference across the membrane, in Pa, A_t, the total membrane area, in m^2, τ, the tortuosity, δ, the membrane thickness, in m and M, the gas molecular weight. i refers either O_2 or N_2.

Based on Equations (3.1), (3.5) and (3.6), pore size and pore size distributions can be obtained with a stepwise reduction (or increase) of relative pressure.

Cao *et al.* used the permporometry technique to study pore size distribution of γ-alumina membranes with a pore radius ranging from about 2 nm to 10 nm [16]. Their results, plotted in Figure 3.9, indicated that the permporometry technique can effectively measure the active pores which show a sharp pore size distribution with an average Kelvin radius, r_K of 9 nm. In Figure 3.9, the long tail of the cumulative oxygen permeability at $r_K > 10$ nm is probably an indication of a small amount of large (larger than 10 nm) pores also existing in the membrane.

3.3.3 Mercury Porosimetry

Mercury porosimetry is one of the most popular and fastest methods for determining pore size and pore size distribution in ceramic membranes. It is based on the fact that mercury is

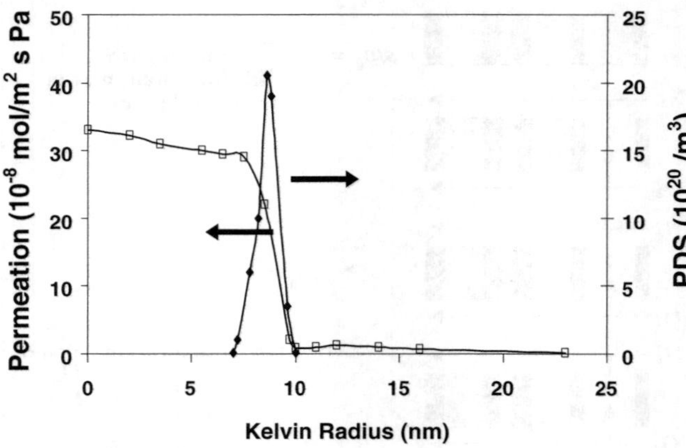

Figure 3.9 Oxygen permeability as a function of Kelvin radius and the pore size distribution (PSD) for a γ (gamma)-alumina membrane determined by permporometry [16]. Reprinted from Journal of Membrane Science, **83** (2), Cao *et al.*, *Permporometry study on . . .* , p. 221–235. Copyright (1993) with permission from Elsevier

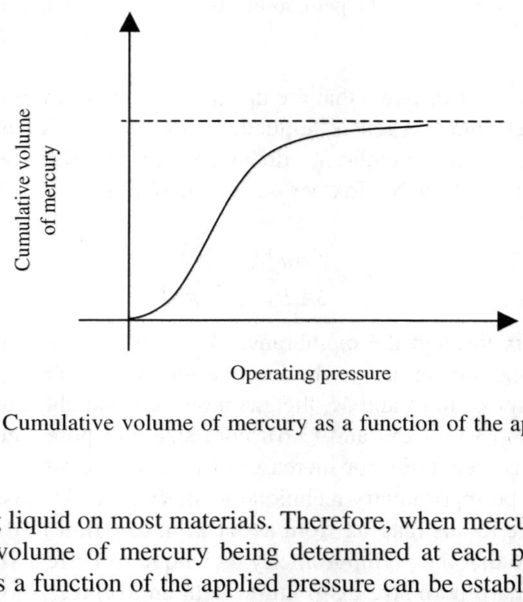

Figure 3.10 Cumulative volume of mercury as a function of the applied pressure

a strongly nonwetting liquid on most materials. Therefore, when mercury is forced into a dry membrane with the volume of mercury being determined at each pressure, a cumulative volume of mercury as a function of the applied pressure can be established, as illustrated in Figure 3.10, from which the pore size distribution can be deduced. The relationship between the operating pressure and pore size of the membrane can be described by the Laplace equation, if the pore in the membrane is assumed to be cylindrical:

$$r_{\mathrm{p}} = -\frac{2\sigma}{p}\cos\theta \tag{3.7}$$

where p is the applied pressure (guage), in Pa, σ, the surface tension at mercury/air, in N m^{-1}, r_{p}, the pore radius, in m and θ, the contact angle.

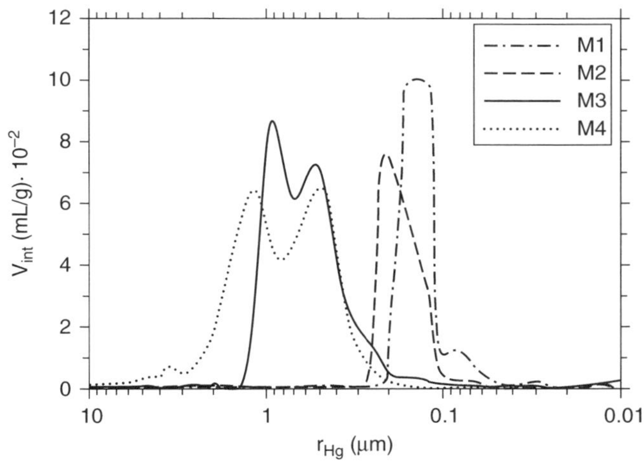

Figure 3.11 Pore size distribution for four ceramic membranes (M1, M2, M3, and M4 (T_s = 1200 °C)) by mercury porosimetry [17]. Reprinted from Journal of Membrane Science, **241** (1), Almandoz *et al.*, *Preparation and . . .*, p. 95–103. Copyright (2004) with permission from Elsevier

Based on Equation (3.7), very high pressures are required for pores in the nanometer range. For example, an operating pressure of 490 MPa is required to measure a ceramic membrane with a pore size of 1.5 nm assuming that the contact angle of the mercury with ceramic materials is at 140°.

It should be noted that Equation (3.7) assumes cylindrical pores in membranes, which is generally not the case for most ceramic membranes and for this reason a morphological constant must be introduced to correct the results. During the experiment, the largest pores will be filled with mercury at a certain minimum pressure. As the pressure increases further, smaller pores are filled until a maximum intrusion value is reached, i.e. all pores are filled. Thus, the pore size distribution of the membrane can be determined because every pressure is related to one specific pore size.

Figure 3.11 shows the pore size distribution curves corresponding to mercury porosimetry measurements for membranes prepared from four different aluminosilicate paste by Almandoz *et al.* [17]. The results indicate that two membranes show a mono distribution, while the other two show a bimodal distribution with lower pore sizes.

In conclusion, both pore size and pore size distribution of membranes can be determined by mercury porosimetry. The disadvantage is that small pore sizes require high pressures, which may damage the membrane pore structures. Also, this method determines all the pores presented in the membrane including dead end pores; this may result in overestimation of the membrane permeation characteristics.

3.3.4 Thermoporometry

Thermoporometry exploits a solid–liquid phase change of a pore penetrant in porous membranes. The relation between the pore size and the extent of undercooling, ΔT, for cylindrical pores was derived by Brun *et al.* [18]:

$$\text{During solidification:} \qquad r_p(\text{nm}) = \left(-\frac{64.67}{\Delta T}\right) + 0.57 \tag{3.8a}$$

$$\text{During melting:} \qquad r_p(\text{nm}) = \left(-\frac{32.33}{\Delta T}\right) + 0.68 \tag{3.8b}$$

where:

$$\Delta T = T - T_o \tag{3.9}$$

In Equation (3.9), T_o is the normal phase transition temperature of the liquid, in °C, and T, the temperature where the phase transition is actually observed when this liquid is held in the pores.

As can be seen from Equation (3.8), the freezing point of water in the membrane pores decreases with decreasing pore size. Therefore, each pore size shows its own specific freezing or melting point.

When melting or solidification takes place, heat is required or released, which can be captured by differential scanning calorimetry (DSC). Figure 3.12 shows a melting thermogram for two membranes with different pore size distributions [19].

When phase change takes place, i.e. from liquid to solid or vice versa, heat is released or is required. The heat of solidification or melting as a function of the undercooling, ΔT for pure water is given as [18]:

$$\text{During solidification:} \qquad Q_a = -5.56 \times 10^{-2} \Delta T^2 - 7.43 \Delta T - 332 \tag{3.10a}$$

$$\text{During melting:} \qquad Q_a = -0.155 \Delta T^2 - 11.39 \Delta T - 332 \tag{3.10b}$$

where Q_a is the energy released or adsorbed per gram of the sample, in J g^{-1} and ΔT, the extent of undercooling, in °C.

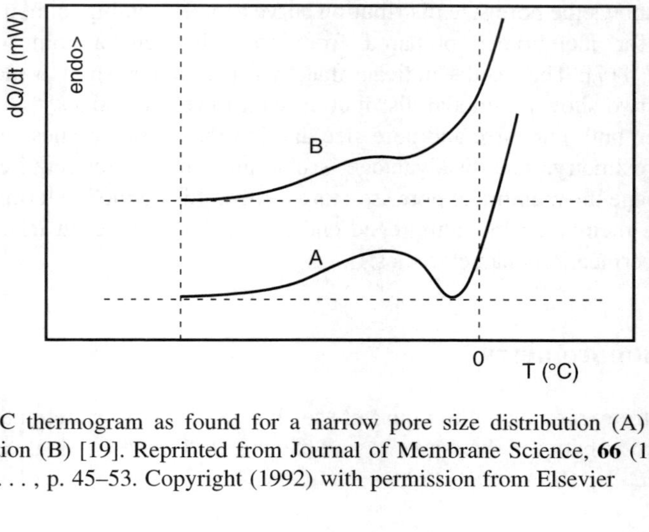

Figure 3.12 DSC thermogram as found for a narrow pore size distribution (A) and for a broad pore size distribution (B) [19]. Reprinted from Journal of Membrane Science, **66** (1), Cuperus *et al.*, *Critical-points in . . .*, p. 45–53. Copyright (1992) with permission from Elsevier

The change of pore volume, dV, corresponding to $d(\Delta T)$ is derived with the following equations:

Since,

$$dV = \frac{1}{\rho} \frac{dQ}{dQ_a} \tag{3.11}$$

where ρ is the water density corresponding to ΔT.

Differentiating Equation (3.8b):

$$dr_p = \frac{32.33}{\Delta T^2} d(\Delta T) \tag{3.12}$$

Combining Equations (3.11) and (3.12) gives:

$$\frac{dV}{dr_p} = \frac{\Delta T^2}{32.33 \rho Q_a} \frac{dQ}{d(\Delta T)} \tag{3.13a}$$

or

$$\frac{dV}{dr_p} = \frac{\Delta T^2}{32.33 \rho Q_a} \frac{\frac{dQ}{dt}}{\frac{d(\Delta T)}{dt}} \tag{3.13b}$$

where: $\frac{dQ}{dt}$ is the heat flow and $\frac{d(\Delta T)}{dt}$ the rate of temperature change.

For a membrane under measurements, a melting thermogram such as that shown in Figure 3.12 can be obtained. With the data given in the thermogram, the pore size, pore volume and their distributions can be calculated from the above mentioned equations.

Figure 3.13 shows the pore size distribution of a γ-alumina ceramic membrane determined by thermoporometry as well as gas adsorption–desorption methods [19]. As can be seen, both curves (both methods) are in good agreement with each other. The alumina samples were also analysed during several successive heating runs (up to eight runs), but differences between the runs were not found, indicating that the membrane is perfectly stable; the structure does not change during the analysis.

In summary, thermoporometry is a very simple method if a DSC apparatus is available. Similar to other methods, assumptions have to be made on the membrane pore geometry so that the pore size and the pore size distribution can be calculated. Again, in this method, all the pores, including dead-end pores, are measured.

3.3.5 Liquid Displacement Techniques

Methods for determining pore sizes and pore size distributions in membranes by monitoring liquid permeation have been available for many years [20]. Various names have been

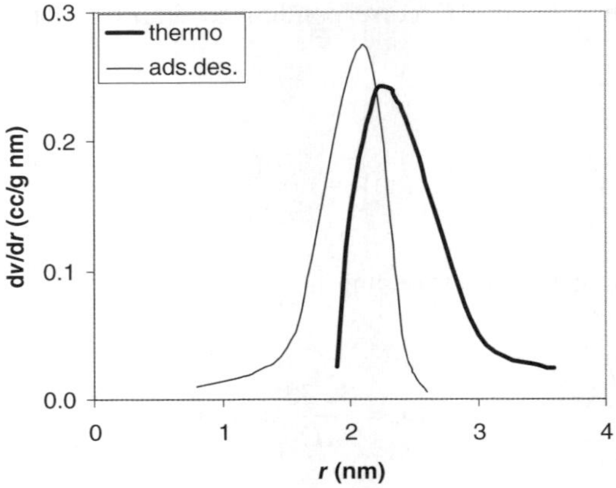

Figure 3.13 Pore size distributions found for alumina membranes using thermoporometry and gas adsorption–desorption measurements (d*v*/d*r*: differential pore volume per g membrane) [19]. Reprinted from Journal of Membrane Science, **66** (1), Cuperus *et al.*, *Critical-points in . . .* , p. 45–53. Copyright (1992) with permission from Elsevier

used in describing the related tests, such as Coulter porosimetry, bubble point method, liquid expulsion permporometry, bubble point with gas permeation, liquid displacement, bi-liquid permporometry, etc. In fact, these techniques are all similar and are based on the same physical principle of displacement of a wetting liquid. In general, the membrane is first wetted with a liquid which is held in the pores by capillary forces. Air or a less wetting liquid used for the displacement is then introduced on one side of the membrane. As the pressure of the air or the less wetting liquid is increased, the initial liquid held in the pores starts to displace from biggest pores until the liquid held in the smallest pores is eventually displaced. Therefore, all the above mentioned techniques can be grouped into two methods: (1) bubble point method, i.e. use of air for the displacement of the initial liquid in the membrane pores; (2) liquid displacement method, i.e. use of a less wetting liquid for the displacement of the initial liquid. These two methods are discussed in the following sections.

3.3.5.1 Bubble Point Method

The bubble point method has been the most widely used technique for determining pore size and pore size distribution of membranes. It is based on the principle that, for a given liquid and pore size with a constant wetting, the pressure required to force an air bubble through the pore is inversly proportion to the size of the pore and can be described by the Laplace equation:

$$r_p = \frac{2\sigma}{\Delta p}\cos\theta \tag{3.14a}$$

where Δp is the pressure difference across the membrane, in Pa, σ the surface tension at the liquid/air interface, in $N\,m^{-1}$, r_p, the pore radius, in m and θ, the contact angle.

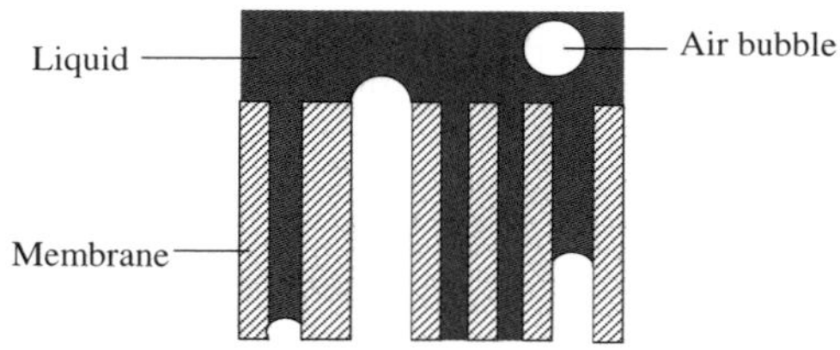

Figure 3.14 Schematic diagram of the principle of the bubble point method

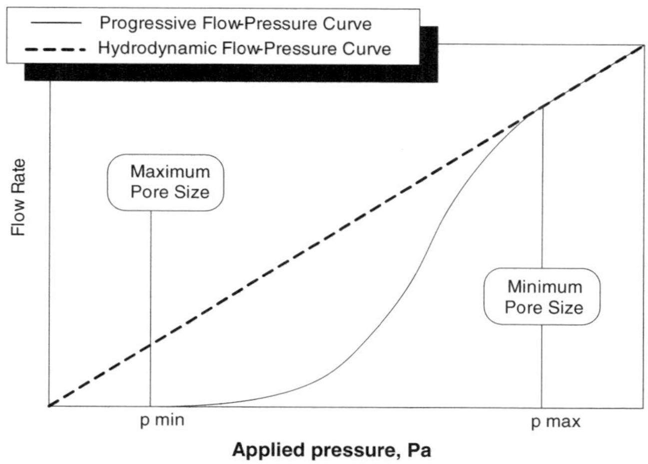

Figure 3.15 Theoretical flow–pressure curve for the bubble point test/progressive displacement test [20]. Reprinted from Journal of Membrane Science, **124** (2), Jakobs and Koros, *Ceramic membrane . . .* , p. 149–159. Copyright (1997) with permission from Elsevier

If the pressure at the downstream side of the membrane is controlled at atmospheric pressure, Equation (3.14a) can be simplified as:

$$r_p = \frac{2\sigma}{p}\cos\theta \tag{3.14b}$$

where p is the applied pressure (gauge), in Pa.

The principle of the bubble point method is illustrated in Figure 3.14. It can be seen that the membrane is wetted by a liquid, which acts as barrier. No flow can occur until the applied pressure reaches the capillary pressure of the largest pores. After increasing the pressure over this limit, the liquid is expelled from the largest pores, allowing the air to permeate. By successively increasing the pressure, smaller and smaller pores are opened for permeation of the air. The ideal flow versus pressure drop curve generated in this fashion is usually S shaped as shown in Figure 3.15 [20].

In Figure 3.15, the dashed and solid lines indicate the relation between the applied pressure and air flow under conditions of dry and wet membranes, respectively. The first air bubble (start of air flow) represents the largest pores measured at a certain minimum pressure. When the air flow measured under wet conditions is equal to that in dry conditions, the smallest pores are determined. Therefore, by use of Figure 3.15 together with the Laplace equation, pore size and the pore size distribution can be determined. This method is suitable for characterization of macropores and can be applied for ultrafiltration membranes with pore sizes up to 50 nm. The bubble point method is a very simple technique for characterizing the pore size and pore size distribution of active pores in membranes by a stepwise increase in the applied pressure. A disadvantage is that different results are obtained when different liquids are used for characterization. In addition, the rate of pressure increase and the pore length may influence the result.

3.3.5.2 Liquid Displacement Method

This method is similar to the bubble point method that was discussed earlier. The difference is that instead of air a liquid is used to displace an initial liquid held in the pores of the membrane. Therefore, these two liquids must be immiscible and the first one must easily wet the membrane, but not the second one. The principle is again based on the Laplace equation:

$$r_p = \frac{2\gamma}{\Delta p}\cos\theta \tag{3.15}$$

where Δp is the pressure difference across the membrane, in Pa, γ, the interfacial tension at the liquid/liquid interface, in Nm^{-1}; r_p, the pore radius, in m and θ, the contact angle between the two liquids and the membrane pore wall.

Again the displacement starts at the largest pores resulting in a flow which can be described by the Hagen–Poisseuille equation if the pore structure is considered to be cylindrical:

$$Q = \frac{n\pi r_p^4}{8\mu l}\Delta p \tag{3.16}$$

where Q is the permeation flow rate $m^3 s^{-1}$, Δp, the pressure drop across the membrane, in Pa, r_p, the pore radius, in m, μ, the liquid viscosity, in Pa s, n, the pore numbers and l, the pore length, in m.

In the case of other pore structures such as pores formed between packed spheres, the Carman–Kozeny equation can be used to describe the flow as a function of pressure drop. The flow can be measured with a mass flow meter. By increasing the pressure, the liquid in smaller pores will be displaced. In this way the flow is obtained as a function of the pressure (or pore radius calculated using Equation (3.15) and from this relation the pore size distribution can be calculated. Figure 3.16 illustrates a typical flow and pressure relation for the liquid displacement method. As can be seen, the maximum and minimum pore sizes as well as pore size distribution can be obtained from the figure and Equation (3.15).

3.3.6 Permeation Method

This method is involved in measurements of membrane permeability to a fluid (either liquid or gas) which leads to an average pore radius. For measurements using liquid, the analysis

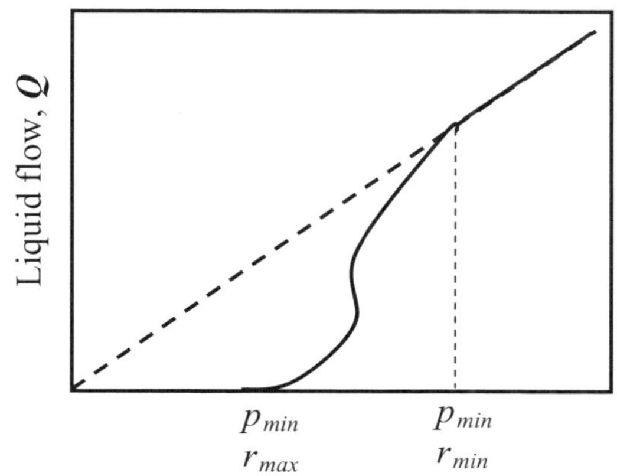

Figure 3.16 Flow versus pressure for a ceramic membrane with a distributed pore sizes

is usually based on either the Hagen–Poiseuille equation or the Carman–Kozeny equation depending on the pore structures of membranes. The former is used to describe capillary pores, while the latter is used to describe membrane pores formed between packed spheres. In terms of gas permeation measurements, the analysis is based on the kinetics of gas theory. For the flux of a liquid through membranes with highly ordered pores, the membrane pore radius can be determined relatively accurately, while for membranes with highly disordered pores, the pore values obtained become questionable, as the flux of a fluid through the membrane is sensitive to all structural aspects of the membrane [21]. Thus, despite of simplicity of the method, the accuracy of the results is dependent on validity of the assumptions made.

3.3.6.1 Liquid Permeation

Ceramic membranes can be considered as a layer of well packed particles (Figure 3.17A) and liquid permeation takes place in interstices between the packed particles. If the interstices are viewed as many tubes with tortuous paths (Figure 3.17B), then, the average radius of the membrane pores can be described by Hagen–Poiseuille equation:

$$J = \frac{\varepsilon \pi r_p^2}{8\mu\delta\tau}\Delta p \tag{3.17}$$

where J is the permeation flux of the liquid, in $m^3 m^{-2} s^{-1}$, Δp, the pressure drop across the membrane, in Pa, r_p, the average pore radius, in m, μ, the liquid viscosity, in Pa s and τ, the tortuosity.

In Equation (3.17), ε is the surface porosity defined as $\varepsilon = A_p/A_t$, the ratio of total pore area ($A_p = n\pi r_p^2$) to the total membrane area, A_t.

When Equation (3.17) is employed for calculation of the average pore size, it becomes uncertain because of the introduction of the membrane surface porosity and tortuosity. Once these two parameters are determined independently, the liquid permeation method can be used with a great advantage in experimental simplicity.

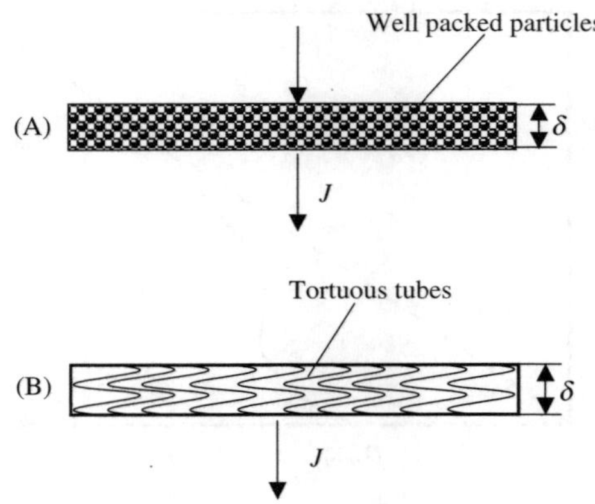

Figure 3.17 Ceramic membranes, (A): well packed particles (B): tortuous tubes

3.3.6.2 Gas Permeation

(i) Without consideration of membrane pore size distribution

This method is very simple. It measures flow rates of a gas through a porous membrane at different operating pressures. Then, average pore size and effective surface porosity of the membrane can be obtained from the measured data [22].

According to Tan *et al.* [23], total gas permeation rate through ceramic membranes can be written in a general nonmechanistic flux equation in terms of Poiseuille flow and Knudsen flow in the porous medium as follows:

$$N_t = P_v A_p \frac{\Delta p}{L_p} + P_k A_p \frac{\Delta p}{L_p} \tag{3.18}$$

where N_t is total permeation flow rate, in mol s^{-1}, A_p is the porous area on the membrane surface, in m^2, L_p, the effective pore length, in m, P_v and P_k are the permeability coefficients of gas transport according to Poiseuille flow and Knudsen flow mechanisms, in mol m m^{-2} Pa^{-1} s^{-1}, respectively and Δp is the pressure difference across the membrane, in Pa. Equation (3.18) is based on the following assumptions: the pores in the ceramic membrane are approximated by cylindrical pores with a modal pore radius, r_p and effective length, L_p.

As the effective pore length of ceramic membranes (tortuous pores) is usually an unknown value and cannot be determined using available methods, the total gas permeation rate through the membranes is usually expressed as:

$$N_t = \bar{P} A_t \Delta p \tag{3.19}$$

where $\bar{P}$ is the gas permeance, in mol m^{-2} Pa^{-1} s^{-1} and A_t is total membrane permeation area, in m^2. The surface porosity of the membrane is defined as:

$$\varepsilon = \frac{A_p}{A_t} \tag{3.20}$$

Substituting Equations (3.19) and (3.20) into Equation (3.18), upon rearranging, the following equation is obtained:

$$\overline{P} = P_v \frac{\varepsilon}{L_p} + P_k \frac{\varepsilon}{L_p} \tag{3.21}$$

When a gas is transferred through a long straight capillary tube with a pore radius of r_p by Poiseuille flow, the molar flux is expressed as [24]:

$$J_v = \frac{r_p^2}{8\mu RT} \overline{p} \frac{\Delta p}{L_p} \tag{3.22}$$

where J_v is molar flux of the gas, in mol m^2 s^{-1}, r_p, the radius of the capillary tube, in m, μ, the viscosity of the gas, in Pa s, R, the gas constant (8.3174 m^3 P$_a$ mol^{-1} K^{-1}), $\overline{p}$, the mean pressure ($\overline{p} = (p_h + p_l)/2$ where p_h and p_l are the upstream and downstream pressures, respectively), in Pa and T is absolute temperature, in K. For the purpose of determination of the permeability coefficient by Poiseuille flow, Equation (3.22) is rewritten as:

$$\frac{J_v L_p}{\Delta p} = P_v = \frac{1}{8\mu} \frac{r_p^2}{RT} \overline{p} \tag{3.23}$$

When transferring a gas through a long capillary tube with a radius of r_p by Knudsen flow, its molar flux is expressed as [24]:

$$J_k = \frac{2}{3} \left(\frac{8RT}{\pi M} \right)^{0.5} \frac{r_p}{RT} \frac{\Delta p}{L_p} \tag{3.24}$$

where J_k is the gas molar flux, in mol m^{-2} s^{-1} and M is the molecular weight, in g mol^{-1}. In order to determine the Knudsen permeability coefficient, Equation (3.24) is rewritten in the following form:

$$\frac{J_k L_p}{\Delta p} = P_k = \frac{2}{3} \left(\frac{8RT}{\pi M} \right)^{0.5} \frac{r_p}{RT} \tag{3.25}$$

Substitution of Equations (3.23) and (3.25) into Equation (3.21) gives:

$$\overline{P} = K_o + P_o \overline{p} \tag{3.26}$$

where:

$$K_o = \frac{2}{3} \left(\frac{8RT}{\pi M} \right)^{0.5} \frac{1}{RT} \frac{r_p \varepsilon}{L_p} \tag{3.26a}$$

$$P_o = \frac{1}{8\mu RT} \frac{r_p^2 \varepsilon}{L_p} \tag{3.26b}$$

An inert gas such as pure N_2 gas can be used as the standard gas for the permeation experiments and the permeation flux through the ceramic membranes prepared can be measured at different pressures. Details of the permeation experiments can be found elsewhere [23]. By plotting the permeance, $\bar{P}$ with mean pressures, $\bar{p}$ according to Equation (3.26), the average pore size can be calculated from the intercept (K_o) and slope (P_o):

$$r_p = \frac{16}{3}\left(\frac{P_o}{K_o}\right)\left(\frac{8RT}{\pi M}\right)^{0.5}\mu \tag{3.27}$$

The effective surface porosity, ε/L_p can also obtained from the slope as follows:

$$\frac{\varepsilon}{L_p} = \frac{8\mu\, RT\, P_o}{r_p^2} \tag{3.28}$$

The average pore size, r_p of Al_2O_3 and yttria stabilized zirconia (YSZ) ceramic hollow fibre membranes has been measured by Tan *et al.* [23] and Liu *et al.* [25] who obtained the average pore sizes of their membranes as ranging from 0.1 to 0.01 μm.

(ii) With consideration of membrane pore size distribution

The above gas permeation method used in characterizing average pore size of membranes may introduce errors when the test conditions were not properly selected or the membrane to be tested shows considerable pore size variation. In this respect, Kong and Li [26] proposed an improved gas permeation method for characterizing porous membranes with incorporation of a log-normal distribution function.

In developing the model equations for describing the permeation of gas through porous membranes, Kong and Li considered that only the Knudsen diffusion, slip flow and viscous flow are the dominant transfer mechanisms. Therefore, the total gas permeation rate through a porous membrane can be written in a general nonmechanistic flux equation in terms of Knudsen flow, slip flow and viscous flow. For gas transport through a long capillary pore (tube) of a membrane with a radius of r_p by Knudsen flow, slip flow and viscous flow, their molar flux can be expressed as the following three equations, respectively [24]:

$$J_k = \frac{2}{3}\left(\frac{8}{\pi MRT}\right)^{0.5} r_p \frac{\Delta p}{L_p} \tag{3.29a}$$

$$J_s = \left(\frac{\pi}{8MRT}\right)^{0.5} r_p \frac{\Delta p}{L_p} \tag{3.29b}$$

$$J_v = \frac{r_p^2}{8\mu RT}\,\bar{p}\,\frac{\Delta p}{L_p} \tag{3.29c}$$

where J_k, J_s and J_v are the gas molar flux under Knudsen flow, slip flow or viscous flow, respectively, in mol m^{-2} s^{-1}, R is the gas constant (8.3174 m^3 Pa mol^{-1} K^{-1}), M is the molecular weight, in g mol^{-1}, Δp is the pressure differential across the membrane, in Pa, L_p is the effective thickness of the membrane, μ, the viscosity of gas, in Pa s and $\bar{p}$, the average pressure across the membrane, in Pa.

For membranes showing distributed pore sizes, a standard normal (Gaussian) distribution or a log-normal distribution function may be possible. Two important parameters, i.e. mean pore radius, r_m and dimensionless standard deviation of the membrane pore size, σ, were used to represent the above two distribution functions. The standard normal distribution and log-normal distribution functions used to represent the pore size distributions are respectively given as [27, 28]:

$$g(r) = \frac{1}{\sqrt{2\pi} r_m \sigma} \exp\left[\frac{-(1-r/r_m)^2}{2\sigma^2}\right] \tag{3.30a}$$

$$g(r) = \frac{1}{\sqrt{2\pi} r}\left[\ln(1+\sigma)^2\right]^{-0.5} \times \exp\left[-\frac{\left(\ln\left(\frac{r}{r_m}\right)(1+\sigma^2)^{0.5}\right)^2}{2\ln(1+\sigma)^2}\right] \tag{3.30b}$$

By considering the pore size distribution, the total permeation flow rate, N_t, through a membrane having a total number of pores, n, can be written as

$$N_t = n\pi\left(\int_0^{0.05\lambda} J_k r^2 g(r) \mathrm{d}r + \int_{0.05\lambda}^{3\lambda} J_s r^2 g(r) \mathrm{d}r + \int_{3\lambda}^{r_{max}} J_v r^2 g(r) \mathrm{d}r\right) \tag{3.31}$$

where $g(r)$ is the pore size distribution function (either standard normal distribution or log-normal distribution), r_{max} is the maximum pore radius of membrane (set at 10 μm for example), and λ is the mean free path of permeate gas which could be expressed as:

$$\lambda = \frac{RT}{\sqrt{2}\pi d^2 N \bar{p}} \tag{3.32}$$

where d is the collision diameter of gas molecular, in m and N is the Avogadro number. Integration limits for the different transport mechanisms depend on the ratio of membrane pore radius, r, to the mean free path of the permeating gas molecules, λ. In this study, the upper limit of Knudsen flow is set at 0.05λ as suggested by Liepmann [29], while the upper limit of slip flow is defined experimentally as 3λ [26].

For a membrane, the gas permeance is commonly defined as:

$$\bar{P} = \frac{N_t}{A_t \cdot \Delta p} \tag{3.33}$$

where A_t is the total membrane area, in m^2.

In addition, the surface porosity of a membrane is defined in Equation (3.20) as:

$$\varepsilon = \frac{A_p}{A_t} \tag{3.20}$$

where A_p is the cross sectional area of the total number of pores, defined as:

$$A_p = n\int_0^{r_{max}} \pi r^2 g(r)\mathrm{d}r \tag{3.34}$$

Substituting Equations (3.31), (3.20) and (3.34) into Equation (3.33) gives:

$$\bar{P} = \frac{\varepsilon}{L_p} \frac{\left(C_1 \int_0^{0.05\lambda} r^3 g(r)\mathrm{d}r + C_2 \int_{0.05\lambda}^{3\lambda} r^3 g(r)\mathrm{d}r + C_3 \int_{3\lambda}^{r_{max}} r^4 g(r)\mathrm{d}r \right)}{\int_0^{r_{max}} r^2 g(r)\mathrm{d}r} \tag{3.35}$$

where

$$C_1 = \frac{2}{3}\left(\frac{8}{\pi MRT}\right)^{0.5} \tag{3.35a}$$

$$C_2 = \left(\frac{\pi}{8MRT}\right)^{0.5} \tag{3.35b}$$

$$C_3 = \frac{\bar{p}}{8\mu RT} \tag{3.35c}$$

The mean pore radius, r_m, effective porosity, ε/L_p, as well as standard deviation, σ, can, therefore, be obtained via regression analysis of gas permeation data, i.e. pressure difference across the membrane, Δp and permeance, $\bar{P}$. The three unknown parameters in Equation (3.35), i.e. mean pore radius, r_m, effective porosity, ε/L_p, and standard deviation, σ, have been searched using Fortran with a library routine [26] via the following steps:

(1) Input the permeation data of a membrane at different pressures (five points).

(2) Calculate the permeance $\bar{P}_c$ at a different operating pressure from the initial values of r_m, ε/L_p and σ.

(3) The sum of square relative error is then calculated as:

$$SSE = \sum_{n=1}^{n} (1 - \bar{P}_c / \bar{P}_e)^2 \tag{3.36}$$

where $\bar{P}_e$ is the experimental permeance.

(4) For a fixed ε/L_p, a set of r_m and σ that minimises SSE is searched by a nonlinear regression routine.

By varying the ε/L_p, the minimum value of SSE is found and hence, the optimum set of ε/L_p, r_m and σ can be obtained.

Table 3.1 The effective porosity, mean pore sizes and standard deviations (SD) of alumina membranes with and without surface modification

	Original Membranes			Surface Modified Membranes		
Sintering Temp.	Effective Porosity (m^{-1})	Mean Pore Size (μm)	SD	Effective Porosity (m^{-1})	Mean Pore Size (μm)	SD
1500 °C	2446	0.025	1.31	3326	0.032	1.31
1400 °C	2873	0.038	1.67	3108	0.039	1.23
1300 °C	6269	0.039	1.45	7450	0.034	1.31
1200 °C	13538	0.053	1.31	13647	0.050	1.31

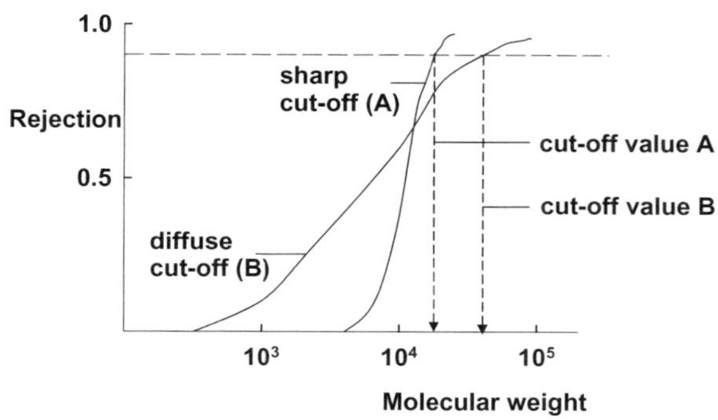

Figure 3.18 Rejection characteristics for a membrane with a sharp cut off compared with those of a membrane with a diffuse cut off [5]. Reprinted from *Basic Principles of Membrane Technology*, 2nd edn, 1996, M. Mulder, Figure IV-27. With kind permission of Springer Science and Business Media

The mean pore size, r_m, its standard deviation, σ and ε/L_p of Al_2O_3 ceramic hollow fibre membranes with or without surface modification has been measured by Koonaphapdeelert and Li [30] and are given in Table 3.1.

3.3.7 Measurements of Solute Rejection

This method utilizes a size exclusion technique to characterize porous membranes. Depending on the pore sizes and surface properties of the membranes to be characterized, different types of mixtures of solute/solvent solutions are required. For example, a mixture of mono- and multivalent salt solution is usually employed for characterizing reverse osmosis (RO) or nanofiltration (loose RO) membranes, while for ultrafiltration and microfiltration membranes macromolecule solutions such as dextrans, insulin or polyethylene glycols in water are often used [31]. Recently, Silva and Livingston used tetraoctylammonium bromide (TOABr) in either methanol or dimethyl methylsuccinate (DMMS) solvent to test solvent resistant nano-filtration membranes [32].

A rejection parameter called 'molecular weight cut off (MWCO)' is often used in solute rejection measurements. It is defined as the molecular weight above which molecules are at least 90 % rejected by the membrane. The measured result is usually represented by a curve of % solute rejection against molecular weight as shown in Figure 3.18 [5].

In characterizing porous membranes using the solute rejection measurements, the main uncertainty is the assumption of the solute rejection which is based only on molecular weight. This may lead to errors in analyses as most macromolecular solutes show flexible shapes and may 'squeeze' through pores that are much smaller than the molecular dimension. Also, surface interaction between the solutes and the membrane surface can occur, which may cause an overestimate of % rejection. Finally, concentration polarization and membrane fouling can have considerable effects on rejection measurements.

Solute rejection of a membrane is defined for each main solute as:

$$R = 1 - \frac{C_p}{C_f} \tag{3.37}$$

or

$$R^o = 1 - \frac{C_p}{C_w} \tag{3.38}$$

where C_f, C_p and C_w are the solute concentration in the feed, in the permeate and at the membrane interface, respectively, in mol m^{-3}. In Equation (3.37), R is the observed solute rejection and the quantity is measured by sampling the feed (and reject) and permeate phases, while in Equation (3.38) R^o is the intrinsic solute rejection and the quantity takes into account the actual solute concentration at the membrane interface, which is impossible to measure. The difference between Equations (3.37) and (3.38) is due to the hydrodynamic resistance, which can be minimized by proper control of process parameters such as the feed solute concentration and the hydrodynamic velocity of the feed solution.

The rejection characteristics of two of alumina membranes tested using polyethylene glycol (PEG) and water solutions were carried out by Leenaars and Burggraaf [33] who obtained a plot for curves of % solute rejection versus molecular weight as shown in Figure 3.19. As can be seen, the cut off value for the Al_2O_3-400 membrane is 2000 and is about 20000 for the Al_2O_3-800 membrane.

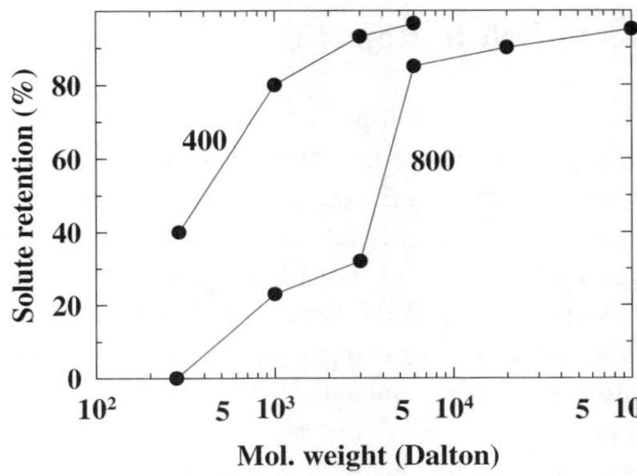

Figure 3.19 Solute rejection as a function of the molecular weight for polyethylene glycol. The left curve is obtained with Al_2O_3,-400 membranes, while the right curve with Al_2O_3-800 membranes [33]. Reprinted from Journal of Membrane Science, **24** (3), Leenaars and Burggraaf, *The preparation and . . .*, p. 261–270. Copyright (1985), with permission from Elsevier

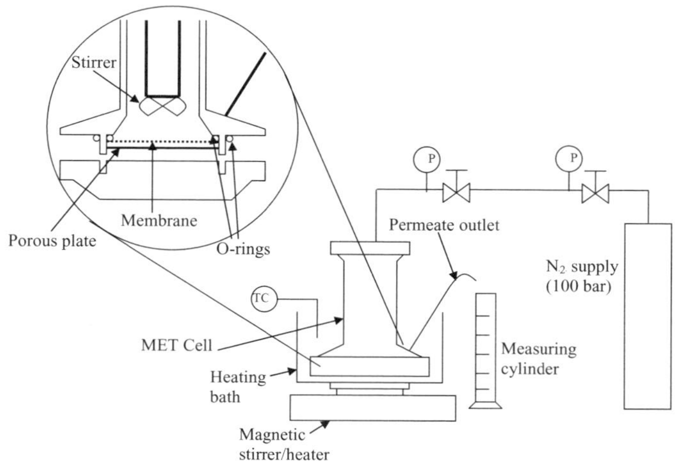

Figure 3.20 Dead end test cell

Solute rejection measurements can be performed in a batchwise stirred cell or in a recalculated crossflow mode. In batchwise operations as illustrated in Figure 3.20, a stirring bar situated close to the top of the membrane surface is used to decrease the concentration polarization or fouling during the measurements. The pressure difference across the membrane is usually achieved by an inert gas blanket. Maximum operating pressures are dependent on the pore size of the membrane to be tested. For example, for an ultrafiltration membrane, the pressure difference would not exceed 9–10 bar, while for a nanofiltration membrane, a pressure difference over 30 bar is not unusual. The membrane cell for solute rejection measurements can accommodate feed volumes from as low as a few μL to several L. The cell design is typically a dead end filtration type. Therefore, the system is simple and the measurements can be accomplished very fast. The crossflow mode operation as illustrated in Figure 3.21 resembles a plant operation. Therefore the operating mode is very popular in industry for a quick estimation of membrane functional properties.

3.4 DENSE CERAMIC MEMBRANES

Dense ceramic membranes for separations are mainly made from solid oxides, a mixture of solid oxides and metals, i.e. cermet or a mixture of two different ceramics with a mixed ionic–electronic conducting property. Because of their unique separation mechanism, i.e. the gas component (usually oxygen or hydrogen) is transported in a dissociated or ionized form rather than conventional molecular diffusion, a membrane with extremely high selectivity to oxygen (or hydrogen) is attainable if the prepared membrane is dense and defect free. Therefore these membranes cannot be characterized by the methods mentioned earlier where the techniques introduced are only suitable for characterizing porous membranes. For mixed ionic–electronic conducting ceramic membranes, the physical structures of the materials are more important. Also, the membranes prepared should be free of defects so that very high oxygen or hydrogen selectivity can be achieved. Therefore, in this respect, the techniques described below are mainly related to gas tightness of membranes, their crystal structures and oxygen or hydrogen permeation characteristics. Characterizations on conductivities of ions and electrons of specific gases of dense ceramic membranes are not included as they are already available in several books.

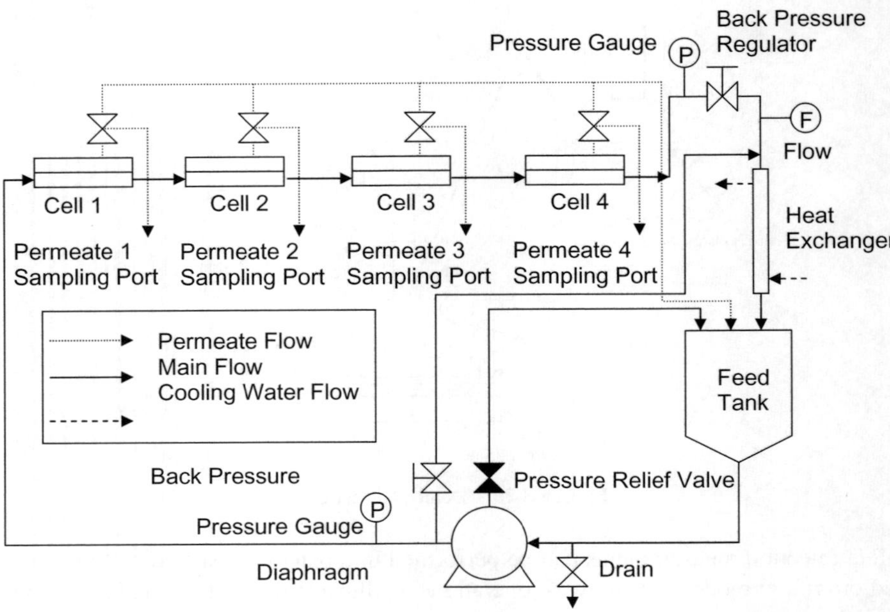

Figure 3.21 Crossflow mode

3.4.1 Leakage Test

Gas tightness of the prepared membranes can be tested using a gas permeation apparatus as shown schematically in Figure 3.22. As can be seen, the membrane, usually hollow fibre type, for the test is glued onto a stainless steel sample holder with an epoxy resin. It is then assembled into a cylinder with a known volume capacity. Nitrogen is used as the test gas. During the test, the pressure change with time of the cylinder is measured with a pressure transducer attached to the test cylinder and is recorded through a panel meter. The gas permeance is, thus, obtained based on the cylinder pressure change with time:

$$\bar{P} = \frac{V}{RT \cdot A_t t} \ln\left(\frac{p_0 - p_a}{p_t - p_a}\right) \tag{3.39}$$

where $\bar{P}$ is the permeance of the test membrane, in mol m^{-2} s^{-2} Pa^{-1}, V, the volume of the test cylinder, in m^3, R, the gas constant, 8.314 J mol^{-1} K^{-1} and T, the measured temperature, in K, p_0 and p_t are the initial and final measured pressures in test cylinder, in Pa, p_a, the atmospheric pressure, in Pa, A_t is the membrane area, in m^2 i.e. $A_t = [2\pi (R_o - R_{in})L]/[\ln(R_o/R_{in})]$ where R_o and R_{in} are respectively the outer and the inner radius of the hollow fibre, in m; L, the length of the hollow fibre membrane, in m and t, the time for measurement, in s. In general, a membrane with a measured permeance of 10 000 times less than the actual permeance of a permeating gas measured at high temperature is considered to be gas tight.

Tan *et al.* [34] conducted a gas tight test for a LSCF membrane they developed. Figure 3.23 illustrates a typical result for one of the hollow fibre membranes, where the pressure in the test cylinder is plotted against the time. The gas permeance of the fibre was calculated

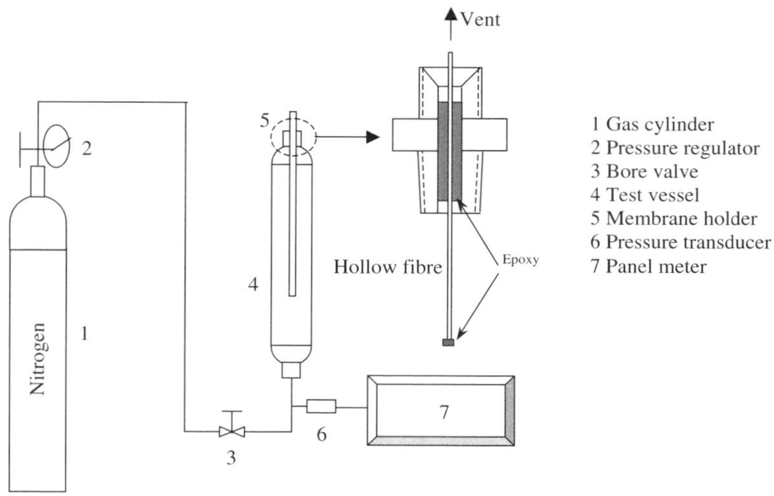

Figure 3.22 Experimental setup for gas tightness test [34]. Copyright (2005) John Wiley & Sons, Ltd, American Institute of Chemical Engineers, **51** (7), Tan, X., Liu, Y. and Li, K., *Mixed conducting ceramic hollow-fiber membranes for air separation*, p. 1991–2000

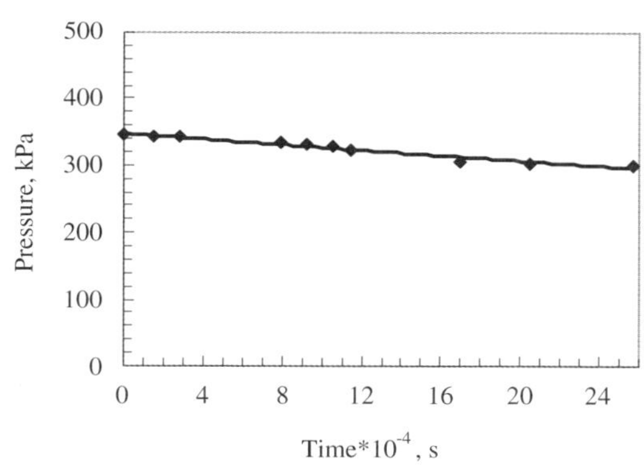

Figure 3.23 Pressure change with time in the gas-tightness test of hollow fibre membranes [34]. Copyright (2005) John Wiley & Sons, Ltd, American Institute of Chemical Engineers Journal, **51** (7), Tan, X., Liu, Y. and Li, K., *Mixed conducting ceramic hollow-fiber membranes for air separation*, p. 1991–2000

according to Equation (3.39) and was found to be 2.82×10^{-10} mol ($m^{-2} Pa^{-1} s^{-1}$). This value is much lower than the oxygen permeance through the hollow fibre at elevated temperatures (estimated to be 2.2×10^{-4} mol ($m^{-2} Pa^{-1} s^{-1}$) at 800 °C [34]. In Figure 3.23, the slight decrease in pressure in the test cylinder is not only due to the negligible gas permeation through the membrane, but also a possible leak from joints of the system. Therefore, the actual permeance due to the leakage of the membrane is even lower than the above calculated result and the membrane can, thus, be considered to be gas tight. The same test has also been carried out by Liu *et al.* [35] for a $SrCe_{0.95}Yb_{0.05}O_{3-\alpha}$ (SCYb) hollow fibre membrane which permeates only protons/hydrogen.

3.4.2 Permeation Measurements

The measurement of oxygen or hydrogen flux through a mixed conducting membrane under realistic conditions is of great importance because it may be the only way to determine the oxygen or hydrogen transport in mixed conducting membranes where the measurements of transport number by the EMF method or other electrical methods are not possible due to the high electronic conductivity of the membrane.

A permeation experiment can be performed so long as a pressure gradient of the gas component to be measured (oxygen or hydrogen) across the membrane is maintained. Because the membranes used for such measurements are usually disc shape with small areas, perfect mixing conditions can always be assumed on both sides of the membrane. Therefore, the partial pressure of the gas on the permeate side has a linear relation with the permeation flux.

Although gas tightness of a membrane can be determined using the leakage test mentioned above, it is obtained at room temperature and may be different under high temperature operation. Therefore, during high temperature permeation experiments, it would be useful to add an inert gas as a marker to the feed side so that it's content in the permeate side would reveal the degree of the leakage. For instance, for an oxygen permeation measurement, air is used as feed gas and Ar as permeate sweep gas; the leakage can be detected if the nitrogen content in the permeate side is considerable.

3.4.2.1 Oxygen Permeation

Measurements of oxygen permeation in dense ceramic membranes can be performed in apparatus schematically shown in Figure 3.24 [36]. As can be seen, the membrane is sealed on the end of a quartz tube.

Helium is supplied to the permeate side of the membrane and the oxygen partial pressure in this compartment can be varied by adjusting the total flow rate. The oxygen partial pressure at the feed side can be varied by mixing air, N_2 and O_2. The oxygen partial pressures of the

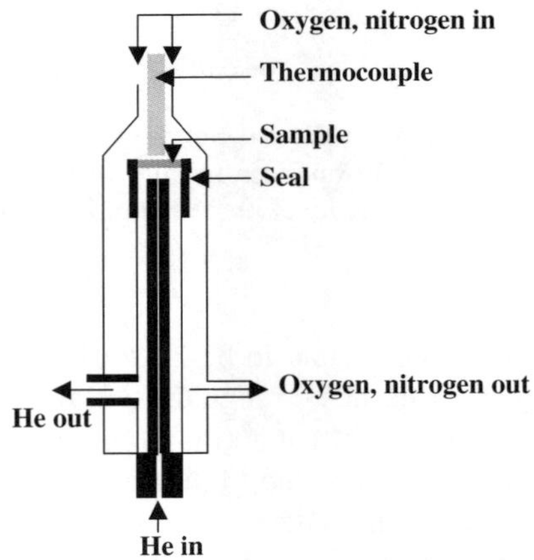

Figure 3.24 The oxygen permeation apparatus [36]. Reprinted from Solid State Ionics, **81** (1–2), Tenelshof *et al.*, *Oxygen-transport through . . .*, p. 97–109. Copyright (1995), with permission from Elsevier

retentate and permeate streams are continuously measured using an oxygen meter. Analysis of the composition of the effluent stream can also be performed by a gas chromatograph, especially for detection of N_2 contents for leakage monitoring. All gas flows are controlled and monitored. The oxygen flux is calculated from:

$$J_{O_2} = \frac{1}{G}\frac{V_t C_{O_2}}{A_t} \tag{3.40}$$

where V_t is the volumetric flow rate of the effluent stream, in $m^3 s^{-1}$, C_{O2}, the oxygen concentration determined by the oxygen sensor meter, in $mol\,m^{-3}$ and A_t, the membrane area, in m^2. G is a dimensionless factor that corrects for the effect of nonaxial diffusion, which occurs due to the fact that the surfaces exposed to air and helium have different surface areas [36].

One of the major problems in the above apparatus is the high temperature sealing. Since the membrane and the quartz tube have different thermal expansion coefficients, a certain degree of leakage from the seal inevitably takes place at high temperature, leading to uncertainties in the experimental results. A different approach was employed by Tan *et al.* [37] who measured oxygen permeation flux with a long hollow fibre membrane placed in a CARBOLITE tube furnace. The design of the experimental apparatus is schematically illustrated in Figure 3.25. In operation, argon used as sweep gas was fed into the fibre lumen, while compressed air was introduced into the furnace tube in order to keep the oxygen concentration outside the membrane constant. The flow rate of the sweep gas was controlled using a mass flow controller. The oxygen concentration in the effluent gas stream was measured online using an oxygen meter. In order to check the oxygen permeation, the flow rate of the effluent was measured with a soap-bubble flow meter, as also shown in Figure 3.25. The oxygen permeation flux was calculated by:

$$J_{O_2} = \frac{V_t x_{O_2}}{A_t} \tag{3.41a}$$

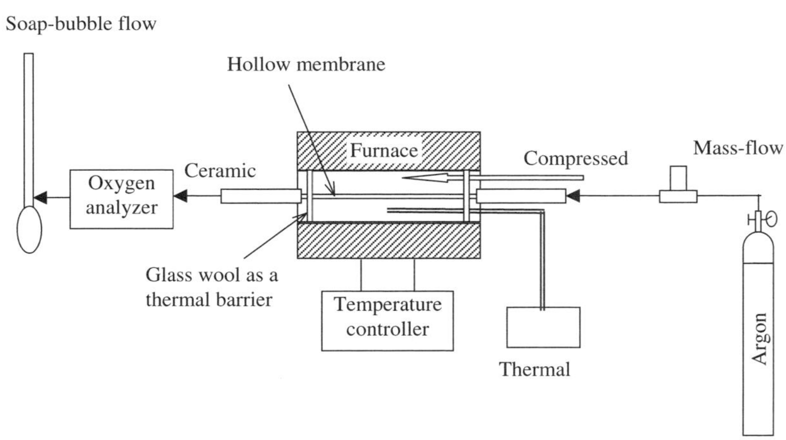

Figure 3.25 Schematic diagram of an apparatus for oxygen permeation measurements [37]. Copyright (2005) American Chemical Society, Industrial and Engineering Chemistry Research, **44**, Tan, X., Liu, Y. and Li, K., *Preparation of LSCF ceramic hollow fibre membrane for oxygen production by a phase-inversion/sintering technique*, p. 61–66

or

$$J_{O_2} = \frac{V_{ar} x_{O_2}}{A_t(1 - x_{O_2})} \tag{3.41b}$$

where V_t and V_{Ar} are the volumetric flow rates of the effluent stream and of the sweep Ar gas, respectively, in $m^3 s^{-1}$; x_{O2}, the oxygen concentration determined by the oxygen sensor meter, in volume percent, i.e. %, A_t, the effective membrane area of the hollow fibre for oxygen permeation, $A_t = [2\pi(R_o - R_i)L]/[\ln(R_o/R_i)]$, in which R_o and R_i are respectively the outer and the inner radius of the hollow, in m and L is the effective heated length of the fibre, in m.

3.4.2.2 Hydrogen Permeation

Measurements for hydrogen permeation in ceramic membranes show some similarities to the apparatus used for oxygen permeation. However, it is more difficult because water vapour sometimes has to be added. In mixed proton conducting membranes, water vapour is, in many cases, the actual source of the protons in the membrane material. Thus, a hydrogen permeable oxide may simply not work in dry gases. Secondly, water vapour determines the oxygen activity, and the membrane may be unstable in dry hydrogen. Thirdly, the water vapour may be used to set up or to balance out an oxygen activity gradient. It may also affect the electronic conductivity. Normally, it is a good practice to operate with wet gases on both sides of the membrane so that conditions are well defined.

Cheng *et al.* [38] conducted a hydrogen permeation measurement using a high temperature permeation system with a gas tight $SrCe_{0.95}Tm_{0.05}O_{3-\alpha}$ membrane. The schematic diagram of this experimental system is shown in Figure 3.26. Permeation experiments were performed with upstream H_2 in a balance of helium and downstream nitrogen or air. The hydrogen concentration in the effluents was detected by a gas chromatography. The H_2 permeation flux was calculated from the rate of water produced at the downstream side as the permeated H_2 reacted with O_2 to produce water. A thermohygrometer was used to measure the water vapour concentration change of the sweep gas. Hydrogen flux, J_{H_2} (mol m^2 s) was calculated from the following equation:

$$J_{H_2} = \frac{p_o}{RT} F_s \frac{\Delta RH}{A_t} \tag{3.42}$$

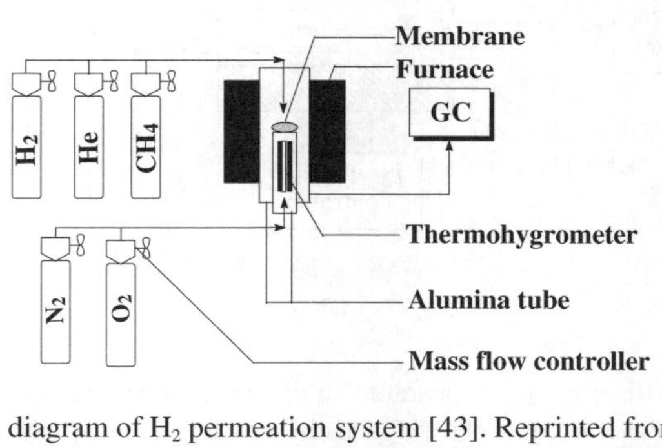

Figure 3.26 Schematic diagram of H_2 permeation system [43]. Reprinted from Solid State Ionics, **120** (1–4), Qi and Lin, *Electrical conducting . . .*, p. 85–93. Copyright (1999) with permission from Elsevier

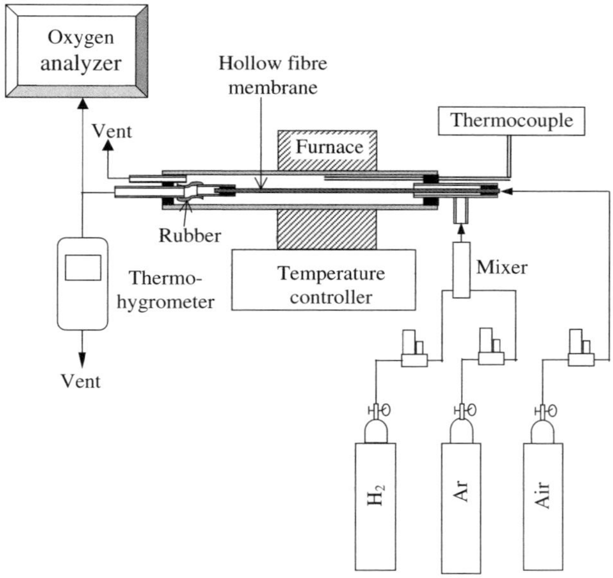

Figure 3.27 Diagram of the experimental setup for hydrogen permeation [35]. Copyright (2006) John Wiley & Sons, Ltd, American Institute of Chemical Engineers Journal, **52** (4), Liu, Y., Tan, X. and Li, K., *$SrCe_{0.95}Yb_{0.05}O_{3-\alpha}$ hollow fiber membrane and its property in proton conduction*, p. 1577–1585

where F_s is the total flow rate of the sweep gas stream, in mol m^{-3}, ΔRH, the difference in the relative humidity between the outlet and the inlet of the sweep gas, A_t, the membrane area, in m^2, p_o, the saturated vapour pressure, in Pa and T, the air at temperature, in K.

Similarly, H_2 permeation can also be measured for hollow fibre membranes. Liu *et al.* [35, 39] developed an experimental apparatus as shown schematically in Figure 3.27. It can be seen that a membrane module containing a long H_2 hollow fibre membrane is fabricated for hydrogen permeation and is placed in a tubular furnace (MTF 10/25/130, CARBOLITE) with a constant temperature zone length that is much shorter than that of the membrane module. Therefore, the two sealing joints, which are glued with normal epoxy resin, can be kept away from the furnace tube. Figure 3.27 further shows that a short soft silicone rubber tube is used to offset the thermal expansion mismatch of the hollow fibre membrane and the module shell at high temperatures. In the hydrogen permeation experiments, a mixture of hydrogen and argon is fed into the shell side, while air was introduced into the lumen side of the module. The gas flow rates are controlled and measured using mass flow controllers. Oxygen concentrations were measured with an oxygen meter. The concentrations of water vapour in the gas streams were measured online in terms of relative humidity (RH) using a thermohygrometer. Obviously, the permeation fluxes of hydrogen through the membrane can be calculated from the formation rates of water vapour in the air stream. Also, a GC is used to check the leakage of the membrane at its operating temperatures.

Figure 3.28 shows hydrogen permeation flux for a SCYb hollow fibre membrane as a function of hydrogen concentration obtained by Liu *et al.* [39]. As can be seen, the hydrogen permeation flux increases with increasing hydrogen concentration in the feed stream and also increases with increasing operating temperature. At a hydrogen feed concentration of 35.9 % and an operating temperature of 950 °C, a hydrogen permeation flux of 0.2 ml (STP)

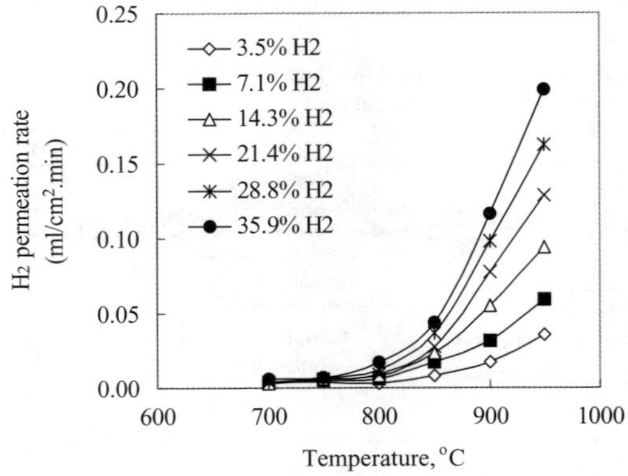

Figure 3.28 Hydrogen permeation flux measured under different hydrogen concentrations in the feed gas and at different temperatures [39]. Copyright (2006), Journal of Membrane Science, **283** (1–2), Liu, Y., Tan, X. and Li, K., *$SrCe_{0.95}Yb_{0.05}O_{3-\alpha}$(SCYb) hollow fibre membrane: Preparation, characterization and performance*, p. 380–385

$cm^{-2}\,min^{-1}$ is obtained, which is in line with literature data, considering the thickness of the membrane used for the measurements [35].

3.4.3 XRD

X-ray diffraction (XRD) is an essential technique for phase identification and crystal structure refinement. It is very often used to examine whether there are any changes in the crystalline structures of membrane materials after being processed into the membranes. Also, stability of ceramic membranes after long term exposure to a reaction environment is usually checked by XRD.

Tan and Li [40] carried out an XRD analysis on a membrane material, a fresh membrane made from the material as well as the membrane after long term exposure to a methane coupling environment using a PW1710 X-ray diffractometer with CuKα radiation. A continuous scan mode was used to collect 2θ data from 10 to 80° with a 0.04 sampling pitch and a 2.4° min^{-1} scan rate. The generator tension and generator current were at 40 kV and 40 mA, respectively.

Figure 3.29 shows the XRD patterns obtained by Tan and Li [40]. It can be seen that the perovskite structure was kept unchanged either after the LSCF powders were made into hollow fibres by the spinning/sintering processes, or after OCM (oxidative coupling of methane) reactions. Since XRD measurement of the reacted membranes was conducted after finishing all the oxidative coupling of methane reactions which lasted for about four weeks, it suggests the LSCF material is very stable in the perovskite structure.

3.4.4 Mechanical Strength

The mechanical strength of a ceramic membrane in either disc shape or in hollow fibre form can be obtained from the three point bend test, which is a well known mechanical strength test. It will accurately give properties such as the Youngs modulus and the tensile strength

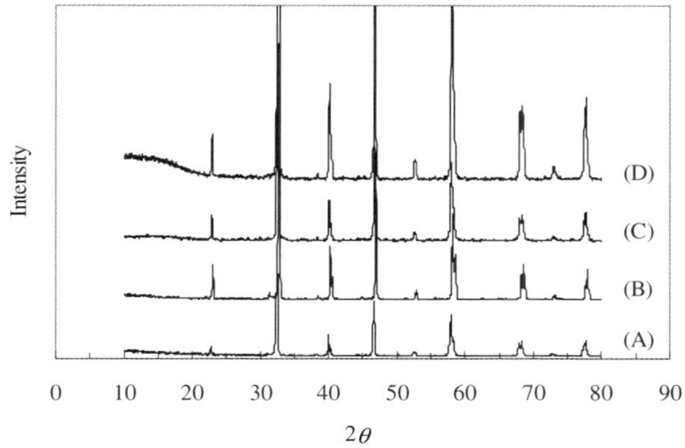

Figure 3.29 XRD patterns of the LSCF raw material and the hollow fibre membrane: (A) membrane material in powder; (B) fresh membrane in powder; (C) reacted membrane in powder; (D) reacted membrane in fibre on the outer surface [40]. Copyright (2006) American Chemical Society, Industrial and Engineering Chemistry Research, **45** (1), Tan, X. and Li, K., *Oxidative coupling of methane in a perovskite hollow-fiber membrane reactor*, p. 142–149

of a material. These properties govern the eventual strength of the membrane and are a function of several factors whilst making the ceramic membranes.

The mechanical strength of a ceramic hollow fibre membranes was measured by Liu *et al.* [35, 39] through a bending test performed using an Intron Model 4466 provided with a load cell for 1 kN. The hollow fibre samples were fixed in a sample holder with a spec gauge length of 15 mm. The crosshead speed was set at 0.02 cm min^{-1}. The load needed to break the hollow fibre was recorded for each sample with which the bending strength σ_F is calculated using the following equation [41]:

$$\sigma_F = \frac{8F_m L D_o}{\pi(D_o^4 - D_i^4)} \tag{3.43}$$

where, F_m is the force at which fracture takes place, L, D_o and D_i are the length, outer diameter and inner diameter of the hollow fibre, respectively. The same test has also been used for other membranes by Tan *et al.* [23] and Liu *et al.* [25].

NOTATION

A_p	Total pore areas, m^2
A_t	Total membrane area, m^2
C_{O_2}	O_2 concentration, mol m^{-3}
C_f	Solute concentration in the feed, mol m^{-3}
C_p	Solute concentration in the permeate, mol m^{-3}
C_w	Solute concentration at the membrane interface, mol m^{-3}
d	Collision diameter of gas molecular, m
d_p	Pore diameter, m
D_i	Inner diameter of a hollow fibre, m
D_o	Outer diameter of a hollow fibre, m
F_m	Force used for test the fibre bending strength, N

F_s	Total flow rate of the sweep gas stream, $mol\,m^{-3}$
G	Dimensionless factor used in Equation (3.38)
J	Permeation flux, $mol\,m^{-2}\,s^{-1}$ or $m^3\,m^{-2}\,s^{-1}$ ($m^3(STP)m^{-2}s^{-1}$ for gases)
l	Pore length, m
L	Length of hollow fibre membrane, m
L_p	Effective pore length, m
n	Number of pores
N	Avogadro number
N_t	Total molar permeation flow rate, $mol\,s^{-1}$
p	Applied pressure, Pa
p_a	Atmospheric pressure, Pa
p_o	Saturated pressure, Pa
p_o	Initial measured pressure, Pa
p_t	Final measured pressure, Pa
$\overline{p}$	Mean pressure, defined as $\overline{p} = (p_1 + p_2)/2$, where p_1 and p_2 are the upstream and downstream pressures, respectively.
Δp	Pressure difference across a membrane, Pa
P	Permeability coefficient, $mol\,m\,m^{-2}\,s^{-1}Pa^{-1}$
$\overline{P}$	Gas permeance, $mol\,m^{-2}\,s^{-1}\,Pa^{-1}$
Q	Permeation flow rate, $m^3\,s^{-1}$
Q_a	Energy released or adsorbed per gram of a sample, $J\,g^{-1}$
r	Local pore radius, m
r_K	Kelvin radius, m
r_p	Pore radius, m
r_m	Mean pore radius, m
r_{max}	Maximum pore radius, m
R	Ideal gas constant
R	Solute rejection
R^o	Intrinsic solute rejection
R_{in}	Inner radius of hollow fibre membrane, cm
R_o	Outer radius of hollow fibre membrane, m
T	Operating temperature, K
T	Phase transition temperature of liquid held in membrane pores, °C, Equation (3.9).
T_o	Normal phase transition temperature of liquid, °C, Equation (3.9)
t	Thickness of the adsorbed layer in the pore wall, m
t	Measured time, s
V	Molar volume of liquid, $m^3\,mol^{-1}$
V	Pore volume, m^3
V	Volume of the test cylinder m^3
V_t	Volumetric flow rate of gas, $m^3(STP)s^{-1}$
x	Concentration, %

Greek Letters

δ	Thickness of a membrane, m
ε	Membrane surface porosity, defined as $\varepsilon = A_p/A_t$
σ	Surface tension of liquid condensate, $N\,m^{-1}$

σ	Dimensionless standard deviation
σ_F	Bending strength, MPa
ρ	Density, kg m^{-3}
μ	Viscosity, Pa s
τ	Tortuosity
θ	Contact angle
γ	Interfacial tension, N m^{-1}
λ	Mean free path of gas

Subscripts

Ar	Argon
i	Oxygen or nitrogen
k	Knudsen flow
v	Poiseuolle flow
O_2	Oxygen
s	Slip flow

REFERENCES

1. Flegler, S. L., Heckman, J. W. and Klomparens, K. L., *Scanning and Transmission Electron Microscopy: An Introduction.* (1995), New York: Oxford University Press.
2. Quate, C. F., *The AFM as a tool for surface imaging.* Surface Science, **300** (1–3): 980–995 (1994).
3. Julbe, A. and Ramsay, J. D. F., *Methods for the characterisation of porous structure in membrane materials*, in *Fundamentals of Inorganic Membrane Science and Technology*, Burggraaf, A. J. and Cot, L., Editors. (1996) Amsterdam: Elsevier Science B.V., p. 67–118.
4. Benfer, S., Arki, P. and Tomandl, G., *Ceramic membranes for filtration applications – preparation and characterization.* Advanced Engineering Materials, **6** (7): 495–500 (2004).
5. Mulder, M., *Basic Principles of Membrane Technology.* 2nd ed. (1996) Dordrecht, The Netherlands: Kluwer Academic Publishers.
6. Hubbard, A. T., *The Handbook of Surface Imaging and Visualization* (1995) CRC Press.
7. Cortalezzi, M. M., Rose, J., Barron, A. R. and Wiesner, M. R., *Characteristics of ultrafiltration ceramic membranes derived from alumoxane nanoparticles.* Journal of Membrane Science, **205** (1–2): 33–43 (2002).
8. Sing, K. S. W., Everett, D. H., Haul, R. A. W., Moscou, L., Pierotti, R. A., Rouquerol, J. and Siemieniewska, T., *Reporting physisorption data for gas solid systems with special reference to the determination of surface-area and porosity.* Pure and Applied Chemistry, **57** (4): 603–619 (1985).
9. Miyata, T., Endo, A., Ohmori, T., Akiya, T. and Nakaiwa, M., *Evaluation of pore size distribution in boundary region of micropore and mesopore using gas adsorption method.* Journal of Colloid and Interface Science, **262** (1): 116–125 (2003).
10. Halsey, G., *Physical adsorption on non-uniform surfaces.* The Journal of Chemical Physics, **16** (10): 931–937 (1948).
11. Avery, R. G. and Ramsay, J. D. F., *Sorption of nitrogen in porous compacts of silica and zirconia powders.* Journal of Colloid and Interface Science, **42** (3): 597–606 (1973).
12. Horvath, G. and Kawazoe, K., *Method for the calculation of effective pore-size distribution in molecular-sieve carbon.* Journal of Chemical Engineering of Japan, **16** (6): 470–475 (1983).

13. Saito, A. and Foley, H. C., *Curvature and parametric sensitivity in models for adsorption in micropores.* American Institute of Chemical Engineers Journal, **37** (3): 429–436 (1991).
14. Elmarraki, Y., Cretin, M., Persin, M., Sarrazin, J. and Larbot, A., *Elaboration and properties of TiO_2-$ZnAl_2O_4$ ultrafiltration membranes.* Materials Research Bulletin, **36** (1–2): 227–237 (2001).
15. Eyraud, C., Betemps, M., Quinson, J. F., Chatelut, F., Brun, M. and Rasneur, B., *Determination of the pore-size distribution of an ultrafilter by – gas-liquid permporometry measurement-comparison between flow porometry and condensate equilibrium porometry.* Bulletin De La Societe Chimique De France Partie I-Physicochimie Des Systemes Liquides Electrochimie Catalyse Genie Chimique, **(9–10)**: I237-I244 (1984).
16. Cao, G. Z., Meijerink, J., Brinkman, H. W. and Burggraaf, A. J., *Permporometry study on the size distribution of active pores in porous ceramic membranes.* Journal of Membrane Science, **83** (2): 221–235 (1993).
17. Almandoz, M. C., Marchese, J., Pradanos, P., Palacio, L. and Hernandez, A., *Preparation and characterization of non-supported microfiltration membranes from aluminosilicates.* Journal of Membrane Science, **241** (1): 95–103 (2004).
18. Brun, M., Lallemand, A., Quinson, J. F. and Eyraud, C., *New method for simultaneous determination of size and shape of pores – thermoporometry.* Thermochimica Acta, **21** (1): 59–88 (1977).
19. Cuperus, F. P., Bargeman, D. and Smolders, C. A., *Critical-points in the analysis of membrane pore structures by thermoporometry.* Journal of Membrane Science, **66** (1): 45–53 (1992).
20. Jakobs, E. and Koros, W. J., *Ceramic membrane characterization via the bubble point technique.* Journal of Membrane Science, **124** (2): 149–159 (1997).
21. Petropoulos, J. H. and Petrou, J. K., *Possibilities of structural characterization of porous solids by fluid flow methods.* Separations Technology, **2** (4): 162–175 (1992).
22. Li, K., Kong, J., Wang, D. L. and Teo, W. K., *Tailor-made asymmetric PVDF hollow fibres for soluble gas removal.* American Institute of Chemical Engineers Journal, **45** (6): 1211–1219 (1999).
23. Tan, X., Liu, S. and Li, K., *Preparation and characterization of inorganic hollow fibre membranes.* Journal of Membrane Science, **188** (1): 87–95 (2001).
24. Present, R. D., *Kinetic Theory of Gases.* (1958) New York: McGraw-Hill.
25. Liu, L., Tan, X. and Liu, S., *Yttria stabilized zirconia hollow fibre membranes.* Journal of the American Ceramic Society, **89** (3): 1156–1159 (2006).
26. Kong, J. and Li, K., *An improved gas permeation method for characterising and predicting the performance of microporous asymmetric hollow fibre membranes used in gas absorption.* Journal of Membrane Science, **182** (1–2): 271–281 (2001).
27. Kreyszig, E., *Advanced Engineering Mathematics.* 7th ed. (1993) New York: John Wiley & Sons, Inc.
28. Zydney, A. L., Aimar, P., Meireles, M., Pimbley, J. M. and Belfort, G., *Use of the log-normal probability density function to analyze membrane pore size distributions: functional forms and discrepancies.* Journal of Membrane Science, **91** (3): 293–298 (1994).
29. Liepmann, H. W., *Gas kinetics and gas dynamics of orifice flow.* Journal of Fluid Mechanics, **10**: 65–79 (1961).
30. Koonaphapdeelert, S. and Li, K., *Preparation and characterization of hydrophobic ceramic hollow fibre membrane.* Journal of Membrane Science, doi:10.1016/j.memsci.2006.12.039 (2007).
31. Ho, W. S. W. and Sirkar, K. K., eds. *Membrane Handbook.* (1992), Van Nostrand Reinhold: New York.
32. Silva, P. and Livingston, A. G., *Effect of solute concentration and mass transfer limitations on transport in organic solvent nanofiltration – partially rejected solute.* Journal of Membrane Science, **280** (1–2): 889–898 (2006).
33. Leenaars, A. F. M. and Burggraaf, A. J., *The preparation and characterization of alumina membranes with ultra-fine pores .4. Ultrafiltration and hyperfiltration experiments.* Journal of Membrane Science, **24** (3): 261–270 (1985).
34. Tan, X., Liu, Y. and Li, K., *Mixed conducting ceramic hollow-fibre membranes for air separation.* American Institute of Chemical Engineers Journal, **51** (7): 1991–2000 (2005).

35. Liu, Y., Tan, X. and Li, K., *$SrCe_{0.95}Yb_{0.05}O_{3-\alpha}$ hollow-fibre membrane and its property in proton conduction,* American Institute of Chemical Engineers Journal, **52** (4): 1577–1585 (2006).
36. Tenelshof, J. E., Bouwmeester, H. J. M. and Verweij, H., *Oxygen-transport through $La_{1-x}Sr_xFeO_{3-\delta}$ membranes .1. Permeation in Air/He gradients.* Solid State Ionics, **81** (1–2): 97–109 (1995).
37. Tan, X., Liu, Y. and Li, K., *Preparation of LSCF ceramic hollow fibre membrane for oxygen production by a phase-inversion/sintering technique.* Industrial and Engineering Chemistry Research, **44**: 61–66 (2005).
38. Cheng, S. G., Gupta, V. K. and Lin, J. Y. S., *Synthesis and hydrogen permeation properties of asymmetric proton-conducting ceramic membranes.* Solid State Ionics, **176** (35–36): 2653–2662 (2005).
39. Liu, Y., Tan, X. and Li, K., *$SrCe_{0.95}Yb_{0.05}O_{3-\alpha}$ (SCYb) hollow fibre membrane: Preparation, characterization and performance.* Journal of Membrane Science, **283** (1–2): 380–385 (2006).
40. Tan, X. and Li, K., *Oxidative coupling of methane in a perovskite hollow-fibre membrane reactor.* Industrial & Engineering Chemistry Research, **45** (1): 142–149 (2006).
41. Cernica, J. N., *Strength of Materials.* 2nd ed. (1977) New York: Holt, Rinchart&Winston, p. 469.
42. de Vos, R. M. and Verweij, H., *Improved performance of silica membranes for gas separation.* Journal of Membrane Science, **143** (1–2): 37–51 (1998).
43. Qi, X. W. and Lin, Y. S., *Electrical conducting properties of proton-conducting terbium-doped strontium cerate membrane.* Solid State Ionics, **120** (1–4): 85–93 (1999).

4

Transport and Separation of Gases in Porous Ceramic Membranes

4.1 INTRODUCTION

Separation of gases using ceramic membranes is a dynamic and growing field. Intense research efforts have been made in developing novel ceramic membranes for use in separations/reactions where harsh feed conditions or high temperature operations precludes the use of existing polymeric membranes.

There are two types of ceramic membranes suitable for gas separations: (1) dense and (2) porous, especially microporous, membranes. The dense ceramic membranes are made from crystalline ceramic materials such as perovskites or fluorites, which allow permeation of only oxygen or hydrogen through the crystal lattice. Therefore, they are mostly impermeable to all other gases, giving extremely high selectivity towards oxygen or hydrogen. Dense ceramic membranes for oxygen and hydrogen permeation is a rapidly growing field. Detailed coverage of these membranes for oxygen or hydrogen separation will be given in Chapters 6 and 7, respectively. Thus, in this chapter, porous membranes for gas separation will be the main objective.

Microporous ceramic membranes with pore sizes less than 2 nm are mainly composed of amorphous silica or zeolites. These membranes are usually prepared as a thin film supported on a macroporous ceramic support, which provides mechanical strength, but offers minimal gas transfer resistances. However, in most cases, some intermediate layers are required between the macroporous support and the top separation layer to bridge the gap between the large pores of the support and the small pores of the top separation layer. In principle, separation properties of a multilayer ceramic membrane, such as permselectivity, are dependent only on the microstructure of the top separation layer. However, they can be compromised if resistances in the intermediate layers and the macroporous support become significant.

Separation of a gas mixture can be achieved based on differences in their molecular masses, size or shape, or on differences in the affinity of the gas molecules to the membrane surface. Therefore, any theoretical models used in analyzing the transport data of gases through a porous ceramic membrane with a distributed pore size must take the following contributions into consideration: (1) viscous flow, (2) slip flow, (3) Knudsen flow and (4) molecular sieving effects.

In this chapter, various ceramic/inorganic membranes for gas separation are reviewed after a brief discussion on performance evaluators of gas separation membranes. Gas transport through a mono pore size ceramic membrane is then outlined, followed by the development of a model which includes different contributions for a porous membrane with a given pore size distribution for the purpose of analysis of gas permeation data. Gas transport through an asymmetric or a multilayered membrane will also be described using a resistance model. Finally, the engineering considerations of gas permeator design for various operating conditions will be provided.

4.2 PERFORMANCE INDICATORS OF GAS SEPARATION MEMBRANES

Figure 4.1 illustrates the simplest description of a membrane system used for gas separation. As can be seen, a membrane is placed in a vessel so that two compartments, i.e. upper and lower compartments, are established. As a gas mixture containing gases *A* and *B* flows along the membrane in the upper compartment, one of the gases permeates preferentially and is enriched in the lower compartment. The fraction of the feed that permeates across the membrane is defined as the permeate and the fraction that fails to pass through, the retentate.

When selecting a membrane for separation of a gas mixture, there are two main parameters that need to be considered.

Firstly, the mole flux across the membrane, J (mol m^{-2} s^{-1}), the amount of material permeated from the feed per square metre of membrane each second. For a binary component system with linear relation of sorption and diffusion characteristics, the mole fluxes for the components *A* and *B* across the membrane can be respectively expressed as:

$$J \times y_A = \frac{P_A}{\delta}(p_{\mathrm{h}} x_A - p_{\mathrm{l}} y_A) \tag{4.1}$$

$$J \times y_B = \frac{P_B}{\delta}(p_{\mathrm{h}} x_B - p_{\mathrm{l}} y_B) \tag{4.2}$$

where J is the total mole flux of the gas mixture permeating through a membrane, in mol m^{-2} s^{-1}, p_{h} and p_{l} are the operating pressures at the retentate and permeate sides, respectively, in Pa, P_A and P_B, the permeabilities of gases *A* and *B*, respectively, in mol m Pa^{-1} m^{-2} s^{-1}, y_A is the mole fraction of the more permeable component in the permeate side, y_B, the mole fraction of the less permeable component in the permeate side, x_A, the mole fraction of the

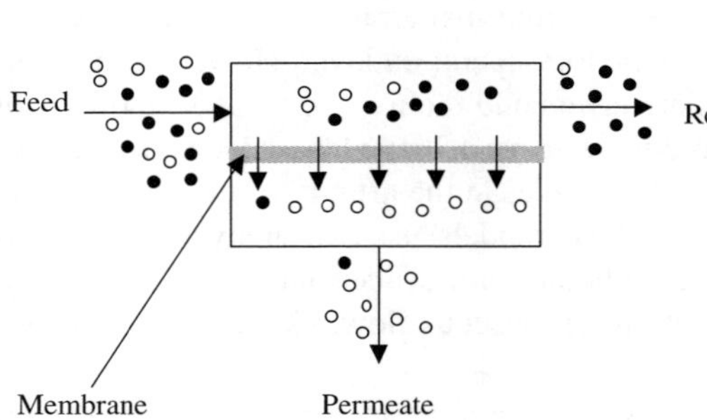

Figure 4.1 A membrane process

more permeable component in the retentate and x_B, the mole fraction of the less permeable component in the retentate, δ is the membrane thickness, in m.

Secondly, the membrane separation factor, which is a measure of the quality of separation that the membrane provides:

$$S_{AB} = \frac{y_A / y_B}{x_A / x_B} \tag{4.3}$$

It can be seen from Equation (4.3) that the higher the value of S_{AB}, the greater the degree of separation offered by the membrane. As $S_{AB} \rightarrow \infty$ so the membrane tends towards being superselective with respect to the component A.

Rearranging Equations (4.1) and (4.2) gives:

$$\alpha = \frac{P_A}{P_B} = \frac{\dfrac{y_A}{y_B}}{\dfrac{(p_h x_A - p_l y_A)}{(p_h x_B - p_l y_B)}} \tag{4.4}$$

where α is the ratio of the permeabilities between gases A and B, defined as the selectivity of the membrane. When the permeate pressure approaches to zero, i.e. $p_l \rightarrow 0$ the selectivity of the membrane for gases A and B is equal to the separation factor:

$$\alpha = \frac{P_A}{P_B} = S_{AB}\big|_{(p_l=0)} \tag{4.5}$$

When looking at membranes to enhance performance, there is often a trade off between the separation factor and flux, performing a modification that increases one leads to a decrease in the other. As an attempt to define the separation ability of a membrane, an index called the separation index (SI, $kg\,m^{-2}\,h^{-1}$), defined as $SI = JS_{AB}$, is often employed. This factor can be useful in comparing similar membranes, however it fails to distinguish the overall performance of the membrane since a membrane with a low separation factor and high flux can have the same SI as one with a high separation factor and a low flux. The particular separation requirements for an individual process need to be assessed and an appropriate membrane selected; just selecting the membrane with the highest SI may not always be the optimum choice for a process.

4.3 CERAMIC MEMBRANES FOR GAS SEPARATION

According to the IUPAC definition, microporous membranes are referred to as those with a pore diameter smaller than 2 nm. There are two main types of microporous membranes used in gas separations, namely crystalline zeolite membranes and XRD amorphous membranes such as silica, carbon, etc. The practically useful crystalline microporous membranes have polycrystalline structures, consisting of many crystallites packed together without any crystallite (grain) boundary gap in the ideal case [1]. Among the zeolite membranes mainly MFI type membranes have been successfully used so far for H_2 separation. Since zeolite, silica and carbon microporous membranes have been developed mainly for gas separations, they will be reviewed in the following section even although the latter two are not considered to be ceramic membranes.

4.3.1 Zeolite Membranes

Zeolites are crystalline microporous aluminosilicate materials with a regular three dimensional pore structure, which is relatively stable at high temperatures. They are currently used as catalysts or catalyst supports for a number of high temperature reactions. Efforts in preparation of polycrystalline zeolite membranes started in the late 1980s, but not until the early 1990s, were MFI type zeolite membranes with very good permeation and separation properties successfully prepared [1]. Since then, zeolite membrane synthesis has become a very active research field and a considerable number of research articles have been published.

The unique properties of zeolite membranes are: (1) their size and shape selective separation behaviour and (2) their thermal and chemical stabilities, which are also the general advantages of ceramic membranes. Due to their 'molecular sieve' function, zeolite membranes can principally discriminate the components of gaseous or liquid mixtures dependent on their molecular size. In order to perform the molecular sieving function, the membranes must have negligible amounts of defects and pinholes of larger than 2 nm. However, since templates are often used in zeolite membrane synthesis, removal of the template which is usually accompanied by shrinkage of the zeolite crystallites, can often create or enlarge the intercrystalline pores [2]. Consequently, all zeolite membranes reported so far contain intercrystalline nonzeolitic pores in addition to the well defined zeolitic pores. For good quality membranes, the nonzeolitic pores should be smaller than 2 nm.

There are basically two approaches to synthesizing zeolite membranes: liquid phase synthesis and the vapour phase transport (VPT) method [3]. The liquid phase synthesis approach is to bring the surface of a porous support in contact with a zeolite synthesis solution (sol or gel) and keep the system under controlled conditions so that the zeolite can nucleate and grow to a continuous film on the support surface. The vapour phase transport method treats a dry precursor gel deposited on a support with saturated steam at a similar temperature. With this method, the membrane thickness may be well controlled since the amount of nutrient for growing zeolite is directly controlled by the amount of gel applied. However, the crystals produced are most likely randomly oriented. Furthermore, since the density of gel is much lower than that of zeolite crystals, the crystallization process produces a large number of intercrystalline voids, which must be filled through a dissolution–recrystallization process. As a result, it is difficult with this method to make a thin and crack free gel film on a porous support.

For the liquid phase synthesis method, the challenges are the intrusion of zeolite into the pores of the support layer and the sealing of the intercrystalline void to make the zeolite layer continuous and pinhole free. Nucleation and deposition of zeolite inside the pores has a negative effect on the permeation properties of the resulting membrane. In order to avoid the penetration of sol into the pores, different approaches, including filling of pores with oil, positioning the support just slightly above the liquid surface and utilizing reverse osmosis pressure, have been applied. On the other hand, in order to produce a thin zeolite film without intercrystalline voids, the crystal seeding technique has been used during the last few years. By decoupling the nucleation step (at high supersaturation) from crystal growth (at low supersaturation), the seeds can grow in low concentration solutions under suppression of secondary nucleation. There are two established methods for attaching the seeds to the surface of the support [4]: (1) by changing the pH of the solution to adjust the zeta potential between the ceramic support (e.g. α-Al_2O_3) and the SiO_2 seeds (e.g. silicalite I) or (2) by adsorbing cationic oligomers/polymers on the support to create a positive surface charge to which the negatively charged zeolite seeds become attached [5].

The channel orientation of the zeolite crystals and seeded growth of an oriented particle monolayer show an important influence on the separation performance of the membrane. For example, MFI type zeolite membranes, which have been mostly studied due to their relatively easy preparation as well as their high temperature (>900 °C) and chemical stabilities, often show a crystallographic orientation of the *c* axis of the zeolite layer perpendicular to the plane of the support surface [6]. This *c* oriented MFI membrane is less favourable to gas permeation. By seeding of the support, MFI zeolite membrane layers of different crystallographic orientations can be obtained. Recently, Lai *et al.* have prepared a *b* oriented MFI silicalite I membrane [7]. They used relative large seeds and attached the seeds as a *b* oriented monolayer to the support surface. By using di- and trimers of tetrapropylammoniumhydroxide (TPAOH), the growth of the seeds in the *b* direction could be enhanced. The resulting polycrystalline films are 1 μm thin and consist of large *b* oriented single crystals with straight channels running down the membrane thickness. This *b* oriented MFI membrane exhibits much higher flux and selectivity compared to the normal *c* oriented membrane.

4.3.2 Silica Membranes

Microporous silica (SiO_2) membranes are prominent representatives of amorphous membranes. The first successful silica membranes for gas permeation/separation with good quality and high flux were prepared in 1989 using a sol-gel method [8] where SiO_2 polymer sols were firstly prepared by acid catalysed hydrolysis of tetraethoxysilane (TEOS) in alcoholic solution. The acid catalyst reduces hydrolysis but enhances polycondensation rates during the sol preparation process resulting in a polymeric sol containing silica particles of fractal structure. Then the porous support, often in a tubular configuration, is dipped in the SiO_2 gel solutions for some time, followed by wiping to remove excess gel. Finally, the deposited silica film is washed with distilled water, dried at 100–200 °C and calcined at around 450 °C. Such processes of dipping and heating are repeated several times so as to control the membrane thickness and compactness as well.

In sol-gel techniques for preparing silica membranes, process parameters strongly affect the flux and selectivity of the resulting membranes and thus must be precisely controlled [9–11]. In particular, the degree of branching of the polymeric sol turns out to be very important for the pore size. In general, lowly branched inorganic polymers lead to gels which can result in membranes with a smaller mean pore size [9, 12]. Spraying hot colloidal sols on hot supports at around 190°C, which is characterized by quick drying may give a narrow pore system for gas separation [13]. By well controlled processing of every membrane fabrication step, SiO_2 membranes with a H_2 permeance at 200 °C of 2×10^{-5} mol m^{-2} s^{-1} Pa^{-1} and permselectivities of H_2/other components of >1000 can be obtained [14].

The concept of template agents, using amphiphilic systems, was used to prepare tailored silica membranes with the sol-gel method [15–17]. The nonionic surfactant introduced into the refluxed TEOS sols can modify the sol-to-gel transition with a template effect or can participate in the formation of gels through a self organization process leading to an organized gel structure. As a result, the resulting silica membranes exhibit an ordered microporosity with the pore size directly related to the size of the surfactant molecules.

It is known that the membrane pore size strongly depends on the particle size in sol solutions. However, if the colloidal particles are very small in size compared to the pore size of the support, these particles can significantly penetrate into pores of the support resulting in deterioration of permeability of the membrane. To avoid this problem, two or more layers with interconnected pores may be applied. Such a membrane may be synthesized by sequentially dipping a porous support in several sols in descending order of sol size, which have been prepared by acid catalysed hydrolysis of tetraethylorthosilicate (TEOS), using different molar ratios of H_2O/TEOS and of HNO_3/TEOS [18, 19]. This membrane has a one to two orders of magnitude of higher gas permeation rate than those previously prepared by the normal sol-gel technique.

A major problem for microporous silica membranes is that these membranes are not stable at high temperatures, especially in humid atmospheres, leading to loss of permeability [1]. This is attributed to the closure of membrane pore channels by densification ($\equiv$Si—OH+$\equiv$Si—OH$\Rightarrow$ $\equiv$Si—O—Si$\equiv$), which is catalysed by moisture, particularly at higher temperatures. In addition, densification can also cause embitterment of silica film and damage the integrity of the film and the support, deteriorating the separation properties of the membrane.

Several methods can be potentially used to improve the stability of sol-gel derived silica membranes. One approach is to dope a small amount of inorganic oxides such as TiO_2, ZrO_2, Fe_2O_3, Al_2O_3, NiO, etc. [20, 21]. Of these metal doped membranes, Ni doped silica membranes showed relatively high H_2 permeance and high stability against water vapour at 35–300 °C. However, the pore sizes of silica–zirconia microporous membranes appear larger than the pure silica membranes [22]. It is difficult to prepare microporous binary membranes with lower silica content or microporous nonsilica membranes.

Another approach is to make the silica membrane hydrophobic by replacing —OH groups on the pore surface with —CH_3 groups using a hydrophobic methyl template covalently bonded to silica in the sol-gel process [23]. It, in principle, can reduce the condensation reaction rate at high temperatures (in an oxygen free atmosphere), improving the membrane's thermal stability. Giessler and coworkers [24] incorporated a short carbon chain cationic surfactant as a template to the sol-gel process for silica hydrophobicity. The resulting membranes showed superior microstructure and hydrophobicity over standard and methyl templated silica, without compromise to pore size alteration. They later successfully applied this technology to practical use in a hydrophobic silica membrane reactor [25]. However, such hydrophobic microporous silica membranes are limited to the applications of low temperature and nonoxidative environments.

Chemical vapour deposition (CVD) is another method used in preparation of microporous silica membranes. It was first reported in 1989 by Gavalas *et al.* [26], who deposited inside the pore wall of a Vycor glass tube using SiH_4 and O_2 or $SiCl_4$ and H_2O as the reactants. Figure 4.2 illustrates the CVD process in preparation of a silica membrane with the aid of evacuation at operating temperatures of 400–500 °C [27].

An important property of the CVD derived silica membranes is that they have pores with diameters less than 1 nm and indicate molecular sieving performance. Therefore, the membranes show enhanced selectivity, but low permeability. In contrast, the sol-gel derived membranes generally possess lower selectivity but higher permeability. In addition, the sol-gel method suffers from a lack of reproducibility (although easy to perform), but the CVD method usually requires substantial capital investment and controlled conditions of deposition.

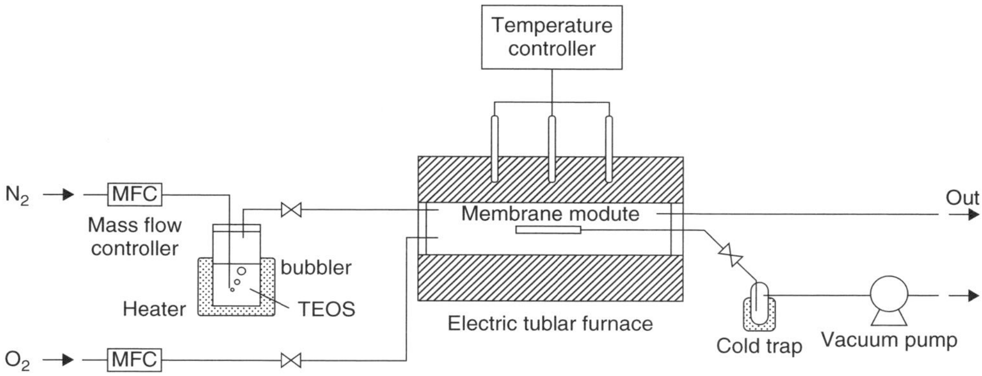

Figure 4.2 Schematic diagram of the apparatus used for CVD [27]. Reprinted from Journal of Membrane Science, **160** (1), Kuraoka *et al.*, *Permeation of...*, p. 31–39. Copyright (1999), with permission from Elsevier

4.3.3 Carbon Membranes

Carbon membranes are highly selective due to their pores of molecular dimensions. They are relatively inexpensive membranes and are prepared basically by carbonizing organic polymers as starting materials at high temperatures under controlled conditions. It is expected that carbonized materials are stable at high temperatures and resist chemical attack.

The challenge for carbon membranes is how to increase the gas permeation rate. One approach is to make the membranes on mesoporous substrates. For example, carbon membranes were prepared by ultrasonic deposition of polyfurfuryl alcohol on a porous inorganic support, followed by pyrolysis at 473–873 K to convert the polymer layer to microporous carbon film [28]. The resulting carbon membranes give an oxygen to nitrogen separation factor of about 30, the largest reported for any membranes at low temperatures. Both kinetic and equilibrium factors may contribute to such a large separation factor. This requires development of a delicate technique for formation of a continuous, defect/pinhole free, thin microporous carbon film on a porous support.

Another approach is using asymmetric hollow fibre membrane precursors [29, 30]. The hollow fibres are preferable because of their high packing density (membrane area per unit volume of vessel) and ease of module construction. An additional advantage is that the carbonization of original polymeric fibres of several hundred micrometers in diameter can be carried out in a continuous process. Carbon hollow fibre membranes, having an asymmetric structure, showed a large permeability and a high selectivity for some gases and light hydrocarbons [31].

4.4 TRANSPORT MECHANISMS

Porous ceramic membranes possess different types of pore structures depending on the formation process. They can be highly connected and tortuous or nonconnected and straight. Pores in ceramic membranes can be classified according to their size, as macropores (>500 Å), mesopores (500–20 Å) or micropores (<20 Å) [32]. Assuming that pores in membranes are composed of a bundle of capillary tubes, transport of gases through the capillary tubes can be described based on the kinetic theory of gases. Three types of mechanism have been proposed for gas transport through capillaries, namely: (1) Knudsen flow (2) slip flow and (3) viscous flow. In the case of uniform sized capillaries, it has been found that, depending on the relative magnitudes of the pore radius, r_p, and mean free path, λ, of the gas, gas molecules pass through the capillary by any one of the mechanisms stated above. Liepmann [33] suggested that when $(r_p/\lambda) < 0.05$, Knudsen type flow is predominant. According to Kong and Li [34], slip flow occurs in the range $(r_p/\lambda) = 0.05$ to 3 and viscous flow occurs when $(r_p/\lambda) > 3$. The mean free path is defined as:

$$\lambda = \frac{RT}{\sqrt{2}\pi d^2 N \bar{p}} \tag{4.6}$$

where R is the gas constant, T, the temperature, in K, d, the collision diameter of gas molecules, in m, N, the Avogadro number; $\bar{p}$, the average pressure across the membrane defined as $\bar{p} = (p_h + p_l)/2$ where p_h and p_l are the upstream and downstream pressures, respectively, in Pa.

4.4.1 Knudsen and Slip Flow

Knudsen mode of gas transport is important when the mean free path of the gas molecules is much greater than the pore size of the membrane, i.e. $(r_p/\lambda) < 0.05$ [33]. In such situations the collisions of the molecules with the pore wall are more frequent than the collisions among molecules. This mechanism is often predominant in macroporous and mesoporous membranes [32, 35–37] and is described by the Knudsen Equation [38] for diffusive flow of molecules in a capillary tube in this regime:

$$J_k = -D_k\left(\frac{dC}{dz}\right) \tag{4.7a}$$

or

$$J_k = -\frac{D_k}{RT}\left(\frac{dp}{dz}\right) \tag{4.7b}$$

where D_k is the Knudsen diffusion coefficient. It is dependent on the mean molecular speed, u and pore radius, r_p and is given by:

$$D_k = \frac{2}{3} u r_p \tag{4.8}$$

An expression for the mean molecular speed can be obtained from the kinetic theory of gases [39, 40]:

$$u = \sqrt{\frac{8RT}{\pi M}} \tag{4.9}$$

where R and T are the gas constant and temperature, respectively, and M is the molecular weight of the gaseous species, in g mol^{-1}. Substituting Equations (4.8) and (4.9) into Equation (4.7b) and then integrating, the permeation rate based on the Knudsen diffusion mechanism can be obtained for a single capillary tube:

$$n_{\mathrm{k}} = \left(\frac{32\pi}{9MRT}\right)^{1/2} r_{\mathrm{p}}^{3}\,\frac{p_{\mathrm{h}} - p_{\mathrm{l}}}{l} \tag{4.10}$$

where n_{k} is the permeation rate, defined as;

$$n_{\mathrm{k}} = \pi r_{\mathrm{p}}^{2} J_{\mathrm{k}} \tag{4.11}$$

For free molecular transport in a porous medium, the Knudsen expression has to be modified according to the structure of the porous medium. This has been carried out by Mason and Malinauskas [41] who accounted for the number of capillaries per unit volume, i.e. porosity, ε, and the complexities of the pore structure, i.e. tortuosity, τ. Thus Equation (4.8) becomes:

$$D_{\mathrm{k}} = \frac{2}{3}\frac{\varepsilon}{\tau} u r_{\mathrm{p}} \tag{4.12}$$

It can be seen from the above Equations that free molecular diffusion is a selective transport mechanism, because the flux is inversely proportional to the molecular mass of a species. This is due to the decrease in mean molecular velocity with increasing molecular weight. The decrease in the flux with increasing temperature is due to the fact that the concentration of an ideal gas (p/RT) decreases linearly with temperature, while the molecular velocity increases with the square root of temperature.

Similarly, the slip flow has been derived as [39]:

$$n_{\mathrm{s}} = \frac{\pi r_{\mathrm{p}}^{3}(p_{\mathrm{h}} - p_{\mathrm{l}})}{Mul} \tag{4.13}$$

where u, the mean speed, is given by Metz [40]:

$$u = \sqrt{8RT/\pi M} \tag{4.14}$$

4.4.2 Viscous Flow

When the mean free path is much smaller than the pore size, i.e. $(r_{\mathrm{p}}/\lambda) > 3$ [34], gas diffusion occurs primarily through molecule–molecule collisions. If a pressure gradient is applied to such a pore regime, bulk or laminar flow occurs. Therefore, the molar flux of gases can be described by Hagen–Poiseuille law as long as the gas velocity near the pore walls equals to zero:

$$J_{\mathrm{v}} = -\frac{r_{\mathrm{p}}^{2}}{8\mu}\frac{p}{RT}\left(\frac{\mathrm{d}p}{\mathrm{d}z}\right) \tag{4.15}$$

At the steady state, the flux through the membrane pores at any position is equal. Therefore, integrating Equation (4.15) for a single capillary tube, the gas permeation rate can be obtained as:

$$n_{\mathrm{v}} = \frac{\pi r_{\mathrm{p}}^4 \bar{p}(p_{\mathrm{h}} - p_{\mathrm{l}})}{8\mu RTl} \tag{4.16}$$

where

$$n_{\mathrm{v}} = \pi r_{\mathrm{p}}^2 J_{\mathrm{v}} \tag{4.17}$$

Considering the membrane pore structure effects, Equation (4.15) is modified as:

$$J_{\mathrm{v}} = -\frac{\varepsilon}{\tau}\frac{r_{\mathrm{p}}^2}{8\mu}\frac{p}{RT}\left(\frac{\mathrm{d}p}{\mathrm{d}z}\right) \tag{4.18a}$$

or

$$J_{\mathrm{v}} = \frac{\varepsilon}{\tau}\frac{r_{\mathrm{p}}^2}{8\mu}\frac{\bar{p}}{RTl}(p_{\mathrm{h}} - p_{\mathrm{l}}) \tag{4.18b}$$

4.4.3 Surface Flow

Surface flow occurs when the permeating species exhibit a strong affinity for the membrane surface and adsorb along the pore walls at sufficiently low temperature and/or high pressure. When the adsorbed molecules are mobile, surface flow can take place. The extent of surface flow is determined by the amount of molecules adsorbed and their mobility. Also, surface flow often occurs in parallel with other transport mechanisms such as Knudsen or viscous flow [32], depending on the pore characteristics of the membranes. Many descriptions of surface diffusion have been proposed. An excellent review on surface diffusion has been given by Uhlhorn and Burggraaf [42] who describe the surface transport models available in literature in the following three categories:

- The hydrodynamic model: developed with the consideration that the adsorbed gas is in a form of liquid film (condensed gas), gliding along the surface under the influence of a pressure gradient [43].
- The hopping model: developed with the assumption that molecules can hop over the surface. The surface flux is calculated by the mean hopping distance and the velocity with which the molecules leave their site. Weaver and Metzner [44] developed a detailed model to calculate the mean hopping distance, while Ponzi *et al.* [45] developed a simpler way to estimate the mean hopping distance.
- The random walk mode: developed based on the two dimensional form of Fick's law and has been used by Ash *et al.* [46] and Okazaki *et al.* [47].

An expression for surface flow derived by Gilliland *et al.* [43] based on the hydrodynamic model from a two dimensional force balance of an adsorbed film is given by:

$$N_{\mathrm{su}} = \frac{A_{\mathrm{p}} R T \rho_{\mathrm{app}}}{1000 \tau^2 C_{\mathrm{R}} S_{\mathrm{s}} l} \int \frac{x^2}{p} \mathrm{d}p \tag{4.19}$$

where N_{su} is the surface flow through a membrane, A_p is the cross sectional area of pores for a given membrane, in m^2, S_s is the specific surface area of the solid over which the adsorbed gas molecules are mobile, in m^2, C_R is the coefficient of resistance, in $kg\,m^{-2}\,s^{-1}$, τ, the tortuosity factor, l, the pore length, in m; ρ_{app}, the apparent density of the membrane, in $kg\,m^{-3}$ and x, the amount of adsorbed gas per unit weight of the membrane, in $kmol\,kg^{-1}$.

4.4.4 Capillary Condensation [37, 48, 49]

Capillary condensation is a form of surface flow because at a certain critical relative pressure, normally determined by the Kelvin equation, pores of ceramic membranes could get completely filled by a gas if it is condensable. Due to the formation of menisci at both ends of the pore, transport can take place through hydrodynamic flow driven by a capillary pressure difference between the two ends. This mechanism of gas transport can be regard as the ultimate limit of the process of adsorption as pressure is increased. Theoretically, capillary condensation can be employed to achieve a very high selectivity, as the formation of the liquid like dense layer of the condensable gas will block and prevent the flow of the noncondensable gas [37, 48].

Once capillary condensation takes place, the transport behaviour changes because of the slow permeation characteristics of liquid compared to that of vapour [49]. Consequently the permeability values drop as soon as all the pores are filled with liquid. A plot of permeability versus relative pressure was made by Uhlhorn *et al.* [48] to illustrate this behaviour. As can be seen in Figure 4.3, the permeability value increases because of

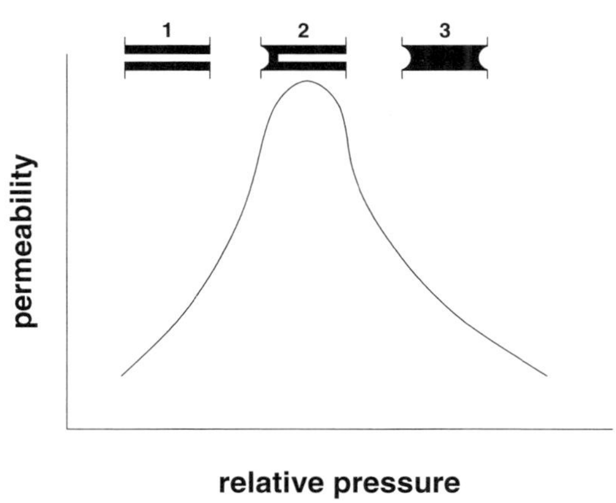

Figure 4.3 Schematic plot of the permeability versus the relative pressure in the case of multilayer diffusion and capillary condensation. The maximum gives the point of capillary condensation [48]. Reprinted from Journal of Membrane Science, **66** (2–3), Uhlhorn *et al.*, *Gas transport . . .* , p. 259–269. Copyright (1992), with permission from Elsevier

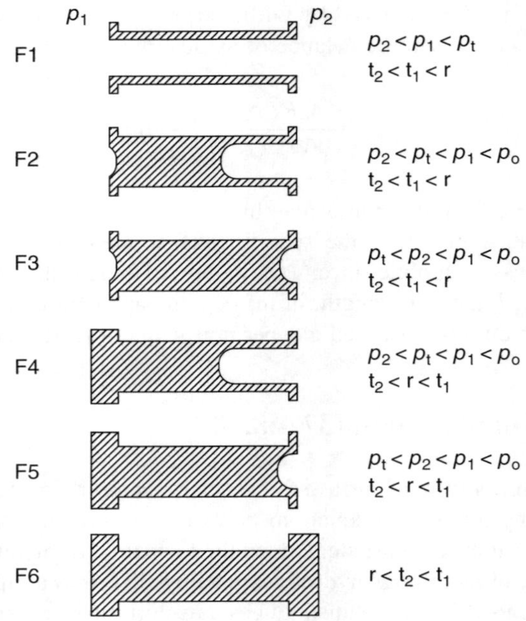

Figure 4.4 Six different flow modes in multilayer diffusion and capillary condensation and their boundary conditions [49]. Reprinted from Journal of Colloid and Interface Science, **110** (2), Lee and Hwang, *The transport of . . .*, p. 544–555. Copyright (1986), with permission from Elsevier

monolayer and subsequent multilayer diffusion and it reaches a maximum once capillary condensation starts. The liquid meniscus is subsequently formed. After this point the permeability decreases due to the liquid transport contribution. Lee and Hwang [49] used six different flow modes to describe the capillary condensation. As illustrated in Figure 4.4, the pressures p_1, and p_2 are the upstream and downstream pressures respectively ($p_1 > p_2$). The capillary condensation pressure p_t, at temperature T is predicted from the Kelvin equation:

$$\frac{\rho RT}{M}\ln\frac{p_t}{p_o} = -\frac{2\sigma\cos\theta}{r_p} \tag{4.20}$$

where ρ is the density of the condensate, in kg m^{-3}, p_o, the temperature dependent saturation vapour pressure for a planar interface, in Pa, σ the surface tension of liquid condensate, in N m^{-1}, θ, the contact angle and r_p the radius of a cylindrical capillary, in m.

The layer thickness of the gas absorbed, t_i, in Figure 4.4 at a certain pressure, p_i and temperature, T, is predicted by:

$$t_i = q(p_i, T)\frac{V_S}{S_W} \tag{4.21}$$

where $q(P,T)$ is the amount adsorbed, in $g\,g^{-1}$, V_S, the specific volume of the adsorbate, in $m^3\,g^{-1}$ and S_W, the specific surface area of the sample, in $m^2\,g^{-1}$. The amount adsorbed can be predicted from the following Equation [50]:

$$\frac{p_r}{q(1-p_r)} = \frac{1}{q_m C} + \frac{(1-C)}{q_m C} p_r \qquad (4.22)$$

where q_m is the monolayer adsorption, in $g\,g^{-1}$, C, the BET constant and p_r, the relative pressure defined as p/p_o. The relative pressure depends on pressure as well as temperature. Equation (4.22) can be applied in a range of temperatures and pressures.

As discussed by Lee and Hwang [49], six possible flow modes as shown in Figure 4.4 for a cylindrical capillary in the case of multilayer diffusion and capillary condensation being available. Mode 1 in Figure 4.4 represents the multilayer diffusion stage. The upstream and downstream pressures are smaller than the capillary condensation pressure. The adsorbate layer thickness is smaller than the pore radius. Mode 2 in Figure 4.4 represents the case where capillary condensation occurs at the upstream side (thus $p_1 > p_t$). A stable meniscus is present at the upstream side (thus $t_1 < r$). The downstream pressure p_2 is, however, still smaller than the capillary condensation pressure. Consequently somewhere inside the pore the capillary pressure p_t is reached ($p_1 > p_t > p_2$). At this point another meniscus occurs, but the matching adsorbate layer thickness (Equation 4.21) is smaller than t_1. As the pore is slowly filled with liquid upon increasing p_2, mode 3 is reached. Thus, capillary condensation exists downstream ($p_2 > p_t$, $t_2 < r$) and the whole pore is therefore filled with liquid. Once the upstream pressure is further increased but the downstream pressure remains smaller than the capillary condensation pressure, mode 4 is obtained. The upstream end of the pore will be filled with bulk condensate ($t_1 > r$). Mode 5 is the limiting case of mode 4. Now the whole pore is filled with condensate ($p_2 > p_t$, $t_2 < r$). Mode 6 finally presents the case where the whole pore is filled with bulk condensate ($t_2 > r$). To each of the above described flow modes, mathematical formulations can be applied and were presented by Lee and Hwang [49].

4.4.5 Configurational or Micropore Diffusion

This type of diffusion has been considered as surface diffusion in the limit where the pore size becomes comparable to the molecular size. In this mechanism, diffusion is perceived as an 'activated' process and separation is a strong function of molecular shape and size, pore size and interactions between the pore wall and gas molecules. This type of mechanism is dominant in microporous zeolite membranes [37].

Gas permeation flux through microporous ceramic membranes occurs by solid state diffusion of molecules. The permeation flux can be derived from Fick's first law:

$$J_m = -\tilde{D}\frac{dC}{dx} \qquad (4.23)$$

where $\tilde{D}$ is the chemical diffusion coefficient, C, the local concentration of molecules and x, the coordinate along the permeation direction.

Normally, microporous diffusion is assumed to be the rate controlling step. Therefore, gas phase diffusional resistance is negligible, and thermodynamic equilibrium can be assumed at the interface. The concentrations of molecules at both membrane surfaces can thus be derived from equilibrium gas phase adsorption data, which in some cases can be described by the Henry's law equation:

$$C = Sp \tag{4.24}$$

where S is a proportionality constant that depends on temperature according to:

$$S = S_o \exp\left(-\frac{\Delta H_s}{RT}\right) \tag{4.25}$$

where S_o is a temperature independent proportionality constant, ΔH_s, the isosteric heat of adsorption, R, the gas constant and T, the absolute temperature. The validity of Henry's law implies that the concentration of molecules absorbed in the microporous solid is small compared to the number of available sites (pores). In the Henry (or Langmuir) law adsorption regime, $\tilde{D}$ is independent of the concentration [51]. Its temperature dependence is given by:

$$\tilde{D} = \tilde{D}_o \exp\left(-\frac{E_D}{RT}\right) \tag{4.26}$$

where $\tilde{D}_o$ is a temperature independent proportionality constant and E_D, the positive mobility energy. According to atomic jump theory [52], E_D represents the energy barrier between two adjacent sorption sites.

Rearrangement of Equations (4.23) to (4.26) and then integration with the condition that $\tilde{D}$ is independent of C yield:

$$J_m = \tilde{D}_o S_o \exp\left(-\frac{\Delta H_s + E_D}{RT}\right)\frac{p_h - p_l}{\delta} = J_o \exp\left(-\frac{E_a}{RT}\right)\frac{p_h - p_l}{\delta} \tag{4.27}$$

where p_h and p_l are the operating pressures at up and down streams, in respectively, Pa and δ is the membrane thickness.

In Equation (4.27) J_o is a temperature independent proportionality constant defined as

$$J_o = \tilde{D}_o S_o \tag{4.28}$$

and E_a is the effective activation energy for permeation defined as:

$$E_a = \Delta H_s + E_D \tag{4.29}$$

4.4.6 Simultaneous Occurrence of Different Mechanisms [53]

The transport mechanisms mentioned above may occur simultaneously due to various reasons. First, the pores on the membrane surface may not be uniform and a distribution of the membrane pore size often exists due to the current fabrication techniques used. As a result, Knudsen diffusion, slip flow and viscous flow can occur simultaneously. Second, when the membrane pores are large and the surface of the membrane shows a strong affinity for the gas molecules to be transported, surface flow together with other diffusional flows coexist. As illustrated in Figure 4.5, for a gas flowing through a single capillary, two distinct phases may be assumed. One is the layer of adsorbed gas molecules which are at the gas–pore wall interface and under a strong interactive force from the pore wall, while the other is the gas molecules that are free from the influence of the pore wall. The former is considered to be a condensed phase and the latter is in a gas phase and under no restriction from the pore wall. Obviously, when the pore diameter is reduced to twice of the adsorbed layer thickness or less, only surface flow would prevail.

If a surface of porous membrane is composed of a bundle of capillary tubes with distributed pore sizes, a Gaussian normal distribution of pore size represented by the function can be assumed:

$$N(r) = \frac{n}{\sqrt{2\pi}\sigma} \exp\left[-\frac{1}{2}\left(\frac{r - r_{\mathrm{m}}}{\sigma}\right)^2\right] \tag{4.30}$$

where r and r_{m} are the pore radius and mean pore radius, respectively, in m, $N(r)$, the normal pore size distribution function, in m^{-1}, n, the total number of pores on a given membrane surface and σ, the standard deviation of the pore size distribution, in m.

As mentioned above, for free gas flowing in a single pore, three flow mechanisms, i.e. viscous flow, slip flow and Knudsen flow, can be observed depending on the size of the pore relative to the mean free path of the permeating gas molecules, defined in Equation (4.6) and for Knudsen flow, n_{k}:

$$n_{\mathrm{k}} = \left(\frac{32\pi}{9MRT}\right)^{1/2} r_{\mathrm{p}}^3 \frac{p_{\mathrm{h}} - p_{\mathrm{l}}}{l} \qquad \text{when } r_{\mathrm{p}}/\lambda < 0.05 \tag{4.10}$$

for slip flow, n_{s}:

$$n_{\mathrm{s}} = \frac{\pi r_{\mathrm{p}}^3 (p_{\mathrm{h}} - p_{\mathrm{l}})}{Mul} \qquad \text{when } 0.05 < r_{\mathrm{p}}/\lambda < 3 \tag{4.13}$$

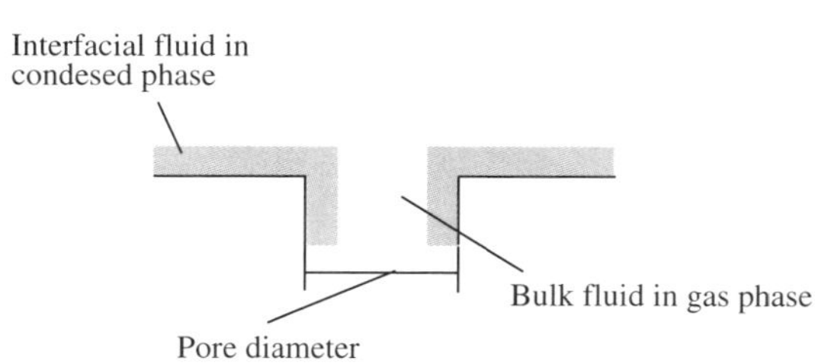

Figure 4.5 Modes of gases flowing through a single capillary

where $u = \sqrt{8RT/\pi M}$;
for viscous flow, n_v:

$$n_v = \frac{\pi r_p^4 \bar{p}(p_h - p_l)}{8\mu RTl} \quad \text{when } r_p/\lambda > 3 \tag{4.16}$$

The total transport of the free gas through all the pores is given by:

$$N_g = N_k + N_s + N_v = \sum_{r_p=0}^{r_p=0.05\lambda} N(r)n_k + \sum_{r_p=0.05\lambda}^{r_p=3\lambda} N(r)n_s + \sum_{r_p=3\lambda}^{r_p=r_{max}} N(r)n_v \tag{4.31}$$

As Gaussian normal distribution of pore size given by Equation (4.30) is assumed, the summation in Equation (4.31) can be replaced by integration to give the following expression:

$$\begin{aligned} N_g = & \frac{n}{l}\left(\frac{32\pi}{9MRT}\right)^{1/2} \frac{p_h - p_l}{\sqrt{2\pi}\sigma} \int_0^{0.05\lambda} r^3 \exp\left[-\frac{1}{2}\left(\frac{r - r_m}{\sigma}\right)^2\right] dr \\ & + \frac{n}{l}\left(\frac{\pi}{Mu}\right) \frac{p_h - p_l}{\sqrt{2\pi}\sigma} \int_{0.05\lambda}^{3\lambda} r^3 \exp\left[-\frac{1}{2}\left(\frac{r - r_m}{\sigma}\right)^2\right] dr \\ & + \frac{n}{l}\left(\frac{\pi\bar{p}}{8\mu RT}\right) \frac{p_h - p_l}{\sqrt{2\pi}\sigma} \int_{3\lambda}^{r_{max}} r^4 \exp\left[-\frac{1}{2}\left(\frac{r - r_m}{\sigma}\right)^2\right] dr \end{aligned} \tag{4.32a}$$

or

$$N_g = \frac{n}{l}(G_1 I_1 + G_2 I_2 + G_3 I_3) \times (p_h - p_l) \tag{4.32b}$$

where:

$$G_1 = \left(\frac{32\pi}{9MRT}\right)^{1/2} \qquad G_2 = \frac{\pi}{Mu} \qquad G_3 = \frac{\pi\bar{p}}{8\mu RT} \tag{4.32c}$$

and

$$I_1 = \frac{1}{\sqrt{2\pi}\sigma} \int_0^{0.05\lambda} r^3 \exp\left[-\frac{1}{2}\left(\frac{r - r_m}{\sigma}\right)^2\right] dr \tag{4.32d}$$

$$I_2 = \frac{1}{\sqrt{2\pi}\sigma} \int_{0.05\lambda}^{3\lambda} r^3 \exp\left[-\frac{1}{2}\left(\frac{r - r_m}{\sigma}\right)^2\right] dr \tag{4.32e}$$

$$I_3 = \frac{1}{\sqrt{2\pi}\sigma} \int_{3\lambda}^{r_{max}} r^4 \exp\left[-\frac{1}{2}\left(\frac{r - r_m}{\sigma}\right)^2\right] dr \tag{4.32f}$$

For condensed gas flowing through the membrane, Gilliland *et al.* [43] derived an expression as shown in Equation (4.19) for surface flow based on a two dimensional force balance on an adsorbed film:

$$N_{\mathrm{su}} = \frac{A_{\mathrm{p}} RT \rho_{\mathrm{app}}}{1000\tau^2 C_{\mathrm{R}} S_{\mathrm{s}} l} \int \frac{x^2}{p} \mathrm{d}p \tag{4.19}$$

Equation (4.19) can be recast to fit the terminology of the present approach to include pore size distribution with the following expressions:

surface flow: N_{su}

$$A_{\mathrm{p}} = \pi \int_0^{r_{\mathrm{Max}}} N(r) r^2 \mathrm{d}r = \pi I_4 \tag{4.33a}$$

$$S_{\mathrm{s}} = 2\pi l \int_0^{r_{\mathrm{Max}}} N(r) r \mathrm{d}r = 2\pi l I_5 \tag{4.33b}$$

In view of Equations (4.33a) and (4.33b), Equation (4.19) becomes:

$$N_{\mathrm{su}} = \frac{RT \rho_{\mathrm{app}}}{2000\tau^2 C_{\mathrm{R}} l^2} \frac{I_4}{I_5} \int_{p_{\mathrm{l}}}^{p_{\mathrm{h}}} \frac{x^2}{p} \mathrm{d}p \tag{4.34}$$

if the adsorption isotherms follow Henry's law form, i.e.:

$$x = k_{\mathrm{H}} p \tag{4.35}$$

Substituting Equation (4.35) into Equation (4.34) and then integrating:

$$N_{\mathrm{su}} = C \frac{I_4}{I_5} \bar{p} (p_{\mathrm{h}} - p_{\mathrm{l}}) \tag{4.36}$$

where:

$$C = \frac{RT \rho_{\mathrm{app}}}{2000\tau^2 C_{\mathrm{R}} l^2} k_{\mathrm{H}}^2 \tag{4.36a}$$

The total gas flow through the membrane is the summation of the free gas flow and condensed flow (surface flow):

$$N_{\mathrm{t}} = N_{\mathrm{g}} + N_{\mathrm{su}} = \frac{n}{l} (G_1 I_1 + G_2 I_2 + G_3 I_3)(p_{\mathrm{h}} - p_{\mathrm{l}}) + C \frac{I_4}{I_5} \bar{p} (p_{\mathrm{h}} - p_{\mathrm{l}}) \tag{4.37}$$

Equation (4.37) indicates the relation between the total permeation rate and operating conditions. Obviously, permeation contributions from the free gas flow must be suppressed as much as possible so that a better degree of separation can be achieved.

4.5 MODIFICATION OF POROUS CERAMIC MEMBRANES FOR GAS SEPARATION

As mentioned above, there are five different mechanisms available for gas transport across porous membranes. The relative contributions of the different mechanisms are dependent on the properties of the membranes and gases, as well as on operating conditions such as temperature and pressure. For ceramic membranes with pore size at 2 nm or slightly greater, Knudsen diffusion is the main dominant mechanism at low pressures and elevated temperatures because capillary condensation and surface diffusion may not be possible at elevated temperatures. Also, the contribution of viscous flow, resulting from a pressure difference across the pores, would be quite small and negligible at low pressures. Finally, micropore diffusion would not take place, because the pore size of 2 nm is much larger than the gas molecules. Thus, it leaves Knudsen diffusion as the only transport mechanism contributing to the separation of gases from gas mixtures at elevated temperatures and low pressures. As discussed in Section 4.4.1, gas permeation by Knudsen diffusion varies inversely with the square root of the molecular weight. Therefore, the selectivity of the membrane for a binary gas mixture is equal to the inverse of the square root of the ratio of the molecular masses.

Gas separation using porous ceramic membranes under the Knudsen flow regime shows very low selectivity for a membrane system. To overcome this deficiency, various surface modification techniques can be employed to promote other modes of gas transport.

Asaeda *et al.* [54] modified an alumina membrane by partially plugging the membrane pores with hydroxide to reduce the pore size to less than 2 nm. Their modified membranes showed a separation factor of above 60 for water–ethanol gaseous mixtures. Using the modified membrane operated under capillary condensation mechanism at 70 °C, the azeotropic points for water–ethanol and water–isoproponal mixtures could be bypassed. Modification of a γ-alumina membrane with impregnation of NaOH was carried out by Sperry *et al.* [55] who used the modified membrane for separation of CH_3OH from a H_2/CH_3OH mixture. Because the CH_3OH was condensed in the pores at temperatures between 373 and 473 K for a certain pressure range and was preferentially removed through the pores, the selectivity of the membrane obtained for CH_3OH/H_2 was as high as 600, i.e. three orders of magnitude higher than that occurring in the Knudsen regime. In both the cases, the increase in selectivity was due to the effect of capillary condensation, which can only take place for condensable components and hence has considerable restrictions in applications.

Modification of porous ceramic membranes to promote surface diffusion has been one of the favourable methods for enhancing the membrane selectivity for gas separations. Keizer *et al.* [56] prepared a membrane consisting of a thin and ultrafine porous (2–4 nm) γ-alumina active layer supported by a substrate with large pores. They observed that surface diffusion of CO_2 in the γ-alumina active layer was found. The surface flow rate depends on surface concentration and surface diffusion coefficient and hence on temperature and pressure. It is of the same order of magnitude as the flow rate through the gas phase. The surface diffusion rates and adsorption energy can be changed by modification of the γ-alumina with MgO. Uhlhorn *et al.* [57] modified both supported and nonsupported γ-alumina membranes with silver to enhance the surface flow of oxygen and hydrogen. Although no surface diffusion of O_2 was found on silver modified γ-alumina membrane, possibly due to sintering of the silver under oxidizing conditions, the hydrogen showed considerable mobility on silver modified γ-alumina membrane. The flux ratio of counterdiffusion measurements, with N_2 as inert gas, was three times higher than that expected. Under a reducing atmosphere, the silver dispersion is maintained, resulting in high surface flow contributions to the overall transport. Sloot *et al.* [58] studied the surface diffusion of H_2S and SO_2 in α-alumina membranes with an

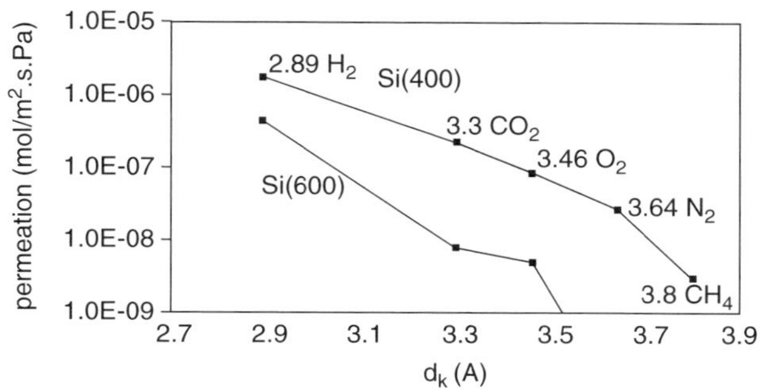

Figure 4.6 Kinetic diameter versus permeance for Si400 and Si600 membranes at 200 °C, $\Delta P = 1$ bar and a mean pressure of 1.5 bar [59]. Reprinted from Journal of Membrane Science, **143** (1–2), Vos and Verweij, *Improved performance of silica membranes for gas separation*, p. 37–51, Copyright (1998) with permission from Elsevier

average pore diameter of 350 μm, impregnated with γ-alumina at temperatures between 446 and 557 K and pressures between 2 and 6 bars. They observed that the contribution of surface diffusion to the total diffusion through the membrane can be as high as about 40 %.

Modification of porous ceramic membranes to show configurational or micropore diffusion characteristics have also been carried out extensively. In these cases, the porous membranes were used as a support where the microporous membrane such as silica was coated. The silica membrane is porous and contains pores of molecular dimensions (>0.5 nm) which can exhibit selectivity according to the size of the molecule. de Vos and Verweij [59] reported that composite microporous silica membranes could be prepared by dip coating a porous γ-alumina membrane in silica sols, followed by drying and calcination at temperatures of 400 °C and 600 °C to obtain the silica membranes Si400 and Si600, respectively, using a clean room technique to prevent mesoscopic defects. Si400 and Si600 membranes prepared are expected to have pore diameters in the range of 3.8–5.5 Å and 3.6–3.8 Å, respectively, with H_2/CH_4 selectivities of more than 500 and H_2 permeance of 2×10^{-6} mol m^{-2} s^{-1} Pa^{-1} and 6×10^{-7} mol m^{-2} s^{-1} Pa^{-1}, respectively. Figure 4.6 shows that the permeance of gases through these microporous silica membranes appears to be strongly related to their kinetic diameters as expected from the micropore diffusion mechanism [59].

Apart from silica membranes, modification of porous ceramic membranes to show configurational or micropore diffusion characteristics have also been carried out using zeolites, which are microporous cystalline alumina silicate with a uniform pore size. As mentioned before, most zeolite membranes have been synthesized using two approaches, i.e. *in situ* synthesis and seeded secondary growth. The *in situ* approach is to bring the surface of a porous support in contact with a zeolite synthesis solution (sol) and keep the system under controlled conditions so that the zeolite can nucleate and grow to a continuous film on the support surface. Many techniques have been reported for growing zeolite films on porous supports using *in situ* synthesis, all of which related to experimental conditions, such as the position of the support and the way the synthesis solution is brought in contact with the support surface, which have to be carefully controlled in order to create local supersaturation near the surface of the support. Dong and Lin illustrated schematically (Figure 4.7) an *in situ* synthesis of P-type zeolite membranes on alumina support in an autoclave [60]. They found that no or poor quality P zeolite films could be formed on the polished surface of the porous alumina disk when it is placed in the vertical or horizontal position, while a continuous and

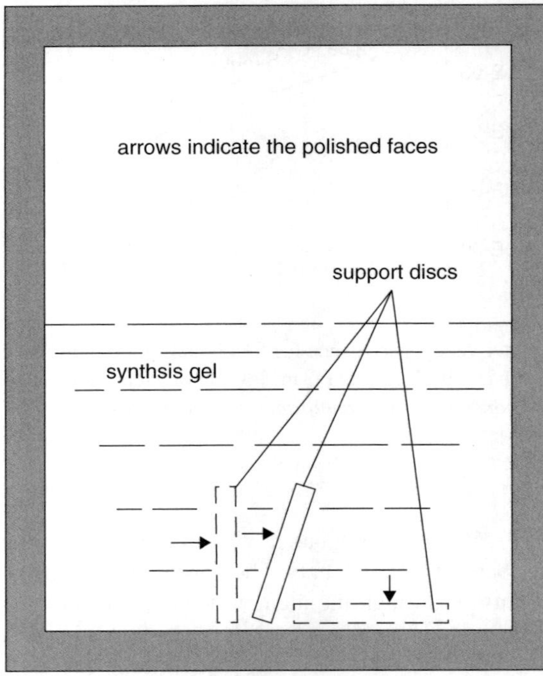

Figure 4.7 Schematic diagram of various support positions in the autoclave [60]. Reprinted with permission from Industrial and Engineering Chemistry Research, **37** (6), Dong and Lin, *In situ synthesis of P-type zeolite membranes on porous alpha-alumina supports*, p. 2404–2409. Copyright (1998) American Chemical Society

high quality P type zeolite film can grow on the alumina support only when it is placed in a slanting position.

Good zeolite membranes should contain negligible amount of defects. However, similar to polymeric membranes, most of the zeolite membranes prepared so far contain intercrystalline nonzeolitic pores in addition to the well defined zeolitic pores. The presence of the intercrystalline pores in a zeolite membrane depending on their size and number could affect the properties of the membrane for the separation of molecules. The presence of nonzeolitic pores means that zeolite membranes often become nonselective for some molecules [61], unless the nonzeolitic pores are small enough (less than 2 nm) and the zeolite material shows good sorption properties towards the molecules to be separated [62].

To minimize the formation of intercrystalline nonzeolitic pores in zeolite membranes, the seeded secondary growth approach may be employed. The basic principle of this approach is first to form a zeolite film or zeolite seeds on a porous support by physical means (such as dip coating or scrubbing the zeolite powder), followed by a conventional crystal growth process as used in the *in situ* approach (but in less stringent conditions). Although this approach requires an additional step of seeding as compared to the *in situ* approach, it offers the advantages of ease of scale up, better reproducibility and the possibility to prepare zeolite films with oriented crystal structure. Gouzinis and Tsapatsis prepared polycrystalline MFI type zeolite membranes using a seeded secondary growth method [63]. The secondary growth was conducted under high temperature synthesis conditions. The oriented silicalite membrane showed good separation for xylene isomers, indicating that the formation of intercrystalline nonzeolitic pores was negligible.

The abovementioned membranes/or intermediate supports were normally prepared via the sol-gel technique which has emerged as one of the most important methods for obtaining a thin, uniform hole free top separation layer and/or intermediate layers on a ceramic support. This technique has been extensively reviewed by Guizard [64]. The quality of the membrane is very much dependent on the preparation procedures and close adherence to a rigid protocol is necessary to obtain membranes of consistent quality.

4.6 RESISTANCE MODEL FOR GAS TRANSPORT IN COMPOSITE MEMBRANES

Microporous ceramic membranes for gas separation are usually composite ones consisting of several layers of one or more different ceramic materials. In general, they have a macroporous support, one or two mesoporous intermediate layers and a microporous top layer. The bottom layer provides mechanical support, while the intermediate layers are used to bridge the gap between the large pores in the bottom layer and the small pores of the top layer where the actual separation occurs.

As mentioned before, the separation properties of a multilayer ceramic membrane, such as permselectivity, are dependent only on the microstructure of the top separation layer. However, it can be compromised if resistances in the intermediate layers become significant. In this section, analyses will be carried out to illustrate the importance of each layer in ceramic composite membranes for gas separation.

For the simplest membrane system for gas separation as illustrated in Figure 4.1, the permeation rate, N_A for a component A through a homogeneous membrane can be written as:

$$N_A = J_A \times A_t = \frac{P_A}{\delta} \times A_t \times (p_h - p_l) = \frac{P_A}{\delta} \times A_t \times \Delta p \tag{4.38}$$

Based on Ohm's law:

$$I = \frac{E}{R} \tag{4.38a}$$

the permeation rate, N_A can be analogous to the current, the driving force, Δp, to the electrical potential and the remainder in Equation (4.38) would be resistance, which can be written as:

$$R_A = \frac{\delta}{P_A A_t} \tag{4.39}$$

Therefore, Equation (4.38) becomes

$$N_A = \frac{\Delta p}{R_A} \tag{4.40}$$

Equation (4.40) is always valid as long as the gas permeation rate is related to the pressure difference in a linear fashion.

4.6.1 Effect of Support Layers

Figure 4.8 shows two membrane layers (for example, top and support layers) of a ceramic membrane with resistances R_1 and R_2 connected in series: 1 (top separation layer) on the top of 2 (support layer). Denoting two different gas permeants by subscripts A and B, the overall resistance R for each gas permeant is:

$$(R)_A = (R_1)_A + (R_2)_A \tag{4.41a}$$

$$(R)_B = (R_1)_B + (R_2)_B \tag{4.41b}$$

Defining the selectivity (ratio of the resistances) for the gas permeants A and B as:

$$\alpha_i = \frac{(R_i)_B}{(R_i)_A} = \frac{(P_i)_A}{(P_i)_B} \tag{4.42}$$

where $i = 1$ and 2. The ratio of the overall resistance for gas permeants A and B, designated as α, becomes:

$$\alpha = \frac{(R)_B}{(R)_A} = \frac{\alpha_1(R_1)_A + \alpha_2(R_2)_A}{(R_1)_A + (R_2)_A} = \frac{\alpha_1 + \alpha_2[(R_2)_A/(R_1)_A]}{1 + [(R_2)_A/(R_1)_A]} \tag{4.43}$$

When the resistance in the support layer is much less than that in the top separation layer, i.e.,

$$\frac{(R_2)_A}{(R_1)_A} \approx 0 \tag{4.43a}$$

Equation (4.43) becomes:

$$\alpha = \alpha_1 \tag{4.44}$$

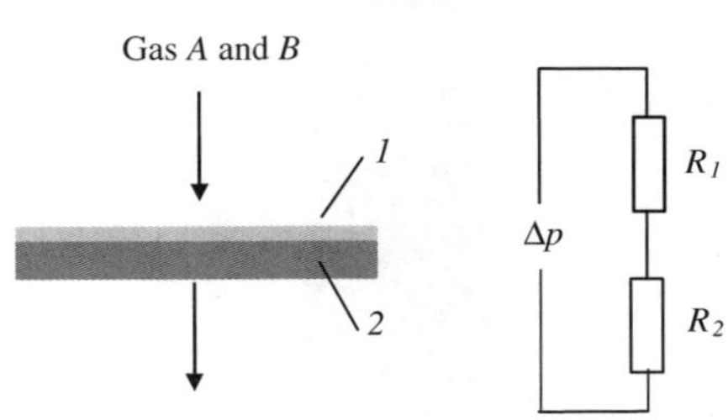

Figure 4.8 Resistances in series

Equation (4.44) indicates that the selectivity of the composite membrane is controlled by the top separation layer whose resistance is far greater than the support layer.

The above derivation indicates (especially Equation (4.43)) that both layers may equally control the selectivity when the resistances from both the top and support layers are nearly equal. Therefore, even when an extremely selective top separation layer is prepared, the overall selectivity could be lowered if the resistance of the support layer is large (i.e. thick layer with small pore sizes and low porosity).

4.6.2 Effect of Nonzeolitic Pores

The gas flow through a zeolite membrane containing intercrystalline pores (nonzeolitic pores) as described in Figure 4.9 can also be analysed using the resistance model.

It can be seen from the Figure 4.9 that the gas flows through both intercrystalline pores and zeolitic pores. The resistance to the flow through the zeolitic pores is R_2, while the resistance to the flow through the nonzeolitic pores is R_3, both of which are connected in parallel. Therefore, the overall resistance for gas A can be written as:

$$R_A = \frac{(R_2)_A \times (R_3)_A}{(R_2)_A + (R_3)_A} \tag{4.45}$$

For gas B:

$$R_B = \frac{(R_2)_B \times (R_3)_B}{(R_2)_B + (R_3)_B} \tag{4.46}$$

The selectivities for the gas permeants A and B are given as:

$$\alpha_i = \frac{(R_i)_B}{(R_i)_A} = \frac{(P_i)_A}{(P_i)_B} \quad \text{(individual)} \tag{4.47a}$$

$$\alpha = \frac{(R)_B}{(R)_A} \quad \text{(overall)} \tag{4.47b}$$

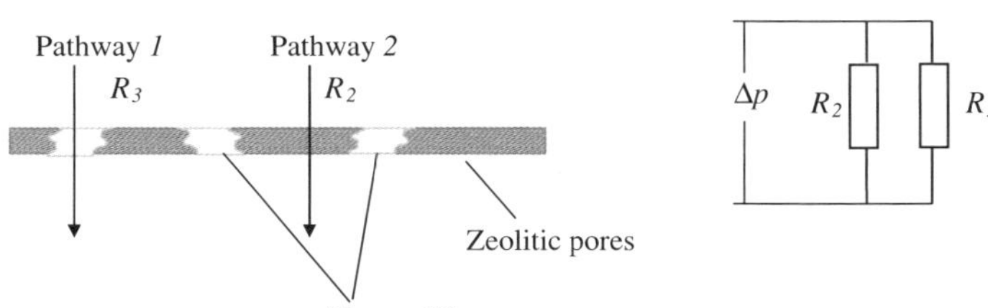

Figure 4.9 Resistances in parallel

Rearranging Equations (4.45) to (4.47), the overall selectivity, α can be written as:

$$\alpha = \frac{R_B}{R_A} = \frac{\dfrac{(R_2)_B \times (R_3)_B}{(R_2)_B + (R_3)_B}}{\dfrac{(R_2)_A \times (R_3)_A}{(R_2)_A + (R_3)_A}} = \frac{\dfrac{\alpha_2 (R_2)_A \times \alpha_3 (R_3)_A}{\alpha_2 (R_2)_A + \alpha_3 (R_3)_A}}{\dfrac{(R_2)_A \times (R_3)_A}{(R_2)_A + (R_3)_A}}$$

$$= \alpha_2 \alpha_3 \frac{(R_2)_A + (R_3)_A}{\alpha_2 (R_2)_A + \alpha_3 (R_3)_A} = \alpha_2 \alpha_3 \frac{1 + [(R_3)_A / (R_2)_A]}{\alpha_2 + \alpha_3 [(R_3)_A / (R_2)_A]} \tag{4.48}$$

Since the resistance of the nonzeolitic pores is much less than that of zeolitic pores, i.e. $R_3/R_2 = 0$, Equation (4.48) reduces to:

$$\alpha \approx \alpha_3 \tag{4.49}$$

Equation (4.49) indicates that the overall selectivity is controlled by the membrane component with nonzeolitic pores (small resistance). Therefore, the gas transport of a zeolite membrane is mostly through the nonzeolitic pores (sometimes called defects) and the selectivity is very low. When the surface area containing the nonzeolitic pores is much reduced, the resistance of R_3 is increased dramatically based on Equation (4.39). This would result in better selectivity as the gas transport is then governed by the zeolitic pores of the membrane.

4.6.3 Effect of Coating

Figure 4.10 illustrates a homogeneous film of relatively high permeability laminated on top of the zeolite membrane discussed in the above case. In addition, the nonzeolitic pores are filled with the highly permeable material.

It can also be seen from Figure 4.10 that it is assumed that the gas flow is strictly vertical to the surface of the membrane. In other words, the gas flow through portion of the homogeneous film that covers the zeolitic pores should pass through the zeolitic pores, and gas flow through the portion of the homogeneous film that covers the nonzeolitic pores should pass through the nonzeolitic pores where the highly permeable material is likely filled. No mixing is assumed between these two flows. The overall resistance for gases A and B is:

$$R_A = \frac{[(R_1)_A + (R_2)_A] \times [(R_1')_A + (R_3)_A]}{[(R_1)_A + (R_2)_A] + [(R_1')_A + (R_3)_A]} \tag{4.50}$$

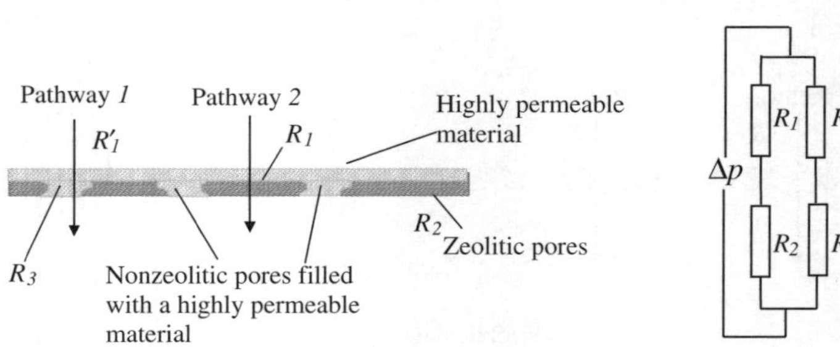

Figure 4.10 Resistances in parallel combination

For gas B:

$$R_B = \alpha R_A = \frac{[\alpha_1(R_1)_A + \alpha_2(R_2)_A] \times [\alpha_1(R_1')_A + \alpha_3(R_3)_A]}{[\alpha_1(R_1)_A + \alpha_2(R_2)_A] + [\alpha_1(R_1')_A + \alpha_3(R_3)_A]} \tag{4.51}$$

Comparing two arms of the parallel resistances, $R_1 + R_2$ and $R_1' + R_3$, the former arm will govern the overall resistance if:

$$R_1' + R_3 \gg R_1 + R_2 \tag{4.52}$$

In the arm $R_1 + R_2$, R_2 will govern the resistance of this series resistance if:

$$R_2 \gg R_1 \tag{4.53}$$

From the above comparison, it is quite obvious that R_2, which corresponds to the resistance of the zeolitic pores, can govern the overall resistance if Equations (4.52) and (4.53) are satisfied simultaneously. This means that the gas flow through nonzeolitic pores in the zeolite membranes can be stopped by the lamination of a top layer which is highly permeable. Equation (4.52) can be satisfied when the area occupied by the nonzeolitic pores, A_3, is very small, i.e. R_3 is extremely large, based on Equation (4.39).

Figure 4.11 illustrates the effect of area ratio A_3/A_2 (A_2, area of zeolitic pores and A_3, area of nonzeolitic pores) on the selectivities of a zeolite membrane and the zeolite membrane coated with a high permeable material for an A–B gas mixture. It is calculated based on the conditions given in Table 4.1 using equations mentioned above. It can be seen that without coating, the selectivity of the zeolite membrane diminishes when $A_3/A_2 > 10^{-6}$ where the coated zeolite membrane still show a very good selectivity towards gases A and B. It, thus, suggests that selectivity of zeolite membranes with small fraction of imperfection (non-zeolitic pores) can be largely restored by coating a thin layer of highly permeable materials.

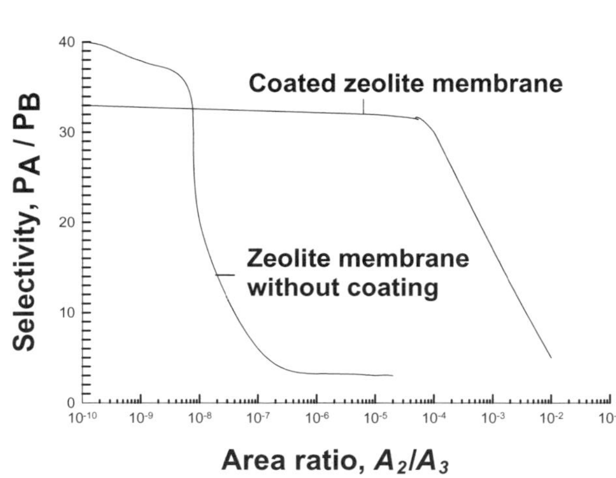

Figure 4.11 Effect of area ratio on selectivity of a zeolite membrane with and without coating

Table 4.1 Data used in Figure 4.11

Material	Permeability A	Coefficient, (Barrer)* B	Thickness, cm
Zeolite membrane	12	0.3	10^{-5}
High permeable coating	520	250	10^{-4}

$* \ 1\,Barrer = 10^{-10}\,\dfrac{cm^3(STP)\,cm}{\sec\,cm^2\,cmHg}$

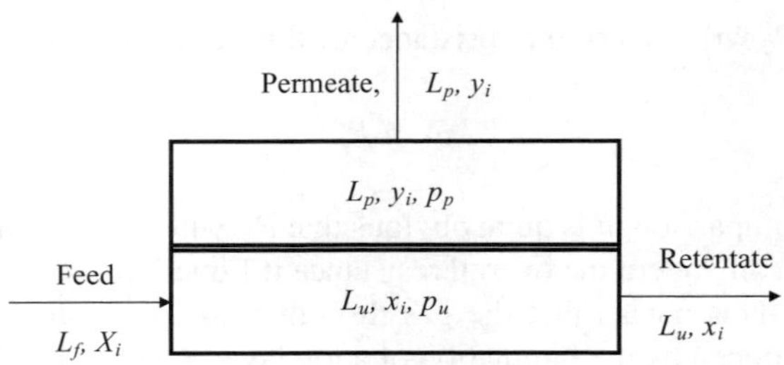

Figure 4.12 Perfect mixing

4.7 SYSTEM DESIGN

4.7.1 Operating Schemes

Gas mixtures are separated by permeation through a ceramic membrane in a device or a module, which is essentially a high pressure vessel containing the membrane in thin sheet form, or in the form of a hollow fibre. In practice, hollow fibre, tubular modules or multichannel monolithic elements are usually employed. From an engineering point of view, membrane modules are elementary separation stages whose function is very similar to those of distillation trays or other separation devices and therefore, each membrane module is considered to be a permeation stage.

For a single permeation stage, the degree of separation achievable, for a specified feed and membrane material, would depend on the temperature, the pressure at the membrane interfaces, the stage cut i.e. fraction of feed allowed to permeate, defined as $\phi = L_p/L_f$ and the flow patterns, all of which are usually termed as operation schemes. The flow patterns, which may include perfect mixing, cross flow, parallel plug flow and one side mixing, have been studied extensively in the past and are discussed below.

4.7.1.1 Perfect Mixing

The perfect mixing condition, as shown in Figure 4.12, in a membrane module is defined as the limiting condition and it is assumed that compositions of the gas mixture on the feed and permeate side of the membrane are uniform and are the same as those in the retentate and permeate streams. The separation of binary gas mixtures under such conditions in a single

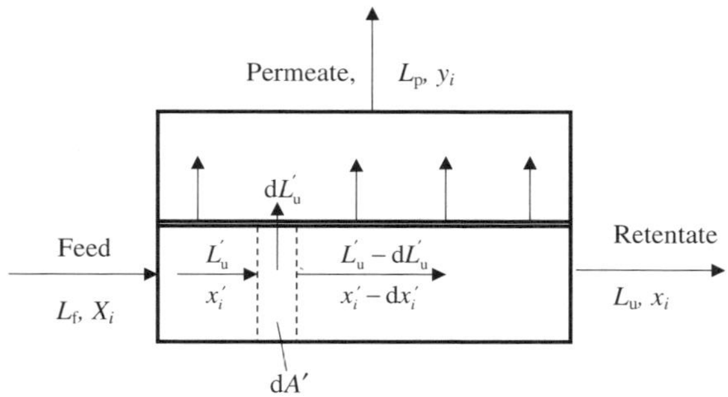

Figure 4.13 Cross flow

stage module was first analysed by Weller and Steiner [65, 66] and was later extended to multicomponent mixtures by Huckins and Kammermeyer [67, 68] and Brubaker and Kammermeyer [69]. A simple iteration method developed by Stern and Walawender [70] was also applied to multicomponent mixtures for the perfect mixing condition.

4.7.1.2 Cross Flow

In this flow pattern as can be seen in Figure 4.13, the feed side stream flows parallel to the membrane, whereas the permeate stream flows perpendicular to, and away from, the membrane. It is assumed that no mixing takes place in either stream, as would be expected if plug flow conditions applied. The composition of the permeate stream at any point near the membrane is determined by the relative rates of the permeating components at that point. The extent of separation achievable in a single stage module and the membrane area requirement with such a flow pattern was studied by Weller and Steiner [65, 66] for a binary gas mixture.

A different method of calculation was proposed by Naylor and Backer [71] for the special case when the stage enrichment factor does not vary with gas composition across the stage. The separation of a multicomponent gas mixture with the crossflow pattern was also studied by Pan and Habgood [72] and Stern and Walawender [70] who proposed an iterative computational method which could be applied to both binary and multicomponent gas mixtures.

4.7.1.3 Parallel Plug Flow

When the longitudinal velocities and the ratio of length to width of the compartments for both the permeate and retentate streams are large, both streams may be in plug flow. In this case, both cocurrent (Figure 4.14a) and countercurrent (Figure 4.14b) flow may have to be considered, depending on the location of the permeate take off. For a single stage module, these flow patterns were studied first by Oishi *et al.* [73] and by Walawender and Stern [74] for a binary gas mixture. They solved the system of ordinary differential equations that comprised the material balance for parallel flow. Later, an improved method of numerical integration coupled with experimental study was carried out by Blaisdell and Kammermeyer [75] who demonstrated experimentally the superior performance of the countercurrent flow patterns in a tubular module.

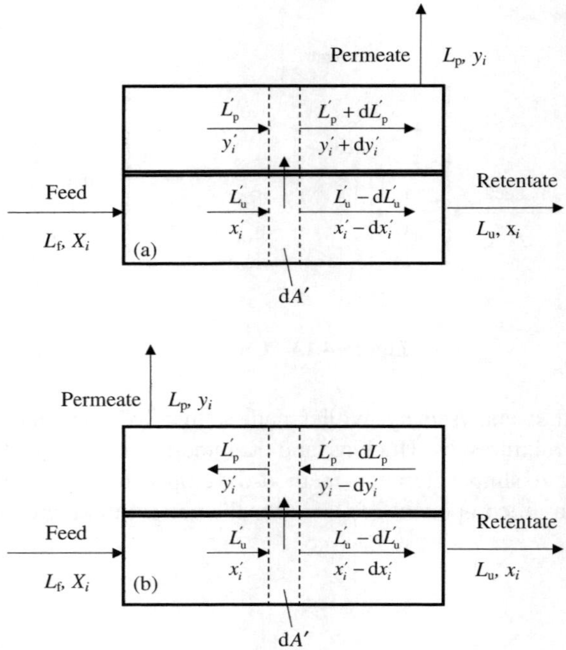

Figure 4.14 Parallel plug flow: (a) courrent mode; (b) countercurrent mode

4.7.2 Design Equations for Membrane Processes in Gas Separation

The models developed in the following sections apply to all types of configuration of membrane modules such as plate and frame, hollow fibre, tubular modules or multichannel monolithic elements, providing that the assumptions listed in the preceeding section are held.

In this section, the models for a single stage system have been provided for a multicomponent mixture with consideration of four different flow patterns, i.e. perfect mixing, crossflow, cocurrent flow and countercurrent flow.

4.7.2.1 Perfect Mixing

In the case of perfect mixing (Figure 4.12), it is assumed that the rate of mixing on the high pressure side of the membrane module is so rapid, compared with the flow rate, that the gas stream has the same composition as the retentate stream at all points along the membrane. The same assumption is made for the low pressure side. The permeation of gases through the membrane is further assumed to be the rate controlling step. Therefore, for a multicomponent gas mixture, the mass conservation over the overall membrane surface area A_t may be written for each component as:

$$L_p y_i = \frac{P_i}{\delta}(p_h x_i - p_l y_i) A_t \qquad i = 1, 2 \ldots\ldots n \tag{4.54}$$

where L_p is permeate flow, in $cm^3(STP)\,s^{-1}$. P_i is the permeability of component i, in Barrer (1 Barrer = $10^{-10}\,cm^3(STP)\,cm/sec\,cm^2\,cmHg$), A_t, the membeane area, in cm^2, δ, the thickness of the membrane, in cm, p_h and p_l are the pressures at feed and retentate sides, respectively, in cmHg and x_i and y_i are the mole fractions of ith component on the retentate and permeate sides, respectively.

The material balance equations for each component are:

$$X_i = (1-\phi)\,x_i + \phi y_i \qquad i = 1, 2 \ldots\ldots n \tag{4.55}$$

$$\sum_{i=1}^{n} x_i = 1 \tag{4.56}$$

$$\sum_{i=1}^{n} y_i = 1 \tag{4.57}$$

where X_i is the mole fraction of the ith component in the feed and ϕ, the fraction of feed allowed to permeate, usually called the 'stage cut', defined as L_p/L_f.

Rearranging Equations (4.54) to (4.57) leads to the following expression:

$$A_t = \frac{L_p y_1 \delta}{P_1\left(\dfrac{p_h}{1-\phi}(X_i - \phi y_1) - p_l y_1\right)} \tag{4.58}$$

$$y_i = \frac{p_h X_i/(1-\phi)}{\dfrac{L_p\delta}{P_i A_t} + \dfrac{\phi}{1-\phi} p_u + p_l} \qquad i = 2, 3 \ldots\ldots n \tag{4.59}$$

Equations (4.58) and (4.59) with Equations (4.55) and (4.57) have been solved by the iteration technique [76]:

(1) A value is assumed for y_1.

(2) The membrane area may be calculated from Equation (4.58).

(3) y_2 to y_n are calculated from Equation (4.59).

(4) $\sum_{i=1}^{n} y_i$ is determined and steps 1 to 3 are repeated until $\sum_{i=1}^{n} y_i = 1$.

(5) x_1 to x_n are calculated from Equation (4.55).

4.7.2.2 Cross Flow

In the case of cross flow with no mixing, the gas on the high pressure side of the stage is assumed to flow parallel to the membrane with plug flow, while the permeated gas on the low pressure side flows perpendicular to the membrane as shown in Figure 4.13.

For a multicomponent gas mixture, mass conservation over the differential membrane area dA' leads to:

$$-\mathrm{d}(L_{\mathrm{u}}'x_i') = \frac{P_i}{\delta}(p_{\mathrm{h}}x_i' - p_{\mathrm{l}}y_i')\,\mathrm{d}A' \qquad i = 1, 2 \ldots\ldots n \tag{4.60}$$

where L_{u}' is the local flow rate of the gas mixture on the high pressure side, in $\mathrm{cm^3(STP)\,s^{-1}}$. An additional assumption is made for the low pressure side of the stage that the permeated stream is swept away by convection from the outgoing membrane surface, and transport by molecular diffusion in this direction is negligible. Therefore, the mole fraction y_i' of the permeated stream can be approximated as:

$$y_i' = \frac{\mathrm{d}(L_{\mathrm{u}}'x_i')}{\mathrm{d}L_{\mathrm{u}}'} \qquad i = 1, 2 \ldots\ldots n \tag{4.61}$$

Rearranging Equations (4.60) and (4.61), a system of ordinary differential equations can be obtained for each component:

$$\frac{\mathrm{d}x_i'}{\mathrm{d}A} = -\frac{1}{L_{\mathrm{u}}'}\left[x_i'\frac{\mathrm{d}L_{\mathrm{u}}'}{\mathrm{d}A'} + \frac{P_i}{\delta}(p_{\mathrm{h}}x_i' - p_{\mathrm{l}}y_i')\right] \tag{4.62}$$

$$\frac{\mathrm{d}L_{\mathrm{u}}'}{\mathrm{d}A'} = -\frac{1}{\delta}\sum_{i=1}^{n} P_i(p_{\mathrm{h}}x_i' - p_{\mathrm{l}}y_i') \tag{4.63}$$

$$\frac{y_i'}{y_{i+1}'} = \frac{P_i}{P_{i+1}}\frac{p_{\mathrm{h}}x_i' - p_{\mathrm{l}}y_i'}{p_{\mathrm{h}}x_{i+1}' - p_{\mathrm{l}}y_{i+1}'} \qquad i = 1, 2 \ldots\ldots n \tag{4.64}$$

With the following condition:

$$\sum_{i=1}^{n} y_i' = 1 \tag{4.65}$$

the system of differential Equations (4.62) and (4.63) with Equations (4.64) and (4.65) can be been solved [76] by a standard Runge Kutta method with the initial conditions $x_i' = X_{\mathrm{f}}\, L_{\mathrm{u}}' = L_{\mathrm{f}}$ and $A' = 0$ to find:

$$x_i' = x_i \qquad y_i = \frac{X_i - (1-\phi)\,x_i}{\phi} \qquad A' = A_{\mathrm{t}}$$

$$\text{at } \phi = \frac{L_{\mathrm{f}} - L_{\mathrm{u}}'}{L_{\mathrm{f}}} \tag{4.66}$$

4.7.2.3 *Cocurrent Flow*

For the cocurrent flow case, as illustrated in Figure 4.14a, the permeate and retentate streams flow in the same direction. Plug flow has been assumed on both sides of the membrane.

For a multicomponent gas mixture, mass conservation over the differential membrane area dA' gives (same as the cross flow case):

$$-d(L_u' x_i') = \frac{P_i}{\delta}(p_h x_i' - p_l y_i')\, dA' \qquad i = 1, 2 \ldots\ldots n \tag{4.60}$$

In addition, at any point in the system downstream from the inlet, it is possible to write the overall material balance:

$$L_f = L_u' + L_p' \tag{4.66}$$

where L_u' and L_p' are the local volumetric flow rates at the high pressure and low pressure compartments, respectively, in $cm^3(STP)\,s^{-1}$. The corresponding material balances for each component are then:

$$L_f X_i = L_u' x_i' + L_p' y_i' \tag{4.67}$$

Rearranging Equation (4.60) yields the same equations as for the cross flow case Equations (4.62) and (4.63):

$$\frac{dx_i'}{dA} = -\frac{1}{L_u'}\left[x_i' \frac{dL_u'}{dA'} + \frac{P_i}{\delta}(p_h x_i' - p_l y_i')\right] \tag{4.62}$$

$$\frac{dL_u'}{dA'} = -\frac{1}{\delta}\sum_{i=1}^{n} P_i(p_h x_i' - p_l y_i') \tag{4.63}$$

Substituting Equation (4.66) into Equation (4.67) yields:

$$y_i = \frac{X_i L_f - x_i' L_u'}{L_f - L_u'} \qquad L_u' \neq L_f \qquad i = 1, 2 \ldots\ldots n \tag{4.68a}$$

$$y_i = L_u' \frac{\left.\dfrac{dx_i'}{dA'}\right|_{A'=0}}{\left.\dfrac{dL_u'}{dA'}\right|_{A'=0}} + x_i' \qquad L_u' = L_f \tag{4.68b}$$

Equation (4.68b) is obtained by application of l'Hopital's rule as $L_u' \rightarrow L_f$. Again, the system of differential Equations (4.62) and (4.63) with Equations (4.68a) and (4.68b) has been integrated [76] by a fourth order Runge Kutta method with initial conditions of $x_i' = X_f$, $L_u' = L_f$, $A' = 0$, to find:

$$x_i' = x_i, \qquad y_i' = y_i \qquad A' = A_t \tag{4.69}$$

$$\text{at } \phi = \frac{L_f - L_u'}{L_f} \tag{4.69a}$$

4.7.2.4 Countercurrent Flow

Countercurrent flow in a single permeation stage is shown in Figure 4.14b. From an analytical point of view, this case is similar to the case of cocurrent flow with the exception that the permeate and retentate streams flow in opposite directions along the membrane.

As for cocurrent flow, mass conservation over the differential membrane area dA' for a multicomponent gas mixture yields Equation (4.60):

$$-\mathrm{d}(L'_\mathrm{u}x'_i) = \frac{P_i}{\delta}(p_\mathrm{h}x'_i - p_\mathrm{l}y'_i)\,\mathrm{d}A' \qquad i = 1, 2 \ldots\ldots n \tag{4.60}$$

and therefore Equations (4.62) and (4.63) are still valid. The overall material balance for this case is

$$L'_\mathrm{u} = L_\mathrm{u} + L'_\mathrm{p} \tag{4.70}$$

with material balances for each component:

$$L'_\mathrm{u}x'_i = L_\mathrm{u}x_i + L'_\mathrm{p}y'_i \qquad i = 1, 2 \ldots\ldots n \tag{4.71}$$

Substituting Equation (4.70) into Equation (4.71) we obtain:

$$y_i = \frac{x'_iL'_\mathrm{u} - (1-\phi)L_\mathrm{f}x_i}{L'_\mathrm{u} - (1-\phi)L_\mathrm{f}} \qquad L'_\mathrm{u} \neq L_\mathrm{u} \qquad i = 1, 2 \ldots\ldots n \tag{4.72}$$

with the condition that:

$$\sum_{i=1}^{n} x_i = 1 \tag{4.73}$$

The system of differential Equations (4.62) and (4.63) with Equations (4.72) and (4.73) has been integrated with initial conditions of:

$$x'_i = X_i \qquad L'_\mathrm{u} = L_\mathrm{f} \qquad A' = 0 \qquad y'_i = y_i = \frac{X_i - (1-\phi)x_i}{\phi} \text{ to find:} \tag{4.74}$$

$$x'_i = x_i, \qquad A' = A_\mathrm{t} \tag{4.74a}$$

$$\text{at } L'_\mathrm{u} = L_\mathrm{f} - \phi L_\mathrm{f} \tag{4.74b}$$

Numerical integration of this system of equations is iterative with an initial guess of x_i in order to find a satisfactory result for $x'_i = x_i$ at $L'_\mathrm{u} = L_\mathrm{f} - \phi L_\mathrm{f}$.

Detailed discussions about the effect of flow patterns on the separation characteristics of the membrane modules and parametric studies on the membrane module performances under various operating conditions will not be carried out here, as they can be found in references [65, 66, 72, 74–76].

NOTATION

A_p Total pore areas, m^2
A_t Total membrane area, m^2 or cm^2
C Constant defined in Equation (4.36a)
C Gas concentration, $mol\,m^{-3}$
C_R Coefficient of resistance given in Equation (4.19), $kg\,m^{-2}\,s^{-1}$
d Collision diameter of gas molecular, m
d_p Pore diameter, m
$\tilde{D}$ Chemical diffusion coefficient, $m^2\,s^{-1}$
$\tilde{D}_o$ Temperature independent proportionality constant
D_k Knudsen diffusion coefficient, $m^2\,s^{-1}$
E_a Effective activation energy for permeation defined as $E_a = \Delta H_s + E_D$
E_D Positive mobility energy
ΔH_s Isosteric heat of adsorption
J Total permeation flux, $mol\,m^{-2}\,s^{-1}$
J_o Temperature independent proportionality constant defined as $J_o = \tilde{D}_o S_o$
k_H Henry's constant, Pa^{-1}
L_f Feed flow, $cm^3(STP)\,s^{-1}$
l Pore length, m
L_p Permeate flow, $cm^3(STP)\,s^{-1}$
L_u Retentate flow, $cm^3(STP)\,s^{-1}$
M Molecular weight
n Permeation flow rate for a single capillary tube, $mol\,s^{-1}$
n Total number of pores
N Avogadro number
N Permeation flow rate, $mol\,s^{-1}$
p Gas pressure, Pa
p_o Saturated pressure, Pa
p_t Capillary condensation pressure, Pa
$\overline{p}$ Mean pressure, defined as $\overline{p} = (p_h + p_l)/2$, where p_h and p_l are the upstream and downstream pressures, respectively.
Δp Pressure difference across a membrane, Pa
P Permeability coefficient, $mol\,m\,m^{-2}\,s^{-1}\,Pa^{-1}$ or Barrer,

$$1 \text{ Barrer} = 10^{-10}\,\frac{cm^3(STP)\,cm}{\sec\,cm^2\,cmHg}$$

q_m Monolayer of gas adsorption, $g\,g^{-1}$
$q\,(P,T)$ Amount of gas adsorbed, $g\,g^{-1}$
r Local pore radius, m
r_p Pore radius, m
r_m Mean pore radius, m
r_{max} Maximum pore radius, m
R Ideal gas constant
R Resistance
S Proportionality constant in Equation (4.24)
S_o Temperature independent proportionality constant
S_{AB} Separation factor
S_S Specific surface area of the solid over which the adsorbed gas molecules are mobile, m^2

S_W	Specific surface area of the sample, $m^2\,g^{-1}$
t	Layer thickness of an adsorbed gas, m
T	Temperature, K
u	Mean molecular speed, $m\,s^{-1}$
V_S	Specific volume of the adsorbate, $m^3\,g^{-1}$
x	Amount of adsorbed gas per unit weight of the membrane, $kmol\,kg^{-1}$
x	Mole fraction in retentate
x	Coordinate along the permeation direction, m
X	Mole fraction of a component in feed
y	Mole fraction in permeate
z	Location coordinate of a capillary tube, m

Greek Letters

α	Selectivity
δ	Thickness of a membrane, m
ε	Membrane surface porosity, defined as $\varepsilon = A_p/A_t$
ϕ	Stage cut, i.e. fraction of feed allowed to permeate defined as L_p/L_f
σ	Surface tension of liquid condensate, $N\,m^{-1}$
σ	Dimensionless standard deviation
ρ	Density of condensate, $kg\,m^{-3}$
ρ_{app}	Apparent density of a membrane, $kg\,m^{-3}$
μ	Viscosity, Pa s
τ	Tortuosity
θ	Contact angle
λ	Mean free path of gas

Superscripts

$'$	Local value

Subscripts

A	More permeable component
B	Less permeable component
g	Free gas flow
k	Knudsen flow
h	Upstream or Retentate
i	Component or a layer in membrane
l	Downstream or Permeate
m	Micropore
s	Slip flow
su	Surface flow
t	Total permeation flow
u	Feed or Retentate
v	Poiseuolle flow

REFERENCES

1. Lin, Y. S., *Microporous and dense inorganic membranes: current status and prospective*. Separation and Purification Technology, **25** (1–3): 39–55 (2001).
2. Dong, J., Lin, Y. S., Hu, M. Z. C., Peascoe, R. A. and Payzant, E. A., *Template-removal-associated microstructural development of porous-ceramic-supported MFI zeolite membranes*. Microporous and Mesoporous Materials, **34** (3): 241–253 (2000).
3. Chiang, A. S. T. and Chao, K. J., *Membranes and films of zeolite and zeolite-like materials*. Journal of Physics and Chemistry of Solids, **62** (9–10): 1899–1910 (2001).
4. Caro, J., Noack, M. and Kolsch, P., *Zeolite membranes: from the laboratory scale to technical applications*. Adsorption – Journal of the International Adsorption Society, **11** (3–4): 215–227 (2005).
5. Lassinantti, M., Jareman, F., Hedlund, J., Creaser, D. and Sterte, J., *Preparation and evaluation of thin ZSM-5 membranes synthesized in the absence of organic template molecules*. Catalysis Today, **67** (1–3): 109–119 (2001).
6. Wong, W. C., Au, L. T. Y., Lau, P. P. S., Ariso, C. T. and Yeung, K. L., *Effects of synthesis parameters on the zeolite membrane morphology*. Journal of Membrane Science, **193** (2): 141–161 (2001).
7. Lai, Z. P., Bonilla, G., Diaz, I., Nery, J. G., Sujaoti, K., Amat, M. A., Kokkoli, E., Terasaki, O., Thompson, R. W., Tsapatsis, M. and Vlachos, D. G., *Microstructural optimization of a zeolite membrane for organic vapor separation*. Science, **300** (5618): 456–460 (2003).
8. Uhlhorn, R. J. R., Veld, M., Keizer, K. and Burggraaf, A. J., *High permselectivities of microporous silica-modified gamma-alumina membranes*. Journal of Materials Science Letters, **8** (10): 1135–1138 (1989).
9. Elferink, W. J., Nair, B. N., DeVos, R. M., Keizer, K. and Verweij, H., *Sol-gel synthesis and characterization of microporous silica membranes. 2. Tailor-making porosity*. Journal of Colloid and Interface Science, **180** (1): 127–134 (1996).
10. Soufyani, M., Jei, T., Bourret, D., Sivade, A. and Larbot, A., *Silicone modified silica membrane. Application to the gas separation*. Separation and Purification Technology, **25** (1–3): 451–457 (2001).
11. Tsai, C.-Y., Tam, S.-Y., Lu, Y. and Brinker, C. J., *Dual-layer asymmetric microporous silica membranes*. Journal of Membrane Science, **169** (2): 255–268 (2000).
12. Schafer, R., Noack, M., Kolsch, P., Stohr, M. and Caro, J., *Comparison of different catalysts in the membrane-supported dehydrogenation of propane*. Catalysis Today, **82** (1–4): 15–23 (2003).
13. Asaeda, M. and Yamasaki, S., *Separation of inorganic/organic gas mixtures by porous silica membranes*. Separation and Purification Technology, **25** (1–3): 151–159 (2001).
14. de Vos, R. M. and Verweij, H., *High-selectivity, high-flux silica membranes for gas separation*. Science, **279** (5357): 1710–1711 (1998).
15. Xomeritakis, G., Naik, S., Braunbarth, C. M., Cornelius, C. J., Pardey, R. and Brinker, C. J., *Organic-templated silica membranes: I. Gas and vapor transport properties*. Journal of Membrane Science, **215** (1–2): 225–233 (2003).
16. Ayral, A., Balzer, C., Dabadie, T., Guizard, C. and Julbe, A., *Sol-gel derived silica membranes with tailored microporous structures*. Catalysis Today, **25** (3–4): 219–224 (1995).
17. Klotz, M., Ayral, A., Guizard, C. and Cot, L., *Synthesis and characterization of silica membranes exhibiting an ordered mesoporosity. Control of the porous texture and effect on the membrane permeability*. Separation and Purification Technology, **25** (1–3): 71–78 (2001).
18. Fujii, T., Yano, T., Nakamura, K. and Miyawaki, O., *The sol-gel preparation and characterization of nanoporous silica membrane with controlled pore size*. Journal of Membrane Science, **187** (1–2): 171–180 (2001).
19. Naito, M., Nakahira, K., Fukuda, Y., Mori, H. and Tsubaki, J., *Process conditions on the preparation of supported microporous SiO_2 membranes by sol-gel modification techniques*. Journal of Membrane Science, **129** (2): 263–269 (1997).

20. Kanezashi, M. and Asaeda, M., *Hydrogen permeation characteristics and stability of Ni-doped silica membranes in steam at high temperature*. Journal of Membrane Science, **271** (1–2): 86–93 (2006).
21. Sekulic, J., Luiten, M. W. J., ten Elshof, J. E., Benes, N. E. and Keizer, K., *Microporous silica and doped silica membrane for alcohol dehydration by pervaporation*. Desalination, **148** (1–3): 19–23 (2002).
22. Urtiaga, A., Casado, C., Asaeda, M. and Ortiz, I., *Comparison of SiO_2-ZrO_2-50% and commercial SiO_2 membranes on the pervaporative dehydration of organic solvents*. Desalination, **193** (1–3): 97–102 (2006).
23. de Vos, R. M., Maier, W. F. and Verweij, H., *Hydrophobic silica membranes for gas separation*. Journal of Membrane Science, **158** (1–2): 277–288 (1999).
24. Giessler, S., da Costa, J. C. D. and Lu, G. Q., *Hydrophobicity of templated silica xerogels for molecular sieving applications*. Journal of Nanoscience and Nanotechnology, **1** (3): 331–336 (2001).
25. Giessler, S., Jordan, L., Diniz da Costa, J. C. and Lu, G. Q., *Performance of hydrophobic and hydrophilic silica membrane reactors for the water gas shift reaction*. Separation and Purification Technology, **32** (1–3): 255–264 (2003).
26. Gavalas, G. R., Megiris, C. E. and Nam, S. W., *Deposition of H_2-permselective SiO_2-films*. Chemical Engineering Science, **44** (9): 1829–1835 (1989).
27. Kuraoka, K., Shugen, Z., Okita, K., Kakitani, T. and Yazawa, T., *Permeation of methanol vapor through silica membranes prepared by the CVD method with the aid of evacuation*. Journal of Membrane Science, **160** (1): 31–39 (1999).
28. Shiflett, M. B. and Foley, H. C., *Ultrasonic deposition of high-selectivity nanoporous carbon membranes*. Science, **285** (5435): 1902–1905 (1999).
29. Itoh, N. and Haraya, K., *A carbon membrane reactor*. Catalysis Today, **56** (1–3): 103–111 (2000).
30. Sznejer, G. and Sheintuch, M., *Application of a carbon membrane reactor for dehydrogenation reactions*. Chemical Engineering Science, **59** (10): 2013–2021 (2004).
31. Rao, M. B. and Sircar, S., *Nanoporous carbon membranes for separation of gas-mixtures by selective surface flow*. Journal of Membrane Science, **85** (3): 253–264 (1993).
32. Karger, J. and Ruthven, D. M., *Diffusion in Zeolites and Other Microporous Solids*. (1992) New York: John Wiley & Sons Inc.
33. Liepmann, H. W., *Gas kinetics and gas dynamics of orifice flow*. Journal of Fluid Mechanics, **10**: 65–79 (1961).
34. Kong, J. and Li, K., *An improved gas permeation method for characterising and predicting the performance of microporous asymmetric hollow fibre membranes used in gas absorption*. Journal of Membrane Science, **182** (1–2): 271–281 (2001).
35. Koros, W. J. and Fleming, G. K., *Membrane-based gas separation*. Journal of Membrane Science, **83** (1): 1–80 (1993).
36. Fain, D. E., *Membrane Gas Separation Principles*. MRS Bulletin, **19** (4): 40–43 (1994).
37. Sotirchos, S. V. and Burganos, V. N., *Transport of gases in porous membranes*. MRS bulletin, **24** (3): 41–45 (1999).
38. Knudsen, M., *Die gesetze der molekularstromung und der innern reibunsstromung der gase durch rohren*. Annals of Physics, **28**: 75–130 (1909).
39. Present, R. D., *Kinetic Theory of Gases*. (1958) New York: McGraw-Hill.
40. Metz, C. R., *Pysical Chemistry*. (1976) New York: McGraw-Hill.
41. Mason, E. A. and Malinauskas, A. P., *Gas Transport in Porous Media: The Dusty Gas Model*. (1983) New York: Elsevier.
42. Uhlhorn, R. J. R. and Burggraaf, A. J., *Gas separations with inorganic membranes*, in *Inorganic Membranes, Synthesis, Characteristics and Applications*, Bhave, R. R., Editor. (1991) New York: Van Nostrand Reinhold, p. 155–176.
43. Gilliland, E., Baddour, R. F. and Russel, J. L., *Rates of flow through microporous solids*. American Institute of Chemical Engineers Journal, **4**: 90–96 (1958).

44. Weaver, J. A. and Metzer, A. B., *The surface transport of adsorbed molecules*. American Institute of Chemical Engineers Journal, **12**: 655–661 (1966).
45. Ponzi, M., Papa, J., Rivarola, J. B. P. and Zgrablich, G., *Surface-diffusion of adsorbable gases through porous-media*. American Institute of Chemical Engineers Journal, **23**: 347–352 (1977).
46. Ash, R., Barrer, R. M. and Pope, C. G., *Flow of adsorbable gases and vapours in a microporous medium*. Proceedings of the Royal Society, **A271**: 1–18 (1963).
47. Okazaki, M., Tamon, H. and Toel, R., *Interpretation of surface flow phenomenon of adsorbed gases by hopping model*. American Institute of Chemical Engineers Journal, **27** (2): 262–270 (1981).
48. Uhlhorn, R. J. R., Keizer, K. and Burggraaf, A. J., *Gas transport and separation with ceramic membranes. Part I. Multilayer diffusion and capillary condensation*. Journal of Membrane Science, **66** (2–3): 259–269 (1992).
49. Lee, K.-H. and Hwang, S.-T., *The transport of condensible vapors through a microporous vycor glass membrane*. Journal of Colloid and Interface Science, **110** (2): 544–555 (1986).
50. Sing, K. S. W., Everett, D. H., Haul, R. A. W., Moscou, L., Pierotti, R. A., Rouquerol, J. and Siemieniewska, T., *Reporting physisorption data for gas solid systems with special reference to the determination of surface-area and porosity*. Pure and Applied Chemistry, **57** (4): 603–619 (1985).
51. Riekert, L., *Sorption, diffusion, and catalytic reaction in zeolites*, in *Advances in Catalysis*, Eley, D. D., Pines, H. and Weisz, P. B., Editors. (1970) New York: Academic Press.
52. Shewmon, P. G., *Diffusion Solids*. (1963) New York: McGraw-Hill.
53. Rangarajan, R., Mazid, M. A., Matsuura, T. and Sourirajan, S., *Permeation of pure gases under pressure through asymmetric porous membranes – membrane characterization and prediction of performance*. Industrial and Engineering Chemistry Process Design and Development, **23** (1): 79–87 (1984).
54. Asaeda, M., Du, L. D. and Fuji, M., *Separation of alcohol/water gaseous mixtures by an improved ceramic membrane*. Journal of Chemical Engineering of Japan, **19** (1): 84–85 (1986).
55. Sperry, D. P., Falconer, J. L. and Noble, R. D., *Methanol–hydrogen separation by capillary condensation in inorganic membranes*. Journal of Membrane Science, **60** (2–3): 185–193 (1991).
56. Keizer, K., Uhlhorn, R. J. R., van Vuren, R. J. and Burggraaf, A. J., *Gas separation mechanisms in microporous modified [gamma]-*Al_2O_3 *membranes*. Journal of Membrane Science, **39** (3): 285–300 (1988).
57. Uhlhorn, R. J. R., Keizer, K. and Burggraaf, A. J., *Gas and surface diffusion in modified [gamma]-alumina systems*. Journal of Membrane Science, **46** (2–3): 225–241 (1989).
58. Sloot, H. J., Smolders, C. A., van Swaaij, W. P. M. and Versteeg, G. F., *Surface diffusion of hydrogen sulfide and sulfur dioxide in alumina membranes in the continuum regime*. Journal of Membrane Science, **74** (3): 263–278 (1992).
59. de Vos, R. M. and Verweij, H., *Improved performance of silica membranes for gas separation*. Journal of Membrane Science, **143** (1–2): 37–51 (1998).
60. Dong, J. H. and Lin, Y. S., In situ *synthesis of P-type zeolite membranes on porous alpha-alumina supports*. Industrial and Engineering Chemistry Research, **37** (6): 2404–2409 (1998).
61. Wegner, K., Dong, J. and Lin, Y. S., *Polycrystalline MFI zeolite membranes: xylene pervaporation and its implication on membrane microstructure*. Journal of Membrane Science, **158** (1–2): 17–27 (1999).
62. Dong, J., Lin, Y. S. and Liu, W., *Multicomponent hydrogen/hydrocarbon separation by MFI-type zeolite membranes*. American Institute of Chemical Engineers Journal, **46** (10): 1957–1966 (2000).
63. Gouzinis, A. and Tsapatsis, M., *On the preferred orientation and microstructural manipulation of molecular sieve films prepared by secondary growth*. Chemistry of Materials, **10** (9): 2497–2504 (1998).
64. Guizard, C., *Sol-gel chemistry and its application to porous membrane processing*, in *Fundamentals of Inorganic Membrane Science and Technology*, Burggraaf, A. J. and Cot, L., Editors. (1996) Amsterdam: Elsevier Science B. V., p. 227–258.

65. Weller, S. and Steiner, W. A., *Separation of gases by fractional permeation through membranes*. Journal of Applied Physics, **21**: 279–283 (1950).
66. Weller, S. and Steiner, W. A., *Engineering aspects of separation of gases: fractional permeation through membranes*. Chemical Engineering Progress, **46**: 585–590 (1950).
67. Huckins, H. E. and Kammermeyer, K., *The separation of gases by means of porous membranes*. Chemical Engineering Progress, **49**: 180–184 (1953).
68. Huckins, H. E. and Kammermeyer, K., *The separation of gases by means of porous membranes*. Chemical Engineering Progress, **49**: 294–298 (1953).
69. Brubaker, D. W. and Kammermeyer, K., *Separation of gases by plastic membranes – permeation rates and extent of separation*. Industrial and Engineering Chemistry Research, **46**: 733–739 (1954).
70. Stern, S. A. and Walawender, W. P., *Analysis of membrane separation parameters*. Separation Science, **4**: 129–159 (1969).
71. Naylor, R. W. and Backer, P. O., *Enrichment calculations in gaseous diffusion: large separation factor*. American Institute of Chemical Engineers Journal, **1**: 95–99 (1955).
72. Pan, C. Y. and Habgood, H. W., *Gas separation by permeation. 1. Calculation methods and parametric analysis*. Canadian Journal of Chemical Engineering, **56** (2): 197–209 (1978).
73. Oishi, J., Matsumura, Y., Higashi, K. and Ike, C., *An analysis of gasous diffusion separating unit*. Journal of the Atomic Energy Society Japan, **3**: 923–928 (1961).
74. Walawender, W. P. and Stern, S. A., *Analysis of membrane separation parameters. II. Countercurrent and cocurrent flow in a single permeation stage*. Separation Science, **7**: 553–584 (1972).
75. Blaisdel, C. T. and Kammerme, K, *Countercurrent and co-current gas separation*. Chemical Engineering Science, **28** (6): 1249–1255 (1973).
76. Li, K., Acharya, D. R. and Hughes, R., *Mathematical-modeling of multicomponent membrane permeators*. Journal of Membrane Science, **52** (2): 205–219 (1990).

5

Ceramic Hollow Fibre Membrane Contactors for Treatment of Gases/Vapours

5.1 INTRODUCTION

The applications of membrane contactors for treatment of gases/vapours have been extensively investigated by many researchers, mostly on laboratory scales [1–8]. There are also several pilot scale plants operating worldwide. For example, a pilot plant for CO_2 removal from flue gas was built and has been studied for years [9]. Another pilot plant that employs a membrane contactor for CO_2 removal is at the Nanko Power plant in Osaka, Japan [10]. A pilot plant for olefin/paraffin separations using a membrane contactor and an aqueous silver nitrate solution as absorbent was built in BP Amerco [11]. However the project was abandoned due to the relatively high and variable cost of the complexing agent and also because of the problem of impurities in the feed stream, such as acetylides, carbon dioxide, hydrogen sulphide, carbonyl sulphide, arsine and ammonia. These impurities reacted with the polymeric membranes employed and had deleterious effects on their performance.

The membranes used in the above studies are all made from polymeric materials such as polypropylene, polyimide, PTFE, etc. and they have considerable disadvantages. It was found that polymeric membranes underwent some changes after having been in contact with absorbing liquid even for a short time. These changes were reported to have deteriorating effects on the system performance. Barbe *et al.* [12] reported the enlargement of the pore entrances of the surface of polypropylene membranes when contacted with water for only 72 hours. Wang *et al.* [13] investigated the impact of amine aqueous solution with and without CO_2 loading on polypropylene porous hollow fibre membranes and found surface morphology and surface roughness changes that might lead to the wetting of the membranes. Dindore *et al.* [4] also reported a significant decrease in membrane performances during a two week operation due to wetting of the membranes induced by membrane morphological changes. In addition, polymeric membranes cannot withstand high temperature; this restricts the membrane contactor to being operated at low temperature only, causing significant energy loss [14]. Also, advantages of using membrane contactors cannot be realised in distillation because

Ceramic Membranes for Separation and Reaction K. Li

of the restrictions in high temperature operations [15, 16]. In this respect, it would be better if the membranes prepared were more chemically resistant and thermally stable.

As ceramic materials are well known for their high chemical and thermal resistances as well as good mechanical strength, they would be a potential alternative as an interface in gas–liquid or liquid–liquid contacting processes [17, 18]. With newly developed ceramic membrane technologies [19], it is now possible to produce ceramic hollow fibre membranes which combine the advantages of both ceramic materials and the hollow fibre configuration [19, 20]. Ceramic hollow fibre membranes can be made to be porous or dense, depending on their application requirements. These hollow fibre membranes have been recently investigated for their applications in many areas, such as gas separation, membrane reactors, solvent recovery etc. [21–24].

One of the limitations that prevent ceramic membranes from being used in some applications is the hydrophilicity of the membranes. Most ceramic membranes are made from metal oxides, which are hydrophilic in nature, some applications such as polar solvent extraction, water ozonation, nonaqueous pervaporation, etc. require hydrophobic membranes [25]. In the case of a membrane contactor, membrane wetting is a major concern as it increases the membrane resistance, $1/k_{Am}$ by many orders of magnitudes and lowers the performance of the contactors. Thus, it is necessary to use membranes with high hydrophobicities in membrane contactors.

The hydrophobicity of ceramic membranes can be promoted by surface modification. One of the possible methods is to graft the surface with silane compounds, such as chloroalkylsilane, fluoroalkylsilanes, etc., which are reported to be solvent resistant and thermally stable [26]. The studies that have been done so far involve only the surface modification of flat sheet and tubular membranes [25–33].

In this chapter, a general review of the progress of the membrane based gas/vapour treatment will be provided first, followed by the discussions on mass transfer of gases through various membrane geometries. Some of the most important theoretical models for gas/vapour treatment in membrane contactors will also be reviewed. Finally, design equations for ceramic hollow fibre membrane modules in gas/vapour treatment will be outlined.

5.2 GENERAL REVIEW

Porous hollow fibre membrane contactors employed in gas/vapour absorption processes have attracted considerable attention in the past and were first studied by Qi and Cussler [34, 35] and Cooney and Jackson [36]. The hollow fibre membrane used by them acts as a fixed interface and keeps the gas/vapour and liquid phases separated while mass transfer of gases/vapours takes place through the membrane. Depending on the membrane material, the physicochemical properties of the liquid and the operating pressures employed, the pores of the membrane can be filled with either gas or liquid, which results in great differences in the mass transfer resistance of the membrane employed [37].

Compared with conventional absorption processes such as bubble columns and packed beds, there are several advantages of using porous hollow fibre contactors for gas/vapour absorption. These include larger interfacial area per unit volume, independent control of gas and liquid flow rates without any flooding, unloading, foaming, etc., and known gas–liquid interfacial area. These advantages have led to a number of investigations on the use of membrane contactors for gas/vapour absorption [6, 38–42]. In most studies, however, emphasis has been focused on the nonwetted operating mode, shown in Figure 5.1(a), whereby the

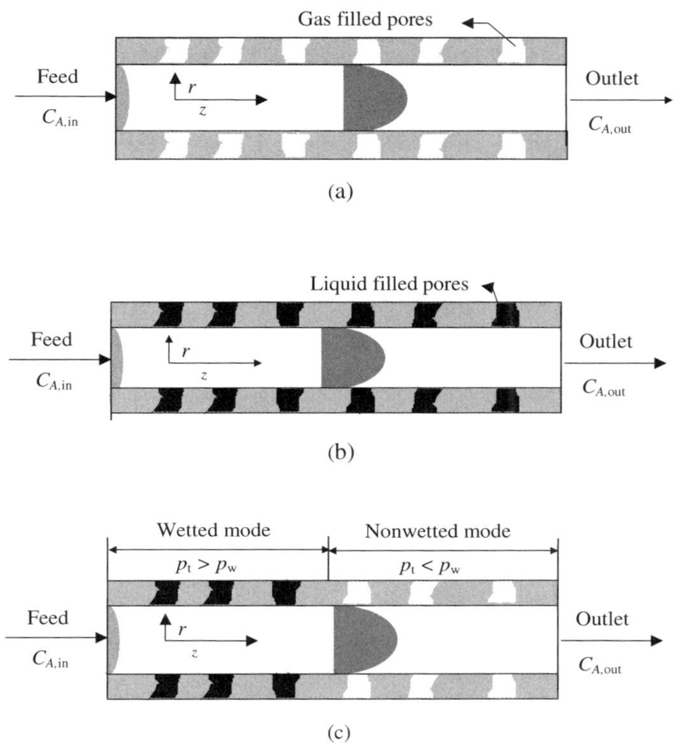

Figure 5.1 Microporous hollow fibre membrane contactor operated under: (a) nonwetted mode; (b) wetted mode; (c) partially wetted mode [45]. Copyright (1997) American Chemical Society, Industrial and Engineering Chemistry Research, **36** (3), Malek, A., Li, K. and Teo, W. K., *Modeling of microporous hollow fibre membrane modules operated under partially wetted conditions,* p. 784–793

membrane pores are filled with gas. The primary reason for such an operating mode is quite obvious, since the usual physical absorption processes used in industries are limited by the mass transfer rates in the liquid phase. The overall mass transfer coefficients in these cases are between 10^{-5} and $10^{-4}\,ms^{-1}$ [43]. Adding the mass transfer resistance of the nonwetted membrane, which is several order of magnitudes smaller than that of liquid, would result in a negligible effect on the overall mass transfer coefficient. Due to the considerably larger interfacial area, the overall mass transfer per unit volume, $K_l a$, for the membrane contactor is, therefore, much higher than that for conventional columns. Thus, membrane based contactors for gas/vapour absorption or distillation offer a promising alternative.

In principle, a membrane process may replace a conventional gas/vapour absorption process. However, the success of membrane contactors over conventional columns would largely depend on the types of gas–liquid systems and hollow fibre membranes used. For example, as mentioned above, for a liquid film resistance controlled system such as CO_2–H_2O, the use of a porous hydrophobic membrane would give a negligible mass transfer resistance to the overall mass transfer provided that the hydrophobic membrane remained 'nonwet', i.e. the membrane pores are filled with gas [6, 38]. In addition to the characteristics of membranes, operating conditions such as solution pressure also play a major role in the overall membrane absorption performance [37, 44]. As illustrated by Poddar *et al.* [44] and Malek *et al.* [45], the operating pressure in the liquid phase should always be kept higher than that in the gas phase in order to prevent bubble formation resulting in loss of gas components and operating stability. However, it has been observed that maintenance of a high pressure at the liquid phase for long term operation could lead to wetting of the porous membrane by the liquid absorbent as shown in Figure 5.1(b) [45]. As a result, the gas transfer rate was found to be greatly reduced due to this wetting of the hydrophobic membrane. It was indicated by Kreulen *et al.* [46] that the performance of a hydrophobic membrane contactor under wetted conditions is inferior compared to conventional processes for gas/vapour absorption.

In order to prevent the membrane from getting wet, the liquid pressure should always be lower than the wetting pressure of the membrane employed [47–49]. For a porous hydrophobic hollow membrane contactor where liquid is fed into the fibre lumen, because of the small size of the fibre diameter, the pressure drop of the liquid in the fibre lumen would inevitably increase with the liquid velocity and length of the fibre used according to the Hagen–Poiseuille law. In this case, the membrane contactor may be partially wetted due to the pressure gradient along the fibre lumen as shown in Figure 5.1(c). Such a phenomena has been observed by Malek *et al.* [45] who found that water droplets are formed at the outer surface of the hollow fibre membrane, especially near the inlet of the membrane module. The overall mass transfer coefficients obtained by Malek *et al.* [45] at different liquid velocities for the O_2–H_2O system exhibit a maximum value which may suggest that the membrane contactor is partially wetted from the inlet of the contactor due to a high liquid pressure in the fibre lumen. A partially wetted membrane contactor may result in a reduction of the overall mass transfer coefficient compared with the nonwetted mode, hence giving a maximum value of the overall mass transfer coefficient as a function of the water velocity.

Porous hydrophobic membranes were also studied for gas film controlled systems such as NH_3 in H_2SO_4, SO_2 in NaOH and H_2S (low concentration) in NaOH by Qi and Cussler [35] and by Kreulen *et al.* [50]. The K_{Ag} values measured by them are in the range of 0.0035 to 0.0073 $m\,s^{-1}$, which are considerably lower than the k_{Ag} values of 0.01–0.1 $m\,s^{-1}$ measured in conventional gas absorbers [51]. Therefore, it is not surprising that membrane resistance controls the absorption of all these gases in the membrane contactors used by Qi and Cussler [35]. The membrane resistance, $1/k_{Am}$, may be reduced if hydrophobic asymmetric membranes are employed for the gas absorption with the systems mentioned above. Li *et al.* [52] studied the removal of H_2S from a gas mixture using an NaOH solution in an asymmetric polysulfone hollow fibre membrane contactor. The experimentally evaluated membrane's coefficient, k_{Am} for the polysulfone is in the range of 0.0125 to 0.025 $m\,s^{-1}$, which is better than the value of 0.0073 obtained from symmetric microporous polypropylene membranes [35].

Membrane based gas–liquid contactors have also been proposed for use in olefin–paraffin separations [53]. As shown in Figure 5.2, absorption of olefin in an aqueous silver nitrate solution occurs in one membrane contactor and desorption of olefin from the aqueous solution is performed via thermal desorption in a separate unit. In the absorption unit, the membrane serves as an interface between the aqueous $AgNO_3$ and the gaseous olefin–paraffin feed mixture. Owing to their high specific surface area (interfacial area per unit volume), membrane contactors promise higher volumetric mass transfer rates than conventional scrubbers, i.e. typically in the range of 500–5000 $m^3\,m^{-2}$ for membrane contactors as compared to 20–500 $m^3\,m^{-2}$ in conventional scrubbers [54]. BP Amoco [11] studied membrane based contactors for olefin separation extensively, using polypropylene hollow fibre membranes and silver nitrate as a carrier up to pilot plant stage. It was found that although the process was technically sound, the operating cost of the process was too expensive, mainly due to short life

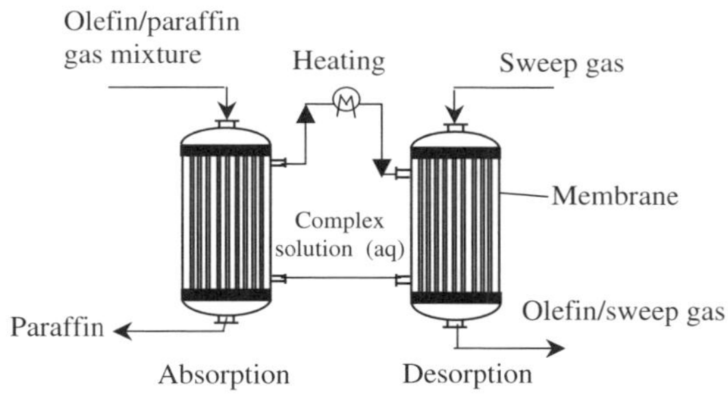

Figure 5.2 Membrane based olefin/paraffin separation

span (limited thermal stability and chemical reaction with silver) of the polymeric hollow fibre membranes and the expensive complexing agent used, although the possibility of other carriers such as molybdenum sulfides was also explored [55].

Poddar *et al.* [44] investigated membrane contactors for absorption of volatile organic compounds (VOCs) from N_2/air into silicone oil or mineral oil. Two types of hollow fibre contactors were used by them. One contained a porous hydrophobic membrane, while the other was packed with composite membranes (a highly VOC permeable nonporous silicone skin coated on a porous substrate). In both the cases the absorbent flowed through the shell side and the spent absorbent was regenerated under vacuum in a separate contactor then recycled. In their study, efficient VOC removal from the gas stream was achieved and the highest efficiency was obtained with toluene, followed by methylene chloride, acetone and methanol. Efficiency was higher with silicone compared to mineral oil because of the higher diffusivities and partition coefficient values of the VOCs in the former. However, the mineral oil provided better long term stability. The absorbent filled porous membrane was responsible for 80–95 % of the mass transfer resistance, while for the skinned membrane most of the resistance was found in the skin and the liquid boundary layer. Mass transfer coefficients were substantially higher than those typically obtained with packed towers. It was suggested that a regenerative membrane based absorption process could eventually replace carbon adsorption in VOC removal applications [44].

5.3 OPERATING MODES AND MASS TRANSFER COEFFICIENTS

In a membrane contactor, gas/vapour absorption is a three step process. It involves transfer of a gas component from the gas/vapour phase to a membrane, through the membrane and then into the liquid phase. Depending on the membrane characteristics, the physicochemical properties of the absorbing liquid and the operating conditions employed, two possible transfer modes, i.e. nonwetted and wetted modes are possible at any point in a membrane. Theoretical treatment of local point mass transfer under these transfer modes as illustrated in Figure 5.3 are presented in detail below.

5.3.1 Nonwetted Mode

The nonwetted transfer mode is often achievable in membrane contactors where the membrane employed is porous and hydrophobic. As shown Figure 5.3(a), an aqueous solution

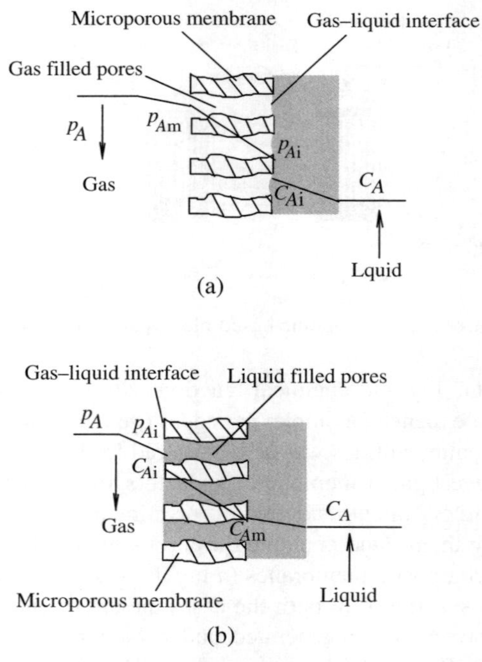

Figure 5.3 Concentration profiles in: (a) nonwetted porous membrane; (b) wetted porous membrane

which does not wet the membrane flows on one side of the membrane, while a gas/vapour mixture containing the gas A flows on the other side of the membrane at a pressure less than that in aqueous phase. The membrane pores remain gas filled so long as the pressure difference between the gas phase and the liquid phase is less than the wetting pressure, p_w of the membrane. The local flux of the gas A transfer, J_A, through the membrane at steady state can, therefore, be expressed as:

$$J_A = k_{Ag}(p_A - p_{Am}) \tag{5.1a}$$

$$= k_{Am,g}(p_{Am} - p_{Ai}) \tag{5.1b}$$

$$= k_{Al}(C_{Ai} - C_A) \tag{5.1c}$$

$$= K_{Ag}(p_A - p_A^*) \tag{5.1d}$$

$$= K_{Al}(C_A^* - C_A) \tag{5.1e}$$

where k_{Ag}, $k_{Am,g}$ and k_{Al} are the individual mass transfer coefficients in gas film, membrane with gas filled pores and liquid film, respectively. p_{Ai} and C_{Ai} are the partial pressure and concentration of A at the interface, respectively, while p_A^* and C_A^* are the equilibrium partial pressure and concentration of the gas A in the gas and liquid phases, respectively, and they can be related by Henry's law:

$$p_{Ai} = H_A C_{Ai} \tag{5.2a}$$

$$p_A^* = H_A C_A \tag{5.2b}$$

$$p_A = H_A C_A^* \tag{5.2c}$$

Because the mass transfers of the gas A through the gas film, the membrane and the liquid film are in series, after rearranging Equations (5.1) and (5.2), the overall mass transfer coefficient, K_{Ag} or K_{Al} in Equations (5.1d) and (5.1e) can be expressed, respectively, as

$$\frac{1}{K_{Ag}} = \frac{1}{k_{Ag}} + \frac{1}{k_{Am,g}} + \frac{H_A}{k_{Al}} \tag{5.3a}$$

and

$$\frac{1}{K_{Al}} = \frac{1}{H_A k_{Ag}} + \frac{1}{H_A k_{Am,g}} + \frac{1}{k_{Al}} \tag{5.3b}$$

5.3.2 Wetted Mode

When the pressure difference between the gas phase and the aqueous phase is greater than the wetting pressure, p_w, of a porous hydrophobic membrane, the membrane becomes wet. As shown in Figure 5.3(b), the aqueous solution fills the pores of the membrane and Equations (5.1) may be modified as:

$$J_A = k_{Ag}(p_A - p_{Ai}) \tag{5.4a}$$

$$= k_{Am,l}(C_{Ai} - C_{Am}) \tag{5.4b}$$

$$= k_{Al}(C_{Am} - C_A) \tag{5.4c}$$

$$= K_{Ag}(p_A - p_A^*) \tag{5.4d}$$

$$= K_{Al}(C_A^* - C_A) \tag{5.4e}$$

Rearranging Equations (5.4a) to (5.4e) together with Equations (5.2a) to (5.2c), the overall mass transfer coefficient based on the gas phase, K_{Ag} or based on the liquid phase, K_{Al} for the completely wetted membranes can be expressed as:

$$\frac{1}{K_{Ag}} = \frac{1}{k_{Ag}} + \frac{H_A}{k_{Am,l}} + \frac{H_A}{k_{Al}} \tag{5.5a}$$

or

$$\frac{1}{K_{Al}} = \frac{1}{H_A k_{Ag}} + \frac{1}{k_{Am,l}} + \frac{1}{k_{Al}} \tag{5.5b}$$

Hollow fibre contactors can also be employed for gas stripping. The operation is identical to the gas/vapour absorption with the exception that the gas species are transferred from the liquid to the gas phase. Therefore, the above equations derived for gas absorption are still useful for gas stripping so long as the sign is reversed in Equations (5.1) and (5.4).

5.3.3 Mass Transfer Coefficients Determined from Experiments

The overall mass transfer and individual coefficients described in Equation (5.3) or Equation (5.5) can be determined experimentally. As long as the concentration of gas A in the liquid is low, i.e. the linear relation of the Henry's law holds, the overall mass transfer coefficient can be obtained using the following equations:

Gas phase:

$$K_{Ag} = \frac{G}{A_t} \frac{p_{A,in} - p_{A,out}}{(p_A - p_A^*)lm} \tag{5.6}$$

where $(p_A - p_A^*)lm$ is the log mean concentration of gas A in the gas phase and is given by:

$$(p_A - p_A^*)lm = \frac{(p_{A,in} - p_{in}^*) - (p_{A,out} - p_{A,out}^*)}{\ln \dfrac{p_{A,in} - p_{A,in}^*}{p_{A,out} - p_{A,out}^*}} \tag{5.7}$$

If the concentration of gas A in the liquid, C_A approaches zero, p_A^* in Equation (5.2b) equals zero. Therefore, Equation (5.6a) can be further simplified as:

$$K_{Ag} = \frac{G}{A_t} \ln \frac{p_{A,in}}{p_{A,out}} \tag{5.8}$$

Liquid phase:

$$K_{Al} = \frac{L}{A_t} \frac{C_{A,out} - C_{A,in}}{(C_A^* - C_A)lm} \tag{5.9}$$

where $(C_A^* - C_A)$ is the log mean concentration of the gas A in the liquid phase and is given by:

$$(C_A^* - C)lm = \frac{(C_{A,out} - C_{A,out}^*) - (C_{A,in} - C_{A,in}^*)}{\ln \dfrac{(C_{A,out} - C_{A,out}^*)}{C_{A,in} - C_{A,in}^*}} \tag{5.10}$$

If the concentration of gas A in the gas phase, p_A approaches zero, C_A^* in Equation (5.2c) equals zero. Therefore, Equation (5.9) can be further simplified as:

$$K_{Al} = \frac{L}{A_t} \ln \frac{C_{A,\text{out}}}{C_{A,\text{in}}} \tag{5.11}$$

In the above equations, the subscripts in and out denote the inlet and outlet of the contactor, respectively. It should be noted that K_{Ag} and K_{Al} are the average rather than the local mass transfer coefficients in the gas and liquid phases, respectively.

In general, the feed concentration and the interfacial area (membrane area) are known. Therefore, measurement of the rate of fluid (G or L) and outlet concentration ($p_{A,\text{out}}$ or $C_{A,\text{out}}$) allows the overall mass transfer coefficient to be calculated using Equation (5.6) or Equation (5.9).

Once the overall mass transfer coefficient is experimentally determined, the individual mass transfer coefficients, especially the membrane coefficient, k_{Am} can be determined using the Wilson plot [6]. This method can not only be used to determine the membrane coefficient, k_{Am}, but also to understand the effect of fluid velocities on individual mass transfer coefficients [38, 56]. Based on the boundary layer theory, the individual coefficients for the gas or liquid boundary layer are proportional to U^{α}, where α is an empirical constant. When plotting K^{-1} versus $U^{-\alpha}$, it gives a straight line; such a plot is known as a Wilson plot.

For a hydrophobic membrane with a high value of H_A, k_{Ag} is independent of the gas velocity, while, k_{Al} is dependent on the liquid velocity and is proportional to U^{α}. The overall mass transfer coefficient can be determined from Equation (5.9) or Equation (5.11). Thus, the Wilson plot of $1/K_{Al}$ versus $1/U^{\alpha}$ can be prepared based on Equation (5.3b). Here the value of α is selected as the one that provides the best straight line through the data points. The membrane's coefficient, $k_{A,m}$ is, thus, obtained from the intercept of the Wilson plot at a condition that the gas velocity is extremely high, i.e. $k_{Ag} \rightarrow \infty$. Figure 5.4 shows the Wilson plot presented by Kartohardjono *et al.* [57] for dissolved oxygen removal from water.

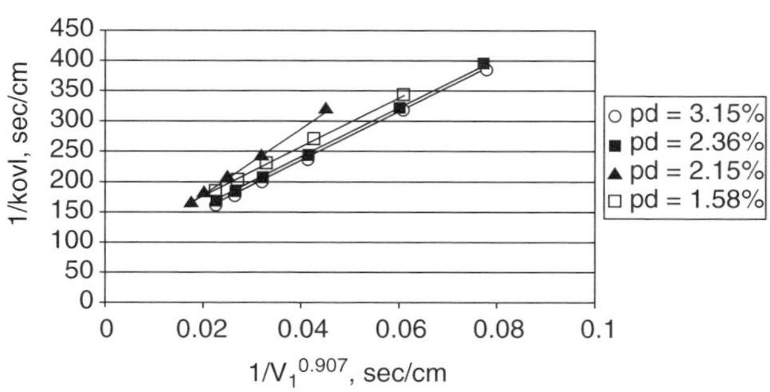

Figure 5.4 Wilson's plot at various packing density of the contactors [57]. Reprinted from Kartohardjono, S., Chen V. and Leiknes T., Dissolved oxygen removal from water by vacuum degassing process using sealed end poly Methyl Penten (PMP) hollow fibre membrane contactor, in The Fifth International Membrane Science & Technology Conference (MSTEC'03), 2003. The University of New South Wales, Sydney, Australia: UNESCO Centre for Membrane Science and Technology, the University of South Wales, Sydney, Australia

5.4 MASS TRANSFER IN HOLLOW FIBRE CONTACTORS

When a hollow fibre membrane contactor is employed for gas/vapour treatment, the gas and liquid can be fed into the fibre lumen and the shell side of the contactor, respectively, or vice versa. The individual mass transfer coefficients in the fibre lumen, across the membrane and in the shell side of the contactor can be evaluated independently.

5.4.1 Mass Transfer in Hollow Fibre Lumens

When a fluid mixture flows through a hollow fibre lumen, the mass conservation of the gas/vapour A in the fluid mixture (either gas or liquid) can be described by a differential steady state material balance equation as below:

$$U\frac{\partial C_A}{\partial z} = D_A\left(\frac{\partial^2 C_A}{\partial r^2} + \frac{1}{r}\frac{\partial C_A}{\partial r}\right)$$

$$2U_{\mathrm{m}}\left(1-\left(\frac{r}{R}\right)^2\right)\frac{\partial C_A}{\partial z} = D_A\left(\frac{\partial^2 C_A}{\partial r^2} + \frac{1}{r}\frac{\partial C_A}{\partial r}\right) \tag{5.12}$$

with boundary conditions:

$$C_A|_{z=0} = C_{A\mathrm{f}} \qquad (0 \le r \le R) \tag{5.13a}$$

$$\left.\frac{\partial C_A}{\partial r}\right|_{r=0} = 0 \qquad (0 \le z \le L) \tag{5.13b}$$

The flux equation at the membrane–fluid interface is:

$$D_A\left(\frac{\partial C_A}{\partial r}\right)_{r=R} = k_{\mathrm{ex}}(C_A - C_{A\mathrm{i}}) \tag{5.13c}$$

where

$$\frac{1}{k_{\mathrm{ex}}} = \frac{1}{k_A^S} + \frac{1}{k_{A\mathrm{m}}} \tag{5.13d}$$

where $1/k_{\mathrm{ex}}$ is the external resistance and is equal to the summation of the membrane resistance, $1/k_{A\mathrm{m}}$ and the resistance in the shell side, $1/k_A^S$. It is obvious that when the resistance of membrane, $1/k_{A\mathrm{m}}$ and the resistance in the shell side of the contactor, $1/k_A^S$ are negligible, Equation (5.7c) can be reduced as:

$$D_A\left(\frac{\partial C_A}{\partial r}\right)_{r=R} = \text{constant} \tag{5.13e}$$

Therefore, analytical solutions for the mass transfer coefficient in the hollow fibre lumen can be derived using the methods given by Leveque and Greatz [43]. The Graetz solution is an infinite series solution, which is obtained by solving Equation (5.12) using the method of separation of variables. This series solution is given as:

$$Sh = 0.5G_Z\theta \tag{5.14a}$$

or

$$\frac{k_A^L \cdot d}{D_A} = 0.5\left(\frac{d_i^2 U_m}{LD_A}\right)\theta \tag{5.14b}$$

where:

$$\theta = \frac{1 - \sum_{j=1}^{\infty} \frac{-4B_j}{\beta_j^2}\left(\frac{d\phi_j}{d\Gamma}\right)_{\Gamma=1} \exp\frac{-\beta_j^2\left(\frac{2L}{d_i}\right)}{\mathrm{Re}\,Sc}}{1 + \sum_{j=1}^{\infty} \frac{-4B_j}{\beta_j^2}\left(\frac{d\phi_j}{d\Gamma}\right)_{\Gamma=1} \exp\frac{-\beta_j^2\left(\frac{2L}{d_i}\right)}{\mathrm{Re}\,Sc}} \tag{5.15a}$$

and

$$\beta_j = 4(j-1) + 8/3 \quad j = 1, 2, 3, 4, \ldots\ldots \tag{5.15b}$$

$$B_j = (-1)^{j-1} \times 2.84606\beta_j^{-2/3} \tag{5.15c}$$

$$\frac{B_j}{2}\left(\frac{d\phi_j}{d\Gamma}\right)_{\Gamma=1} = 1.01276\beta_j^{-1/3} \tag{5.15d}$$

The Leveque solution, on the other hand, is more restricted since it was derived based on the assumption that concentration gradients in the fluid are limited to a thin layer near the fibre wall. This assumption also means that $\partial C_{Ai}/\partial z$ can be taken as constant along the fibre length. One important consequence of this assumption is that the model is only applicable for Gz numbers exceeding 400; that is, for short fibres or large liquid flow rates. The Leveque equation is given as:

$$Sh = 1.615Gz^{1/3} \tag{5.16a}$$

or

$$\frac{k_A^L \cdot d}{D_A} = 1.615\left(\frac{d_i^2 U_m}{LD_A}\right)^{1/3} \tag{5.16b}$$

It is clear from the derivation of the analytical solutions of Graetz and Leveque that these two models are only applicable when the resistance of membrane, $1/k_{Am}$ and the resistance in the shell side of the contactor, $1/k_A^S$ are negligible.

It should be noted that Equation (5.16) overestimates experimentally determined mass transfer coefficients at low flows. This is attributed to nonuniform flow caused by polydispersity of the hollow fibre diameters as suggested by Wickramasinghe *et al.* [58] who used a Gaussian distribution to correct the effect of the nonuniformity of the fibre radii. Therefore, the mass transfer coefficient with the incorporation of the Gaussian distribution function for the fibre radii becomes:

$$\langle k \rangle = k\left[1 - 9kV/QR_o + 7\varepsilon_o^2 + \ldots\right] \tag{5.17}$$

where k is the mass transfer coefficient for a uniform distribution of fibre radii, V the average volume occupied by one fibre, R_o the average fibre radius, and ε_o is the standard deviation of fibre radii divided by the mean. Equation (5.17) was developed for low flow conditions of Graetz number less than four; at these low flows, the quantity in brackets is less than one. Since the distribution of fibre diameters leads to low flows in some fibres, it is obvious that the nonuniformity of the fibre diameters would lead to a reduction in the mass transfer coefficient. Wickramasinghe *et al.*'s study was verified experimentally upon stripping oxygen from water using porous polypropylene hollow fibres. Correlations inferred from the experimental data agreed with the Leveque equation at high water flow rates, but gave lower mass transfer coefficients at low water flow rates, which is corrected well with Equation (5.17).

Nonuniformity of fluid flow in the fibre lumen was also demonstrated using high speed photography and dye tracer by Park and Chang [59] who determined that the distribution depends on the inlet manifold type (cylindrical or conical), manifold height, fibre length, fibre inner diameter, shell diameter, fibre packing density and Reynolds number. Nearly uniform flow was achieved for certain shell and tube geometries with long manifolds at low Reynolds numbers. Experimental data agreed fairly well with model predictions.

5.4.2 Mass Transfer Across Membranes

For nonwetted porous membranes, i.e. gas filled pores, the membrane coefficient, $k_{Am,g}$, in ms^{-1}, can be evaluated independently using the pore structure properties [35, 46]:

$$k_{Am,g} = \frac{D_{A,e}\varepsilon}{\delta\tau_m} \tag{5.18}$$

where ε is the membrane surface porosity, defined as $\varepsilon = A_p/A_t$, A_p, the cross sectional area of the pores and A_t, the total membrane area, respectively, in m, δ is the thickness of the membrane, in m and τ_m is the tortuosity, which corrects the membrane pore geometry.

In Equation (5.18), $D_{A,e}$ is the diffusion coefficient of gas in the membrane pores governed by both the continuum and Knudsen diffusion coefficients, and is expressed by [46]:

$$\frac{1}{D_{A,e}} = \frac{1}{D_A} + \frac{1}{D_{AK}} \tag{5.19}$$

where D_A is the continuum gas diffusion coefficient, while D_{AK} is the Knudsen diffusion coefficient, which can be expressed as:

$$D_{AK} = \frac{2r_p}{3}\sqrt{\frac{8RT}{\pi M_A}} \tag{5.20}$$

where r_p is the membrane pore radius, M_A is the molecular weight of the gas, R is the gas constant (8.314 Pa m^3 mol^{-1} K^{-1}), and T is the absolute temperature, in K.

Combining Equations (5.18) to (5.20), the membrane's coefficient can be expressed as:

$$k_{Am,g} = \frac{A_p}{\delta \tau_m A_t} \cdot \frac{D_A r_p}{\alpha + r_p} \tag{5.21}$$

where α is defined as:

$$\alpha = \frac{3}{2} \cdot \frac{D_A}{\sqrt{8RT/\pi M_A}} \tag{5.22}$$

Inspection of Equation (5.21) indicates that the contribution of Knudsen diffusion to the membrane's coefficient declines as the membrane pore radius, r_p, increases. At $r_p >> \alpha$, the membrane's coefficient, $k_{Am,g}$ is governed solely by the continuum diffusion coefficient, D_A.

5.4.2.1 *Influence of Pore Size Distribution on Membrane Coefficient*

Porous membranes often show a distribution of pore sizes. Thus, the pore size of the membrane varies with its pore size distribution function. If an average pore radius were used in obtaining the $k_{Am,g}$ values, a substantial error would occur. To avoid this, a distribution function of the membrane pore size should be considered. The log-normal distribution function should be employed since it has a distinct advantage of defining only pore radius values greater than zero, unlike the Gaussian normal distribution where the pore radius ranges from $-\infty$ to $+\infty$ [60]. Moreover, a number of studies have shown that the log-normal distribution function could reasonably represent the real pore size distribution of ultrafiltration membranes in most cases [61, 62]. In order to represent the real values of mean pore radius, r_m, and standard deviation, σ, of the membranes, the following form of log-normal distribution function can be employed:

$$g(r) = \frac{1}{\sqrt{2\pi} r}\left[\ln(1+\sigma^2)\right]^{-0.5} \times \exp\left[-\frac{\left(\ln\left(\frac{r}{r_m}\right)(1+\sigma^2)^{0.5}\right)^2}{2\ln(1+\sigma^2)}\right] \tag{5.23}$$

where r_m is the population mean pore radius [61] and σ is the dimensionless standard deviation of the membrane pore.

By considering the pore size distribution, the actual membrane coefficient, $\bar{k}_{Am,g}$ is then expressed as:

$$\bar{k}_{Am,g} = \frac{\varepsilon \bar{D}_{A,e}}{\delta \tau_m} \tag{5.24}$$

where:

$$\bar{D}_{A,e} = D_A \frac{\int_0^\infty \frac{r^2}{\alpha + r} \cdot \exp\left[-\frac{\left(\ln\left(\frac{r}{r_m}\right)(1+\sigma^2)^{0.5} \right)^2}{2\ln(1+\sigma^2)} \right] dr}{\int_0^\infty r \cdot \exp\left[-\frac{\left(\ln\left(\frac{r}{r_m}\right)(1+\sigma^2)^{0.5} \right)^2}{2\ln(1+\sigma^2)} \right] dr} \tag{5.25}$$

The full derivation of Equation (5.25) is given in Appendix 5.1. Equation (5.25) has been represented graphically by Li *et al.* [7] as shown in Figure 5.5. Therefore, with a given stan-

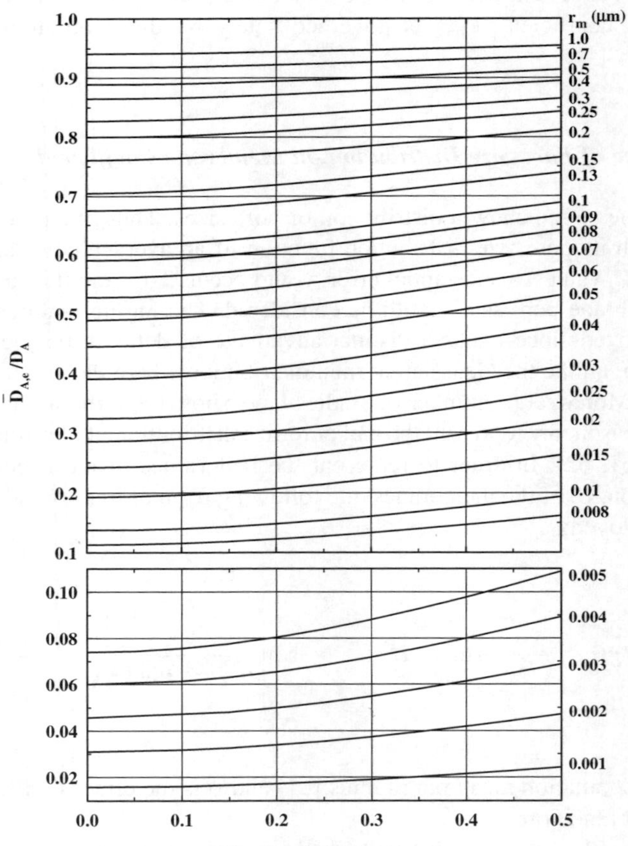

Figure 5.5 Effects of standard deviation of pore size distribution on relative diffusion coefficient at various membrane pore radii [7]. Copyright (2000) Elsevier, Chemical Engineering Science, **55** (23), Li, K., Kong, J. and Tan, X., *Design of hollow fibre membrane modules for soluble gas removal*, p. 5579–5588

dard deviation σ and the mean pore size r_m, the membrane coefficients, $\bar{k}_{Am,g}$ can be easily obtained from Equation (5.24) using Figure 5.5.

Figure 5.6 examines the effect of pore radius on the membrane coefficient, $k_{Am,g}$ for different effective surface porosities under the uniform pore size condition. It can be seen that as the pore radius becomes greater than 1 μm, the Knudsen diffusion effect becomes negligible and the membrane coefficient is governed only by the continuum diffusion coefficient.

As the pores of the membrane become nonuniform, the membrane coefficient evaluated, based on the assumption of the uniform pore size, may deviate from the measured value. Thus the effect of the pore size distribution on the membrane coefficient must be considered. As illustrated in Figure 5.7 where the standard deviation is plotted against the relative error,

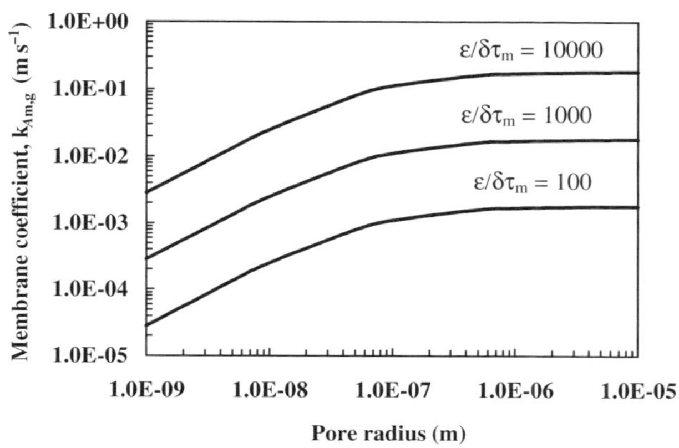

Figure 5.6 Effects of membrane pore radius on membrane coefficient at various effective porosities [7]. Copyright (2000) Elsevier, Chemical Engineering Science, **55** (23), Li, K., Kong J. and Tan, X., *Design of hollow fibre membrane modules for soluble gas removal*, p. 5579–5588

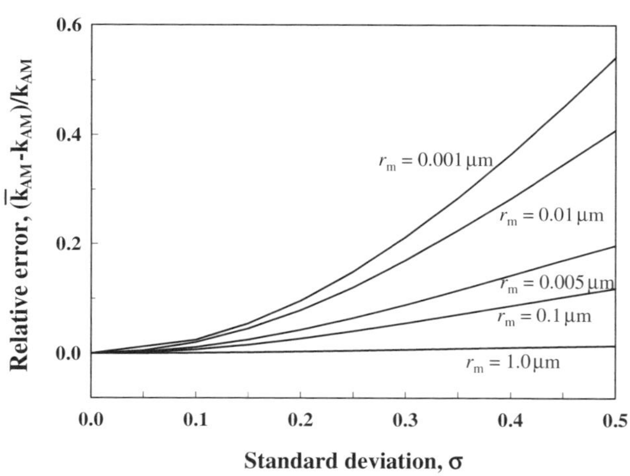

Figure 5.7 Effects of standard deviation of pore size distribution on membrane coefficient at various membrane pore radii [7]. Copyright (2000) Elsevier, Chemical Engineering Science, **55** (23), Li, K., Kong J. and Tan, X., *Design of hollow fibre membrane modules for soluble gas removal*, p. 5579–5588

defined as $(\bar{k}_{Am,g} - k_{Am,g})/k_{Am,g}$, the error increases as the average pore size is decreased. This is expected: as the average pore size decreases, the membrane coefficient is progressively dominated by the Knudsen diffusion coefficient. Under the Knudsen diffusion regime, the pore size distribution greatly affects the membrane coefficient. For example, at $r_m = 0.001\ \mu m$, the error is as high as 30 % at $\sigma = 0.35$. Figure 5.7 further illustrates that the error remains less than 5 % when the standard deviation of the membrane is at 0.15 or less. It thus follows that the effect of pore size distribution on membrane coefficient is only significant when the standard deviation is high and the average pore size of the membrane is small.

Li *et al.* also conducted absorption experiments for H_2S using 22 hollow fibre membrane contactors with different pore size characteristics [7]. The membrane coefficients obtained from the 22 membrane contactors were compared with the calculated membrane coefficients $\bar{k}_{Am,g}$ obtained from either Equation (5.21) (without considering the pore size distribution) or Equation (5.24) (considering the pore size distribution) using known pore structure parameters such as mean pore size, standard deviation and effective porosity [7]. It can be seen from Table 5.1 and Figure 5.8 that for the same membrane, the calculated $\bar{k}_{Am,g}$ values fall into the same order of magnitude compared to the $k_{Am,g}$ values obtained from absorption experiments. Also the $k_{Am,g}$ values predicted from the model incorporated with the log-normal distribution function have a better agreement with the experimental results for most of the cases. The data shown in Table 5.1 and Figure 5.8 suggests that the pore structure parameters

Table 5.1 Comparison of membrane coefficients

No.	Membrane coefficient, $k_{Am,g} \times 10^2$ (calculated) (uniform pores) ($m\,s^{-1}$)	Membrane coefficient, $\bar{k}_{Am,g} \times 10^2$ (calculated) (distribution pores) ($m\,s^{-1}$)	Membrane coefficient, $k_{Am,g} \times 10^2$ (absorption) ($m\,s^{-1}$)
1	0.0248	0.025	0.0225
2	0.0118	0.013	0.0161
3	0.0062	0.006	0.0235
4	1.27	1.40	0.71
5	0.69	0.77	0.78
6	0.64	0.70	1.18
7	0.81	1.15	1.84
8	2.62	2.80	4.09
9	6.71	7.16	7.20
10	0.75	0.96	0.95
11	1.04	1.20	1.21
12	1.23	1.70	1.73
13	1.17	1.32	1.20
14	0.59	0.80	0.83
15	0.73	0.87	0.92
16	0.84	1.07	1.27
17	0.51	0.72	0.81
18	0.55	0.86	1.02
19	0.71	0.85	0.86
20	0.72	0.92	0.90
21	0.57	0.87	1.03
22	0.54	0.88	1.10

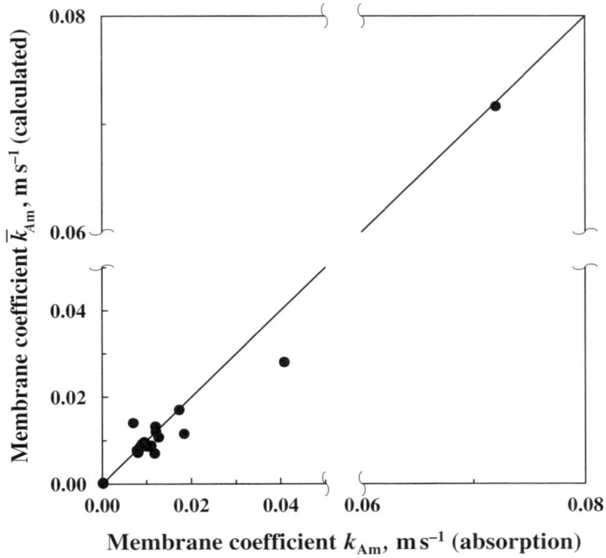

Figure 5.8 Comparison of membrane coefficient evaluated from absorption and calculated methods [7]. Copyright (2000) Elsevier, Chemical Engineering Science, **55** (23), Li, K., Kong, J. and Tan, X., *Design of hollow fibre membrane modules for soluble gas removal*, p. 5579–5588

of porous membranes determined from an improved gas permeation method [62] can be directly employed to estimate the membrane coefficient for various permeant gases. The accuracy of the estimation would largely depend on the availability of the pore structure parameters, which can be obtained fairly accurately using the improved gas permeation method [62].

5.4.3 Mass Transfer in Shell Side of a Contactor

Hollow fibres in the shell of a contactor are usually packed randomly, which complicates the mass transfer in the shell side of the module. As shown by Costello *et al.* [42], the shell side flow is likely to be characterized by regions of low local packing, through which a disproportionately high amount of flow passes, and regions of high local packing which are, by comparison, starved of flow. For a uniform packing in a triangular array, concentration profiles of the gas/vapour A in a fluid flowing in the shell side of a contactor can be described by considering a series of identical sectors as defined in Figure 5.9. The differential material balance equation for gas A at steady state condition is:

$$U\frac{\partial C_A}{\partial z} = D_A\left(\frac{1}{r}\frac{\partial}{\partial r}\left(r\frac{\partial C_A}{\partial r}\right) + \frac{1}{r^2}\frac{\partial^2 C_A}{\partial \theta^2}\right) \tag{5.26}$$

where U is the local velocity and C_A is the concentration of gas A in the fluid, which is a function of z, r and θ. A finite difference solution to an equation of this form developed by Miyatake and Iwashita [63] can be adopted to calculate the change of concentration profiles

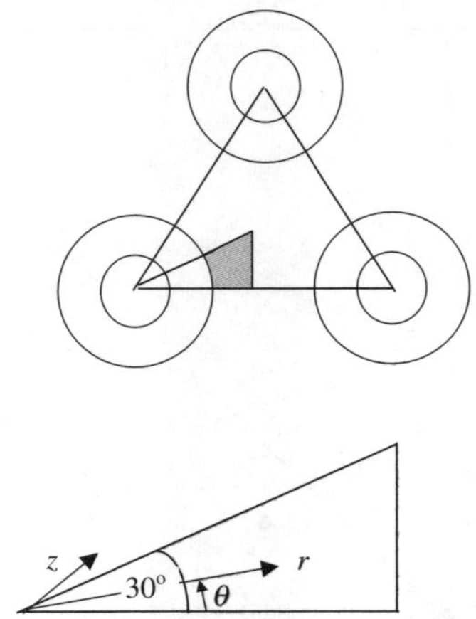

Figure 5.9 Analytical system and coordinates

Table 5.2 Shell side mass transfer correlations

Reference	Correlations	Range of Reynolds number
Yang and Cussler [38]	$\mathrm{Sh}=1.25\left(\mathrm{Re}\frac{d_e}{z}\right)^{0.93} S_C^{0.33}$	$0.1<\left(\mathrm{Re}\frac{d_e}{z}\right)<1000$
Prasad and Sirkar [56]	$\mathrm{Sh}=5.8(1-\phi)\left(\frac{d_e}{z}\right)\mathrm{Re}^{0.6} S_C^{0.3}$	$0<\mathrm{Re}<500$
Ahmed and Semmens [39]	$\mathrm{Sh}=0.0104\mathrm{Re}^{0.806}S_c^{0.33}$	$500<R_e<15\,000$
Costello *et al.* [42]	$\mathrm{Sh}=(0.53-0.58\phi)\mathrm{Re}^{0.53}S_c^{0.33}$	—

$d_e = (4 \times \text{flow area})/(\text{total fibre circumference})$

of the gas/vapour A in the shell side down the length of the contactor leading to a calculation of the mass transfer coefficient, so long as assumptions of steady state operation, insignificant axial diffusion, constant fluid properties and constant concentration of gas A at the interface are valid.

For randomly packed hollow fibre modules, the shell side mass transfer coefficient has been obtained empirically. Several correlations describing the mass transfer coefficient at the outside of the hollow fibres are listed in Table 5.2. The correlation given by Yang and Cussler [38] was developed using the experimental data of deaeration obtained from two hydrophobic hollow fibre modules with different packing fractions. The correlation developed by Prasad and Sirkar [56] includes the packing fraction, ϕ, as a parameter in the correlation and shows a smaller variation with the Reynolds number. The correlation proposed by Costello *et al.* [42] also includes the packing fraction, ϕ, but the factor d_e/z was eliminated as their experimental results indicate that the length of the fibre has no effect on the mass transfer coefficient. The correlation given by Ahmed and Semmens [39] eliminates both packing fraction and the factor d_e/z as their experimental data obtained from oxygenation experiments indicates independence of both the factors from the mass transfer coefficient. Shell side mass transfer has

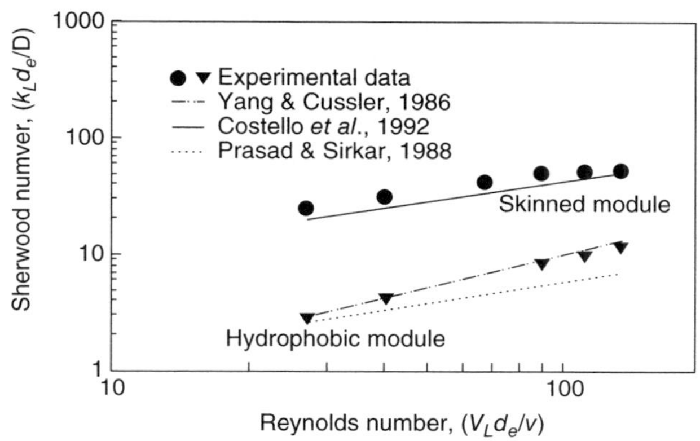

Figure 5.10 Sherwood number versus Reynolds number for the experimental data obtained for CO_2—H_2O system

also been studied by Li *et al.* [6] and Li and Teo [64] who employed both porous hydrophobic and ultrathin skinned hollow fibre modules for CO_2 absorption. Their experimental data, plotted in Figure 5.10, together with the correlations given in Table 5.2 indicate that data from the ultrathin skinned hollow fibre contactor fits well with the correlation of Costello *et al.* [42], but deviates from the correlations given by Yang and Cussler [38], and Prasad and Sirkar [56]. For the porous hydrophobic membrane module, the opposite results were obtained, i.e. the data fits correlations of Yang and Cussler, and Prasad and Sirkar fairly well, but deviates from Costello *et al.* From the comparisons between the experimental data and the correlations in Figure 5.10, it may suggest that these correlations may be correct at a particular operating condition and can be selectively employed for design purposes. The different correlations developed may represent different degrees of back mixing in the different contactor geometries.

5.4.4 Nonwetted, Wetted and Partially Wetted Conditions in a Hollow Fibre Contactor

When an aqueous solution is fed into the hollow fibre lumens of a contactor, the pressure drop of the liquid in the fibre lumens will inevitably be increased with the liquid velocity and length of the fibre used according to the Hagen–Poiseuille law. If porous membrane contactors are employed, three different operating modes, i.e. nonwetted, wetted or partially wetted modes, as shown in Figures 5.1(a)–5.1(c) in a hollow fibre contactor, can be realized depending on the characteristics of the hollow fibre contactor and the operating conditions.

When the porous hollow fibre membrane is totally wetted, the general conservation equations remain the same, except that the interface concentration and the fibre wall concentration are generally not the same and are related by Equation (5.27) assuming that gas phase mass transfer resistance in the shell side is negligible.

$$D_A\left(\frac{\partial C_A}{\partial r}\right)_{r=R} = k_{\mathrm{Am,l}}(C_{A\mathrm{i}} - C_A) \tag{5.27}$$

This problem cannot be solved analytically. Malek *et al.* [45] introduced an orthogonal collocation technique to obtain the concentration profile of the gas in liquid in the hollow fibre lumen down the length of the contactor leading to a calculation of the mass transfer coefficient.

As was mentioned at the beginning of this chapter, in actual hydrophobic hollow fibre modules, it is likely that the membrane contactor is partially wetted. The extent of wetting of the hollow fibre contactor is as yet an unknown function of the operating conditions. Aqueous solutions such as water may not penetrate through the hydrophobic hollow fibres due to the presence of a finite wetting pressure, p_w, acting along the direction opposite to that of penetration. The penetration of water through the pores of the hollow fibres, i.e. wetting of the fibres, however, becomes possible if the hydraulic pressure drop across the fibre wall exceeds the wetting pressure. Therefore, the wetting phenomena of a hollow fibre contactor can be related to the hollow fibre wetting pressure, p_w, described by the Laplace equation:

$$p_w = -2\sigma \cos\theta / r_p \tag{5.28}$$

The fibre wall is considered to be wetted for portions of the fibre length where the hydraulic pressure drop across the fibre wall exceeds the wetting pressure (Figure 5.1c). The shell side pressure can generally be taken as atmospheric pressure, while the tube side gauge pressure can be calculated using the Hagen–Poiseuille equation as follows:

$$p_t = \frac{32U_m \mu (1-z)}{d_i^2} \tag{5.29}$$

where, z is the length of fibre measured from the inlet. Based on Equation (5.29), the pressure profile in the fibre lumen can be calculated. As shown in Figure 5.11 [45], the pressure in the fibre lumen is dependent on both the liquid velocity and length of the fibre. Hence, the partially wetted model solves for mass transfer using the wetted hollow fibre model, where the

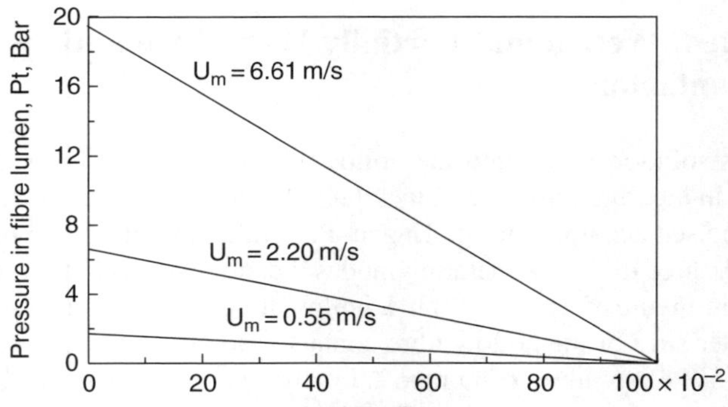

Figure 5.11 Pressure drop in hollow fibre lumen, fibre length, $L = 500$ mm, fibre ID, $d_i = 0.56$ mm [45]. Copyright (1997) American Chemical Society, Industrial and Engineering Chemistry Research, **36** (3), Malek, A., Li, K. and Teo, W. K., *Modeling of microporous hollow fibre membrane modules under partially wetted conditions*, p. 784–793

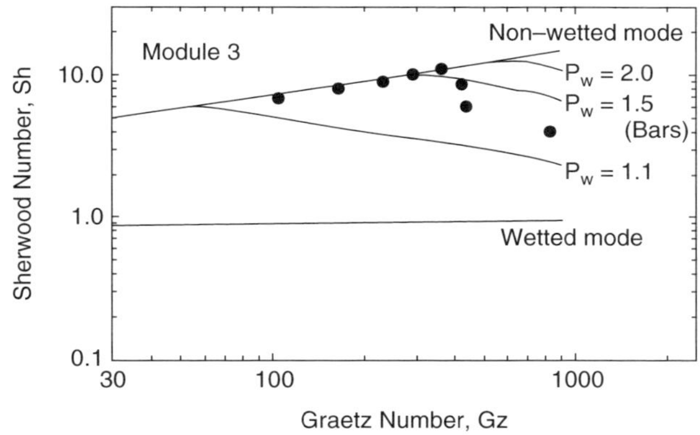

Figure 5.12 Comparison between experimental results and theoretical calculations; filled circles are the experimental data [45]. Copyright (1997) American Chemical Society, Industrial and Engineering Chemistry Research, **36** (3), Malek, A., Li, K. and Teo, W. K., *Modeling of microporous hollow fibre membrane modules operated under partially wetted conditions*, p. 784–793

wall pressure drop is larger than the wetting pressure, p_w, and switches to the nonwetted model otherwise. Again, an orthogonal collocation technique was employed to obtain the concentration profile of the gas in the hollow fibre lumen down the length of a contactor leading to the calculation of the overall mass transfer coefficient [45].

The above theoretical consideration on the partial wetting of a hollow fibre contactor was successfully adopted by Malek *et al.* [45] to explain an unusual experimental result obtained by Tai [65] who studied the removal of dissolved oxygen from water using hollow fibre contactors containing hydrophobic porous membranes. Water containing saturated oxygen was fed into the fibre lumen, while purified nitrogen acting as purge gas was introduced into the shell side. During the course of the experiments, Tai [65] observed that water droplets were formed at the outer surface of the hollow fibre membrane, especially near the inlet of the membrane module and overall mass transfer coefficients obtained at different liquid velocities exhibited a maximum value which could not be explained by existing mass transfer models. Conventionally, for hollow fibre contactors operated in nonwetted mode, increase of the water flow would generally increase the overall mass transfer coefficients for the O_2–H_2O system. The existence of a maximum value of the overall mass transfer coefficient is because of wetting of the contactor from the inlet due to the high liquid pressure in the fibre lumen. The partially wetted membrane contactor may result in a reduction of the overall mass transfer coefficient compared with the nonwetted mode, hence giving a maximum value of the overall mass transfer coefficient. Figure 5.12 shows the theoretical results together with the experimental data given by Malek *et al.* [45]. It can be seen that the observed experimental data shows a maximum value of the mass transfer coefficient at a Graetz number around 350. The experimental data agrees exceptionally well with the theoretical results in the lower Graetz number region, i.e. nonwetted mode. As the Graetz number increases, the partial wetting of the module becomes inevitable, resulting in a decrease in the mass transfer coefficient. Overall, the experimental data is in fairly good agreement with the simulation results obtained at $p_w = 1.5$ bars. When partial wetting of the fibre takes place in the contactor, the experimental results deviate from the theoretical results, however, the general trends are sufficiently clear to merit a meaningful comparison between the simulation and experimental results.

5.5 EFFECT OF CHEMICAL REACTION

Chemical reaction is often employed in gas absorption processes to enhance the mass transfer rate of gases to be removed. The reaction can be instantaneous, fast or slow and can be irreversible or reversible. The enhanced gas transfer rate is due to enhancement of the mass transfer coefficient and an increased driving force. Figures 5.13 and 5.14 illustrate mass transfer with various chemical reactions for a nonwetted operating mode.

5.5.1 Instantaneous Reaction

For an instantaneous reaction, gas *A* reacts with an aqueous solution containing a solute *B* as follows:

$$A + bB \rightarrow R \tag{5.30}$$

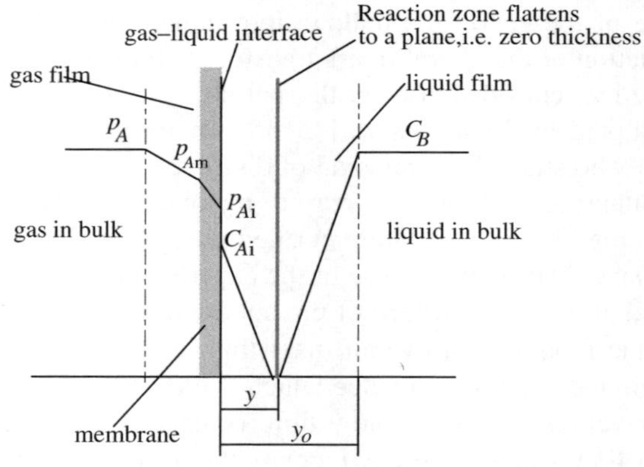

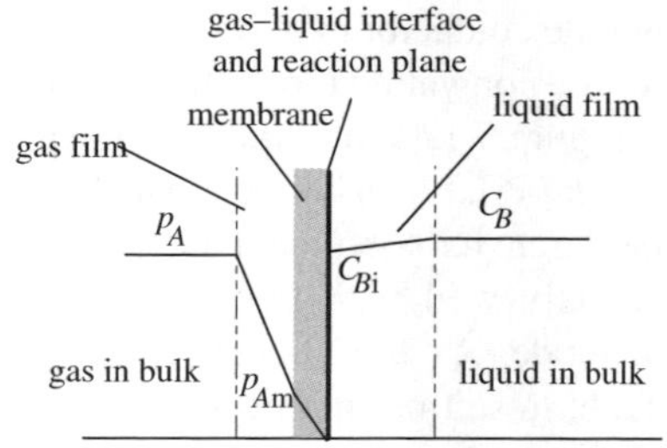

Figure 5.13 Mass transfer with chemical reaction: (a) instantaneous reaction; (b) instantaneous reaction with high C_B

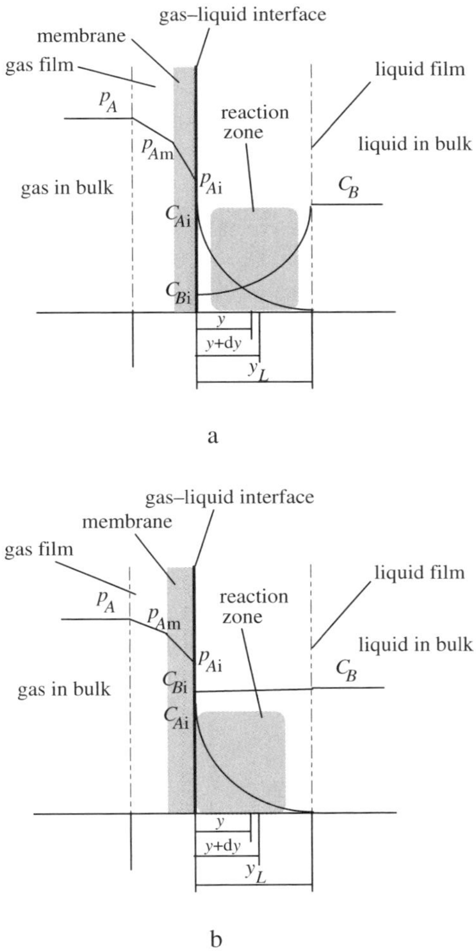

Figure 5.14 Mass transfer with chemical reaction: (a) fast second order reaction; (b) pseudo first order reaction

As the reaction takes place within the liquid film with the reaction zone flattening to a plane, as shown in Figure 5.13a, the rate of gas A transfer can, therefore, be evaluated with the Hatta approach as below:

$$N_A = \frac{\dfrac{D_{Bl}C_B H_A}{D_{Al}b} + p_A}{\dfrac{1}{k_{Ag}} + \dfrac{1}{k_{Am,g}} + \dfrac{H_A}{k_{Al}}} \tag{5.31}$$

If the resistances in the gas phase and in the membrane are negligible, i.e. $p_A = p_{Ai}$, Equation (5.31) is simplified as:

$$N_A = \left(\frac{D_{Bl}C_B}{bD_{Al}C_{Ai}} + 1\right) k_{Al} C_{Ai} \tag{5.32}$$

When the concentration of reactant B in the liquid becomes very high, i.e., C_B is high, the reaction zone is expected to shift to the gas–liquid interface, as shown in Figure 5.13b, and $k_{Al} \rightarrow \infty$; Equation (5.31) reduces to:

$$N_A = k_{ex} p_A \tag{5.33}$$

where:

$$\frac{1}{k_{ex}} = \frac{1}{k_{Ag}} + \frac{1}{k_{Am,g}} \tag{5.34}$$

5.5.2 Fast Reaction

For a fast second order reaction of $A + bB \rightarrow R$, the rate of gas A transfer can be written as:

$$N_A = E k_{Al} C_{Ai} \tag{5.35}$$

Equation (5.35) is written with an assumption that the concentration of gas A in the liquid is negligible, which is admissible as for a fast reaction, the reaction still takes place within the liquid film illustrated in Figure 5.14(a). In Equation (5.35), E is the enhancement factor due to the chemical reaction, which can be evaluated from the classical solution of gas absorption accompanied by a second order chemical reaction [66]:

$$E = \frac{M_H \sqrt{\dfrac{E_i - E}{E_i - 1}}}{\tanh M_H \sqrt{\dfrac{E_i - E}{E_i - 1}}} \tag{5.36}$$

where:

$$E_i = 1 + \frac{D_{Bl} C_B H_A}{b D_{Al} p_{Ai}} \tag{5.37a}$$

$$M_H = \frac{\sqrt{D_A C_B k}}{k_{Al}} \tag{5.37b}$$

where k is the reaction rate constant. At a high value of C_B as shown in Figure 5.14b, the fast second order reaction approaches to pseudo first order [66] and $E \rightarrow M_H$. Under such conditions, the overall mass transfer coefficient can be evaluated by:

$$\frac{1}{K_{Ag}^R} = \frac{1}{k_{Ag}} + \frac{1}{k_{Am,g}} + \frac{H_A}{\sqrt{D_A C_B k}} \tag{5.38}$$

As can be seen from Equation (5.38), the overall mass transfer coefficient, K_{Ag}^R is independent of the liquid velocity if a high concentration of B is employed for the mass transfer.

5.6 DESIGN EQUATIONS

The design equations for membrane based gas absorption closely follow conventional gas absorption processes. Depending on the membrane characteristics and arrangement of gas and liquid flow, the gas–liquid interface can be either in the hollow fibre lumen or at the outer surface of the hollow fibre. In the following sections, the design equations are derived for a hollow fibre module containing n hollow fibres of internal diameter (ID), d_i, external diameter (OD), d_o and hollow fibre length, L. As shown in Figure 5.15, the flow consists of a gas mixture, containing gas A and inert gases flowing in the hollow fibre lumen, while an aqueous solution is introduced countercurrently into the shell side of the contactor. Therefore, the gas–liquid interface is at the outer surface of the hollow fibre for a totally nonwetted operation. The differential mass transfer equation for the gas A absorbed can be written as:

$$-G_m \mathrm{d}Y_A = (n\pi d_o) \times J_A \times \mathrm{d}z \tag{5.39a}$$

The detailed derivation of Equation (5.39a) is given in Figure 5.16. In Equation (5.39a), Y_A is the local mole ratio of gas A and inert gases in the gas phase at a hollow fibre length of z and G_m, the inert gas flow rate. Substituting Equation (5.1d) into Equation (5.39a) gives:

$$-G_m \mathrm{d}Y_A = (n\pi d_o) \times K_{Ag}(p_A - p_A^*) \times \mathrm{d}z \tag{5.39b}$$

The local partial pressure of gas A in the bulk phase and partial pressure of gas A in equilibrium with the liquid phase are given by the following:

$$p_A = p \times y_A = p\left(\frac{Y_A}{1+Y_A}\right) \tag{5.40a}$$

$$p_A^* = p \times y_A^* = p\left(\frac{Y_A^*}{1+Y_A^*}\right) \tag{5.40b}$$

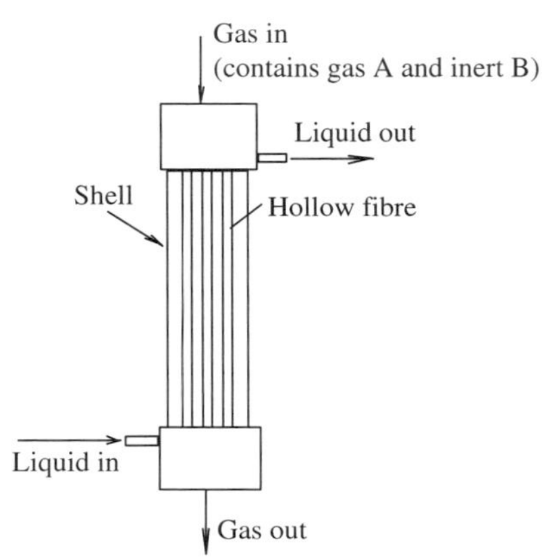

Figure 5.15 Membrane based gas absorption, countercurrent operation

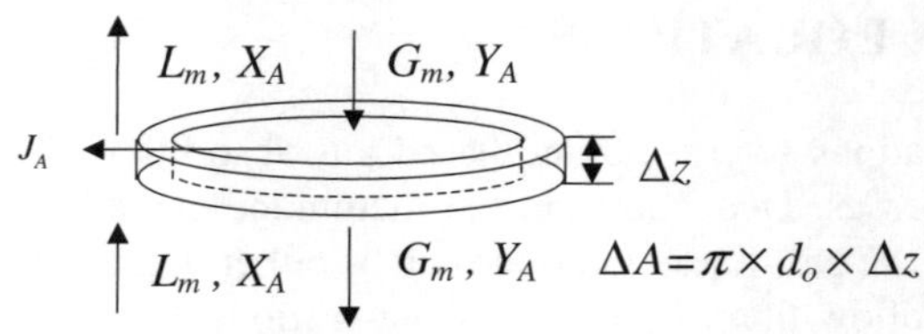

$(G_m Y_A)_{in}$ (input) + 0 (generation) = $((G_m Y_A)_{out} + J_A(\pi d_o \Delta z))$ (output) + 0 (accumulation) + 0 (consumption)

$$\frac{(G_m Y_A)_{in} - (G_m Y_A)_{out}}{\Delta z} = J_A \times \pi \times d_o$$

⇓ Writing in differential form

$$-\frac{G_m \mathrm{d}\, Y_A}{dz} = J_A \times \pi \times d_o$$

⇓ for n hollow fibres

$$-G_m \mathrm{d}\, Y_A = J_A \times (\pi \times d_o \times n) \times dz$$

Figure 5.16 Derivation of Equation (5.39a) from first principles

where, y_A represents local mole fraction of gas A. Substituting Equations (5.40a) and (5.40b) into Equation (5.39) and after integration, we obtain:

$$L = \frac{G_\mathrm{m}}{K_{A\mathrm{g}} n \pi d_\mathrm{o} p} \int_{Y_{A,\mathrm{out}}}^{Y_{A,\mathrm{in}}} \frac{dY_A}{\left(\frac{Y_A}{1+Y_A} - \frac{Y_A^*}{1+Y_A^*}\right)} \tag{5.41}$$

When the aqueous solution is in excess, or an instantaneous or fast reaction takes place between the reactant B and the dissolved gas A in liquid, the concentration of the gas A in the liquid approaches zero resulting in $Y_\mathrm{A}^* = 0$. Therefore Equation (5.41) is reduced, after integration, to:

$$L = \frac{G_\mathrm{m}}{K_{A\mathrm{g}} n \pi d_\mathrm{o} p} \int_{Y_{A,\mathrm{out}}}^{Y_{A,\mathrm{in}}} \frac{\mathrm{d}Y_A}{\left(\frac{Y_A}{1+Y_A}\right)} = \frac{G_\mathrm{m}}{K_{A\mathrm{g}} n \pi d_\mathrm{o} p}\left((Y_{A,\mathrm{in}} - Y_{A,\mathrm{out}}) + \ln\left(\frac{Y_{A,\mathrm{in}}}{Y_{A,\mathrm{out}}}\right)\right) \tag{5.42}$$

If the concentration of gas A in the gas mixture is dilute, i.e. $y_\mathrm{A} = Y_A$ and $y_\mathrm{A}^* = Y_\mathrm{A}^*$, Equation (5.41) can be simplified as:

$$L = \frac{G}{K_{A\mathrm{g}} n \pi d_\mathrm{o} p} \int_{y_\mathrm{out}}^{y_{A,\mathrm{in}}} \frac{dy_A}{\left(y_A - y_\mathrm{A}^*\right)} \tag{5.43}$$

For dilute concentrations, the material balance of the gas A from inlet of the module to any point of the module is:

$$L(x_{A,\text{in}} - x_A) = G(y_{A,\text{out}} - y_A) \tag{5.44}$$

where L is the liquid flow rate. If the entering aqueous solution is free from the gas A, then $x_{A,\text{in}} = 0$ and

$$x_A = \frac{G}{L}(y_A - y_{A,\text{out}}) \tag{5.45}$$

For the dilute solution, the Henry's law holds:

$$y_A^* = mx_A \tag{5.46}$$

where

$$m = H_A \frac{C_\text{T}}{p} \tag{5.47}$$

Substituting Equations (5.45) and (5.46) into Equation (5.43), upon rearrangement, the following expression may be obtained:

$$\begin{aligned} L &= \frac{G}{K_{A\text{g}} n\pi d_\text{o} p} \int_{y_{A,\text{out}}}^{y_{A,\text{in}}} \frac{\text{d}y_A}{y_A - \dfrac{mG}{L}(y_A - y_{A,\text{out}})} \\ &= \frac{G}{K_{A\text{g}} n\pi d_\text{o} p \left(1 - \dfrac{mG}{L}\right)} \ln\left(\left(1 - \frac{mG}{L}\right)\frac{y_{A,\text{in}}}{y_{A,\text{out}}} + \frac{mG}{L}\right) \end{aligned} \tag{5.48}$$

Again, in a condition that the absorbing liquid is in excess, or an instantaneous or fast reaction takes place between the reactant B and the dissolved gas A in liquid, Equation (5.48) reduces to:

$$L = \frac{G}{K_{A\text{g}} n\pi d_\text{o} p} \ln\left(\frac{y_{A,\text{in}}}{y_{A,\text{out}}}\right) \tag{5.49}$$

Finally, the design equation can be written as:

$$L = HTU \times NTU \tag{5.50}$$

where HTU is the height of transfer unit defined as:

$$HTU = \frac{G_\text{m}}{K_{A\text{g}} n\pi d_\text{o} p} \tag{5.51}$$

In Equation (5.50), the number of transfer unit, NTU is defined as:

$$NTU = \int_{Y_{A,\text{out}}}^{Y_{A,\text{in}}} \frac{dY_A}{\left(\dfrac{Y_A}{1+Y_A} - \dfrac{Y_A^*}{1+Y_A^*}\right)} \tag{5.52}$$

or for dilute gas A in the gas mixture:

$$NTU = \int_{y_{\text{out}}}^{y_{A,\text{in}}} \frac{dy_A}{(y_A - y_A^*)} \tag{5.53}$$

The above design equations were derived for nonwetted operation with gas flowing in the hollow fibre lumen and liquid flowing in the shell side. For any other operating modes or for gas stripping processes, the design equations can be easily derived in exactly the same manner shown above.

NOTATION

A_p	Total membrane pore area, m^2
A_t	Total membrane area, m^2
b	Stoichiometric coefficient
C	Concentration in liquid, $mol\,m^{-3}$
d	Hollow fibre diameter, m
d_e	Equivalent diameter, m
d_p	Pore diameter, m
D_A	Continuum diffusion coefficient, $m^2\,s^{-1}$
D_{AK}	Knudsen diffusion coefficient, $m^2\,s^{-1}$
$D_{A,e}$	Diffusion coefficient of a gas in membrane pores, $m^2\,s^{-1}$
$\overline{D}_{Ae}$	Diffusion coefficient by considering membrane pore size distribution, $m^2\,s^{-1}$
E	Enhancement factor
$g(r)$	Log-normal distribution function, m^{-1}
G	Flow rate of a gas mixture, $mol\,s^{-1}$
G_m	Inert gas flow rate, $mol\,s^{-1}$
Gz	Graetz number
H_A	Henry's constant, $atm\,m^3\,mol^{-1}$
k	Reaction rate constant in Equations (5.37b) and (5.38)
$\overline{k}_{Am,g}$	Membrane coefficient with consideration of pore distribution, $m\,s^{-1}$
k	Mass transfer coefficient in liquid phase, $m\,s^{-1}$, or in gas phase, $mol\,s^{-1}\,m^{-2}\,atm^{-1}$ or in membrane, $m\,s^{-1}$
K	Overall mass transfer coefficient in liquid phase, $m\,s^{-1}$ or in gas phase, $mol\,s^{-1}\,m^{-2}\,atm^{-1}$
J	Mass transfer flux, $mol\,m^{-2}\,s^{-1}$
L	Fibre length, m
L	Flow rate of liquid, $mol\,s^{-1}$ or $m^3\,s^{-1}$
L_m	Inert liquid flow rate, $mol\,s^{-1}$

m	Henry's constant, defined in Equation (5.46)
M	Molecular weight of gas
M_H	Hatta number defined in Equation (5.37b)
n	Number of hollow fibres in a module
N	Rate of mass transfer, mol s^{-1}
p	Pressure, atm or Pa
Δp	Pressure difference across the membrane, atm or Pa
p_w	Wetting pressure, atm or Pa
p_t	Gauge pressure in fibre lumen at a given length of the fibre, atm
r	Radial coordinate, m
r	Local pore radius, m
r_p	Pore radius, m
r_m	Mean pore radius, m^{-1}
R	Gas constant (8.314 Pa m^3 mol^{-1} K^{-1})
R	Fibre inner radius, m
Re	Reynolds number
R_o	Fibre outer radius, m
Sh	Sherwood number
U	Velocity in fibre, m s^{-1}
U_m	Average velocity in fibre, m s^{-1}
x	Mole fraction of gas A in aqueous solution
X	$x/(1 - x)$, mole ratio
y	Mole fraction of gas A
Y	$y/(1 - y)$, mole ratio
z	Axial coordinate, m
z	Contactor length given in Table 5.2

Symbols

α	Constant defined in Equation (5.21)
ε	Porosity
δ	Membrane thickness, m
τ	Tortuosity
μ	Liquid viscosity
ϕ	Packing fraction
σ	Dimensionless standard deviation of the membrane pore size
σ	Surface tension
θ	Contact angle

Subscripts

A	Gas A
B	Reactant B in aqueous solution
ex	External
f	Feed
g	Gas phase

i Gas–liquid interface
i Hollow fibre ID
in Contactor inlet
l Liquid phase
m Membrane
o Hollow fibre OD
out Contactor outlet
p pore
T Total

Superscripts

S Shell side
L Hollow fibre lumen
R Reaction
* Equilibirium

APPENDIX 5.1

For a binary gas mixture, the rate of gas transferred through a single capillary pore (tube) of a membrane with a radius *r* can be expressed as:

$$q_A = k_A A \frac{\Delta p}{RT} \qquad \text{(A1)}$$

where q_A is the gas molar flow rate, in mol s^{-1}, R is the gas constant (8.3174 m^3 Pa mol^{-1} K^{-1}), Δp is the pressure difference across the membrane, in Pa, A is the cross sectional area of the pore, defined as πr^2, in m^2, and k_A is the mass transfer coefficient, in m s^{-1}, defined as:

$$k_A = \frac{D_{A,\mathrm{e}}}{\delta\tau_\mathrm{m}} = \frac{1}{\delta\tau_\mathrm{m}}\cdot\frac{D_A r}{\alpha + r} \qquad \text{(A2)}$$

where α has been defined in Equation (5.21).

For a membrane with N_t number of pores, by considering the pore size distribution, the total flow rate, Q_t through the membrane can be written as:

$$Q_\mathrm{t} = \int_0^\infty N_\mathrm{t} q_A g(r)\mathrm{d}r = \frac{N_\mathrm{t}\Delta p \pi D_A}{\Re T \delta\tau_\mathrm{m}}\int_0^\infty \frac{r^3}{\alpha + r} g(r)\mathrm{d}r \qquad \text{(A3)}$$

where $g(r)$ is the log-normal distribution function defined in Equation (5.23).

For a porous membrane, the mass transfer coefficient, $\bar{k}_{A\mathrm{m}}$ is commonly defined as:

$$\bar{k}_{A\mathrm{m}} = \frac{Q_\mathrm{t}}{A_\mathrm{t}}\frac{RT}{\Delta p} \qquad \text{(A4)}$$

where A_t is the total membrane area, in m^2.

In addition, the surface porosity of a porous membrane can be written as:

$$\varepsilon = \frac{A_\mathrm{p}}{A_\mathrm{t}} \tag{A5}$$

where A_p is the cross sectional area of the total number of pores, defined as:

$$A_\mathrm{p} = N_\mathrm{t} \int_0^\infty \pi r^2 g(r) \mathrm{d}r \tag{A6}$$

Substituting Equations (A3), (A5) and (A6) into Equation (A4) gives Equations (A7a) and (A7b) below:

$$\bar{k}_{Am} = \frac{\varepsilon \bar{D}_{A,\mathrm{e}}}{\delta \tau_\mathrm{m}} \tag{A7a}$$

where:

$$\bar{D}_{A,\mathrm{e}} = D_A \frac{\displaystyle\int_0^\infty \frac{r^2}{\alpha + r} \cdot \exp\left[- \frac{\left(\ln\left(\frac{r}{r_\mathrm{m}}\right)(1+\sigma^2)^{0.5} \right)^2}{2\ln(1+\sigma^2)} \right] \mathrm{d}r}{\displaystyle\int_0^\infty r \cdot \exp\left[- \frac{\left(\ln\left(\frac{r}{r_\mathrm{m}}\right)(1+\sigma^2)^{0.5} \right)^2}{2\ln(1+\sigma^2)} \right] \mathrm{d}r} \tag{A7b}$$

REFERENCES

1. Wang, R., Li, D. F. and Liang, D. T., *Modeling of CO_2 capture by three typical amine solutions in hollow fibre membrane contactors.* Chemical Engineering and Processing, **43** (7): 849–856 (2004).
2. Wang, D., Teo, W. K. and Li, K., *Selective removal of trace H_2S from gas streams containing CO_2 using hollow fibre membrane modules/contractors.* Separation and Purification Technology, **35** (2): 125–131 (2004).
3. Yeon, S.-H., Lee, K.-S., Sea, B., Park, Y.-I. and Lee, K.-H., *Application of pilot-scale membrane contactor hybrid system for removal of carbon dioxide from flue gas.* Journal of Membrane Science, **257** (1–2): 156–160 (2005).
4. Dindore, V. Y., Brilman, D. W. F., Feron, P. H. M. and Versteeg, G. F., *CO_2 absorption at elevated pressures using a hollow fibre membrane contactor.* Journal of Membrane Science, **235** (1–2): 99–109 (2004).
5. Kim, Y.-S. and Yang, S.-M., *Absorption of carbon dioxide through hollow fibre membranes using various aqueous absorbents.* Separation and Purification Technology, **21** (1–2): 101–109 (2000).
6. Li, K., Tai, Mona, S. L. and Teo, W. K., *Design of a CO_2 scrubber for self-contained breathing systems using a microporous membrane.* Journal of Membrane Science, **86** (1–2): 119–125 (1994).

7. Li, K., Kong, J. and Tan, X., *Design of hollow fibre membrane modules for soluble gas removal.* Chemical Engineering Science, **55** (23): 5579–5588 (2000).
8. Rangwala, H. A., *Absorption of carbon dioxide into aqueous solutions using hollow fibre membrane contactors.* Journal of Membrane Science, **112** (2): 229–240 (1996).
9. Feron, P. H. M. and Jansen, A. E., *The production of carbon dioxide from flue gas by membrane gas absorption.* Energy Conversion and Management, **38** (Supplement 1): S93–S98 (1997).
10. Simmonds, M. H. P., *Post combustion technologies for CO_2 capture: a techno-economic overview of selected options* in *The 7th International Conference on Greenhouse Gas Control Technology.* (2004) Vancouver, Canada.
11. Kwasniewski, V. J., Calamur, N., Kaminsky, M. P., Mahoney, J. A., Scouten, C. G. and Wilsak, R. A., *Unsaturated Hydrocarbon Separation and Recovery Process.* (1999) US patent 5,863,420.
12. Barbe, A. M., Hogan, P. A. and Johnson, R. A., *Surface morphology changes during initial usage of hydrophobic, microporous polypropylene membranes.* Journal of Membrane Science, **172** (1–2): 149–156 (2000).
13. Wang, R., Li, D. F., Zhou, C., Liu, M. and Liang, D. T., *Impact of DEA solutions with and without CO_2 loading on porous polypropylene membranes intended for use as contactors.* Journal of Membrane Science, **229** (1–2): 147–157 (2004).
14. Li, J.-L. and Chen, B.-H., *Review of CO_2 absorption using chemical solvents in hollow fibre membrane contactors.* Separation and Purification Technology, **41** (2): 109–122 (2005).
15. Zhang, G. L. and Cussler, E. L., *Distillation in hollow fibres.* American Institute of Chemical Engineers Journal, **49** (9): 2344–2351 (2003).
16. Chung, J. B., DeRocher, J. P. and Cussler, E. L., *Distillation with nanoporous or coated hollow fibres.* Journal of Membrane Science, **257** (1–2): 3–10 (2005).
17. Cot, L., Ayral, A., Durand, J., Guizard, C., Hovnanian, N., Julbe, A. and Larbot, A., *Inorganic membranes and solid state sciences.* Solid State Sciences, **2** (3): 313–334 (2000).
18. Burggraaf, A. J., *Fundametals of membrane top layer synthesis and processing*, in *Fundamentals of Inorganic Membrane Science and Technology*, Burggraaf A. J. and Cot, L., Editors. (1996) Amsterdam: Elsevier Science B.V., p. 259–329.
19. Tan, X., Liu, S. and Li, K., *Preparation and characterization of inorganic hollow fibre membranes.* Journal of Membrane Science, **188** (1): 87–95 (2001).
20. De Jong, J., Benes, N. E., Koops, G. H. and Wessling, M., *Towards single step production of multilayer inorganic hollow fibres.* Journal of Membrane Science, **239** (2): 265–269 (2004).
21. Way, J. D. and Roberts, D. L., *Hollow fibre inorganic membranes for gas separations.* Separation Science and Technology, **27** (1): 29–41 (1992).
22. Liu, S. and Gavalas, G. R., *Oxygen selective ceramic hollow fibre membranes.* Journal of Membrane Science, **246** (1): 103–108 (2005).
23. Liu, Y. T., Tan, X. Y. and Li, K., *Nonoxidative methane coupling in a $SrCe_{0.95}Yb_{0.05}O_3$-alpha (SCYb) hollow fibre membrane reactor.* Industrial and Engineering Chemistry Research, **45** (11): 3782–3790 (2006).
24. Liu, S. M., Teo, W. K., Tan, X. Y. and Li, K., *Preparation of PDMSvi-Al_2O_3 composite hollow fibre membranes for VOC recovery from waste gas streams.* Separation and Purification Technology, **46** (1–2): 110–117 (2005).
25. Picard, C., Larbot, A., Sarrazin, J., Janknecht, P. and Wilderer, P., *Ceramic membranes for ozonation in wastewater treatment.* Annales de Chemie-Science des Materiaux **26** (2): 13–22 (2001).
26. Alami Younssi, S., Iraqi, A., Rafiq, M., Persin, M., Larbot, A. and Sarrazin, J., *[gamma]Alumina membranes grafting by organosilanes and its application to the separation of solvent mixtures by pervaporation.* Separation and Purification Technology, **32** (1–3): 175–179 (2003).
27. Alami-Younssi, S., Kiefer, C., Larbot, A., Persin, M. and Sarrazin, J., *Grafting [gamma]alumina microporous membranes by organosilanes: characterisation by pervaporation.* Journal of Membrane Science, **143** (1–2): 27–36 (1998).

28. Caro, J., Noack, M. and Kolsch, P., *Chemically modified ceramic membranes.* Microporous and Mesoporous Materials, **22** (1–3): 321–332 (1998).
29. Dafinov, A., Garcia-Valls, R. and Font, J., *Modification of ceramic membranes by alcohol adsorption.* Journal of Membrane Science, **196** (1): 69–77 (2002).
30. Hyun, S. H., Jo, S. Y. and Kang, B. S., *Surface modification of [gamma]alumina membranes by silane coupling for CO_2 separation.* Journal of Membrane Science, **120** (2): 197–206 (1996).
31. Krajewski, S. R., Kujawski, W., Dijoux, F., Picard, C. and Larbot, A., *Grafting of ZrO_2 powder and ZrO_2 membrane by fluoroalkylsilanes.* Colloids and Surfaces A: Physicochemical and Engineering Aspects, **243** (1–3): 43–47 (2004).
32. Picard, C., Larbot, A., Guida-Pietrasanta, F., Boutevin, B. and Ratsimihety, A., *Grafting of ceramic membranes by fluorinated silanes: hydrophobic features.* Separation and Purification Technology, **25** (1–3): 65–69 (2001).
33. Picard, C., Larbot, A., Tronel-Peyroz, E. and Berjoan, R., *Characterisation of hydrophilic ceramic membranes modified by fluoroalkylsilanes into hydrophobic membranes.* Solid State Sciences, **6** (6): 605–612 (2004).
34. Qi, Z. and Cussler, E. L., *Microporous hollow fibres for gas absorption: I. Mass transfer in the liquid.* Journal of Membrane Science, **23** (3): 321–332 (1985).
35. Qi, Z. and Cussler, E. L., *Microporous hollow fibres for gas absorption: II. Mass transfer across the membrane.* Journal of Membrane Science, **23** (3): 333–345 (1985).
36. Cooney, D. and Jackson, C., *Gas absorption in a hollow fibre device.* Chemical Engineering Communications, **79**: 153–163 (1989).
37. Karoor, S. and Sirkar, K. K., *Gas absorption studies in microporous hollow fibre membrane modules.* Industrial and Engineering Chemistry Research, **32** (4): 674–684 (1993).
38. Yang, M. and Cussler, E. L., *Designing hollow-fibre contactors.* American Institute of Chemical Engineers Journal, **32**: 1910–1916 (1986).
39. Ahmed, T. and Semmens, M. J., *Use of sealed end hollow fibres for bubbleless membrane aeration: experimental studies.* Journal of Membrane Science, **69** (1–2): 1–10 (1992).
40. Kreulen, H., Smolders, C. A., Versteeg, G. F. and van Swaaij, W. P. M., *Microporous hollow fibre membrane modules as gas–liquid contactors. Part 1. Physical mass transfer processes: a specific application: mass transfer in highly viscous liquids.* Journal of Membrane Science, **78** (3): 197–216 (1993).
41. Kreulen, H., Smolders, C. A., Versteeg, G. F. and van Swaaij, W. P. M., *Microporous hollow fibre membrane modules as gas–liquid contactors Part 2. Mass transfer with chemical reaction.* Journal of Membrane Science, **78** (3): 217–238 (1993).
42. Costello, M. J., Fane, A. G., Hogan, P. A. and Schofield, R. W., *The effect of shell side hydrodynamics on the performance of axial flow hollow fibre modules.* Journal of Membrane Science, **80** (1): 1–11 (1993).
43. Skelland, A. H. P., *Diffusional Mass Transfer.* (1974) New York: John Wiley & Sons, Inc.
44. Poddar, T. K., Majumdar, S. and Sirkar, K. K., *Membrane-based absorption of VOCs from a gas stream.* American Institute of Chemical Engineers Journal, **42** (11): 3267–3282 (1996).
45. Malek, A., Li, K. and Teo, W. K., *Modeling of microporous hollow fibre membrane modules operated under partially wetted conditions.* Industrial and Engineering Chemistry Research, **36** (3): 784–793 (1997).
46. Kreulen, H., Smolders, C. A., Versteeg, G. F. and Van Swaaij, W. P. M., *Determination of mass transfer rates in wetted and non-wetted microporous membranes.* Chemical Engineering Science, **48** (11): 2093–2102 (1993).
47. Kim, B. S. and Harriott, P., *Critical entry pressure for liquids in hydrophobic membranes.* Journal of Colloid and Interface Science, **115** (1): 1–8 (1987).
48. Callahan, R. W., *Novel uses of micrporous membranes: a case study* in *New Membrane Materials and Processes for Separation.* American Institute of Chemical Engineers Symposium Series, **84** (261): 54–63 (1988).

49. Sirkar, K. K., *Other new membrane processes*, in *Membrane Handbook*, Winston W. S. and Sirkar, K. K. Editors. (1992) New York: Van Nostrand Reinhold, p. 891–992.
50. Kreulen, H., Versteeg, G. F., Smolders, C. A. and Van Swaaij, W. P. M., *Selective removal of H_2S from sour gas with microporous membranes. Part I. Application in a gas–liquid system.* Journal of Membrane Science, **73** (2–3): 293–304 (1992).
51. Westerterp, K. R., van Swaaij, W. P. M. and Beenackers, A. A. C. M., *Chemical Reactor Design and Operation.* (1984) New York: Wiley.
52. Li, K., Wang, D., Koe, C. C. and Teo, W. K., *Use of asymmetric hollow fibre modules for elimination of H_2S from gas streams via a membrane absorption method.* Chemical Engineering Science, **53** (6): 1111–1119 (1998).
53. Nymeijer, K., Visser, T., Assen, R. and Wessling, M., *Super selective membranes in gas–liquid membrane contactors for olefin/paraffin separation.* Journal of Membrane Science, **232** (1–2): 107–114 (2004).
54. Iversen, S. B., Bhatia, V. K., Dam-Johansen, K. and Jonsson, G., *Characterization of microporous membranes for use in membrane contactors.* Journal of Membrane Science, **130** (1–2): 205–217 (1997).
55. Koval, C. R., Lopez, L. L., Kaul, B. B., Renshaw, S., Green, K. and Dubois, M. R., *Kinetic-studies of olefin binding to sulfido sites in dinuclear molybdenum complexes.* Organometallics, **14** (7): 3440–3447 (1995).
56. Prasad, R. and Sirkar, K. K., *Dispersion-free solvent-extraction with microporous hollow-fibre modules.* American Institute of Chemical Engineers Journal, **34** (2): 177–188 (1988).
57. Kartohardjono, S., Chen, V. and Leiknes, T., *Dissolved oxygen removal from water by vacuum degassing process using sealed end poly Methyl Pentene (PMP) hollow fibre membrane contactor* in *The Fifth International Membrane Science and Technology Conference (MSTEC'03).* (2003) The University of New South Wales, Sydney, Australia: UNESCO Centre for Membrane Science and Technology,
58. Wickramasinghe, S. R., Semmens, M. J. and Cussler, E. L., *Mass-transfer in various hollow fibre geometries.* Journal of Membrane Science, **69** (3): 235–250 (1992).
59. Park, J. K. and Chang, H. N., *Flow distribution in the fibre lumen side of a hollow-fibre module.* American Institute of Chemical Engineers Journal, **32** (12): 1937–1947 (1986).
60. Zydney, A. L., Aimar, P., Meireles, M., Pimbley, J. M. and Belfort, G., *Use of the log-normal probability density function to analyze membrane pore size distributions: functional forms and discrepancies.* Journal of Membrane Science, **91** (3): 293–298 (1994).
61. Mochizuki, S. and Zydney, A. L., *Theoretical analysis of pore size distribution effects on membrane transport.* Journal of Membrane Science, **82** (3): 211–227 (1993).
62. Kong, J. and Li, K., *An improved gas permeation method for characterising and predicting the performance of microporous asymmetric hollow fibre membranes used in gas absorption.* Journal of Membrane Science, **182** (1–2): 271–281 (2001).
63. Miyatake, O. and Iwashita, H., *Laminar-flow heat-transfer to a fluid flowing axially between cylinders with a uniform surface-temperature.* International Journal of Heat and Mass Transfer, **33** (3): 417–425 (1990).
64. Li, K. and Teo, W. K., *An ultrathin skinned hollow fibre module for gas absorption at elevated pressures.* Chemical Engineering Research and Design, **74** (A8): 856–862 (1996).
65. Tai, M. S. L., *Dissolved oxygen removal using microporous hollow fibre membrane modules* in *Department of Chemical Engineering.* (1995) Singapore: National University of Singapore.
66. Danckwerts, P. V., *Gas Liquid Reactions.* (1970) New York: McGraw-Hill.

6

Mixed Conducting Ceramic Membranes for Oxygen Separation

6.1 INTRODUCTION

Research into ionic conducting materials and membranes has been in progress for long time. Initial work was mainly focused on the development of solid oxide fuel cells and sensors [1, 2]. Interest in membranes prepared from these types of materials for oxygen separation from air arose after the 1980s when a number of new materials such as $SrCo_{0.8}Fe_{0.2}O_{3-\delta}$ and $La(Co,Fe)O_3$ were developed [3]. These new materials show mixed conducting properties as they are not only capable of conducting ions, but also electrons. Similar to conventional ionic conducting materials, these materials also possess a unique separation mechanism, i.e. oxygen or hydrogen is transported through the membranes in a dissociated or ionized form rather than the conventional molecular diffusion. Therefore, an extremely high selectivity (theoretically up to 100 %) can be achieved. The electronic conductivity of mixed conducting materials plays an important role, as it acts as an internal short circuit such that the mixed conducting membranes do not need electrodes and an external circuit. If a mixed ionic oxygen and electronic conducting membrane is placed under an oxygen chemical potential gradient, oxygen anions permeate from the high oxygen chemical potential side to the low oxygen chemical potential side while overall charge neutrality is maintained by a counterbalancing flux of electrons. Figure 6.1 shows schematically the operating principle of mixed conducting materials and the difference to that of purely ionic conducting materials. Because the external circuit is unnecessary, fabrication and operation of the mixed conducting membranes is much simplified. Thus, significant cost reduction can be expected and so mixed conducting materials have attracted considerable research interest in recent years.

In the past ten years efforts in efficient utilization of energy and reduction of emissions have indirectly stimulated research in mixed conducting membranes. One of the most important applications of mixed conducting membranes is high purity oxygen production [1, 4, 5]. Significant advances have been achieved through several major initiatives that have brought the technology close to the point of commercial readiness.

Ceramic Membranes for Separation and Reaction K. Li

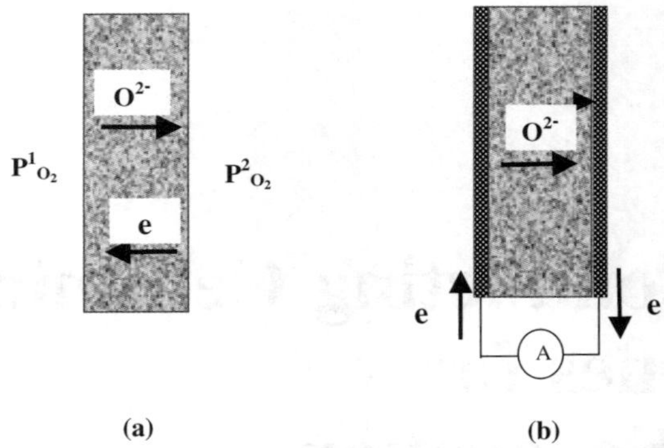

Figure 6.1 Different principles for oxygen permeable membranes: (a) mixed ionic–electronic conductor. (b) solid electrolyte cell (oxygen pump) [128]. Copyright (2006) Taylor & Francis, Catalysis Reviews–Science and Engineering, **48** (2), Liu, Y., Tan, X. and Li, K., *Mixed conducting ceramics for catalytic membrane processing*, p. 145–198

6.2 FUNDAMENTALS OF MIXED CONDUCTING CERAMIC MATERIALS

The performance of mixed conducting ceramic membranes is strongly dependant on the properties of its materials, which are mainly governed by the material compositions and structures. In order to discuss the relationship between the materials and the membrane performance, basic knowledge and understanding of the composition, structure and properties of the materials are required.

6.2.1 Structure of Perovskite-type Materials

It has been found that most mixed conducting materials are perovskites which were originally named after a mineral oxide of $CaTiO_3$. The basic structure of this mineral was first thought to be cubic although it was later found to be orthorhombic and the name perovskite has been retained for this type of structure. Such a structure makes the material very stable at high temperatures. A simple cubic structure emphasizing the coordination environment about the A site cation is shown in Figure 6.2. It can be seen that the A site cation is coordinated to 12 oxygen ions forming a cuboctahedral coordination environment while the B site cation is coordinated to six oxygen ions with an octahedral geometry. This structure provides an indication of chemical composition and the structural unit formed when preparing the materials. An ideal perovskite consists of ABO_3 units, but the chemical composition can vary depending on the valencey of the A and B site cations. Combinations such as $A^{1+}B^{5+}O_3$, $A^{2+}B^{4+}O_3$ and $A^{3+}B^{3+}O_3$, are commonly seen. In most cases the A site is generally occupied by 2+ large alkali earth metals such as barium, lanthanum or strontium, and the B site by 4+ smaller first row transition elements such as cerium, cobalt or iron. Although the same structure is maintained, the properties can be very different. For example, the $BaCoO_3$ and $SrCoO_3$ exhibit oxygen conduction, but the $BaCeO_3$ and $SrCeO_3$ are proton conductors.

In a perovskite structure, the A site cation is normally larger than the B site cation because of the different coordination environment. Hayashi *et al.* [6] examined a correlation between

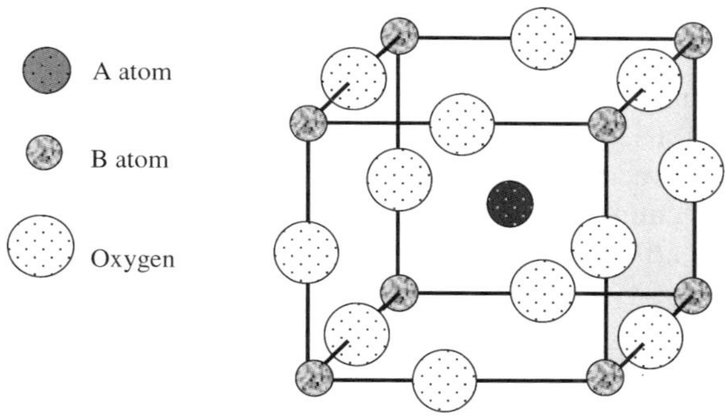

Figure 6.2 Structure of ABO_3, emphasizing the coordination number of the A cation at the body centre [128]. Copyright (2006) Taylor & Francis, Catalysis Reviews–Science and Engineering, **48** (2), Liu, Y. Y., Tan, X. Y. and Li, K., *Mixed conducting ceramics for catalytic membrane processing*, p. 145–198

the electrical conductivity and structurally related parameters such as the tolerance factor and the specific free volume of the lattice. The tolerance factor is defined for indicating the deformation from the cubic lattice as:

$$\text{Tolerance factor} = (r_A + r_O)/\sqrt{2}\,(r_B + r_O) \tag{6.1}$$

where r_A, r_B and r_O are the ionic radii of the A site cation, B site cation and oxygen. High conductivity is found to occur in compositions with large specific free volumes and with tolerance factors of around 0.96. The following two guides were suggested [6]:

(1) A site cation should have a large ionic radii in order to achieve the large free volume.

(2) B site cation should be chosen for adjusting the tolerance factor to around 0.96.

Yamamura *et al.* [7] studied the effect of the radii of A and B site cations on the oxygen ion conductivity with a lanthanum and neodymium series of materials using molecular dynamic simulations. Their results suggested that the simulation was able to reproduce the effects of the tolerance factor and the ionic radius of A site cations on the ion conductivity. A new perovskite oxygen ion conductor $(Sm_{0.8}Ca_{0.2})AlO_{2.9}$ with a conductivity of $0.037\,S\,cm^{-1}$ at 800 °C was developed by them using this method, which could be very useful for the development of cheap and optimal materials.

A perovskite with the ideal structure shown in Figure 6.2 does not show the capability to conduct oxide ions; there must be a certain amount of imperfections or defects produced due to the nonstoichiometry for the conduction or diffusion to take place [8, 9]. The basic concepts of defect chemistry have been adapted to describe the conduction properties of mixed conducting ceramic materials. Kroger and Vink [10] first employed notations to describe the defect chemical reactions [2, 8]. For example, the oxide ZrO_2 doped with Y_2O_3 to create mobile oxygen vacancies is described in terms of defect chemical notation as $Y_2O_3(ZrO_2) \rightarrow 2Y'_{Zr} + V_O^{\bullet\bullet} + 3O_O$ [2]. The subscript usually shows the site of the defect in the Kroger–Vink notation, the effective defect charge is written as a superscript dot (·) for a positive excess charge, and as a prime (′) for a negative excess charge, and the vacancy is written as *V*. Ion defects may take the form of vacancies (vacant lattice sites), interstitial ions (placed at normally unoccupied sites), impurities or dopants with charges different from those expected from the overall stoichiometry. Electron defects may arise in the form of ions present with charges deviating from the normal lattice ions, as well as a consequence of the transition of electrons from normally filled energy levels to normally empty levels. In those cases where the electron is missing, i.e. when there is an electron deficiency, this is usually called a hole (or electron hole). The vacancies and electron holes could be considered the point defects where the ions and electrons move. As a mixed ion and electron conductor, its overall conductivity performance arises from contributions from both ionic and electronic defects.

6.2.2 Doping Strategies

Doping is the principal method to tailor the physical properties of mixed conducting materials through formation of nonintegral stoichiometry phases or solid solutions by homogeneous doping with appropriate elements [11]. The earliest and most successful example is ZrO_2 doped with various rare earth oxides [2], which was previously thought to be a preliminary doping strategy. Fluorite type structural ZrO_2 can form an extended solid solution with oxides of lower valence cations such as CaO, Y_2O_3 and other rare earth oxides. Doping with Y_2O_3 creates mobile oxygen vacancies and stabilizes the cubic structure of ZrO_2, and the tetragonal monoclinic phase transition can be avoided at high temperature (so-called stabilized ZrO_2) [12].

For the perovskite structured oxides ABO_3, it is often seen that a lower valence dopant B′ is introduced on the B site to produce $AB_{1-x}B'_xO_{3-\delta}$ and thus form defects. The symbol δ expresses the amount of vacancies (defects). Ion vacancies provide a pathway for ions, therefore the oxygen flux can be increased by increasing the number of vacancies [13].

A wide range of compounds with the perovskite structure can also be produced with substitutions occurring for either the A atom, the B atom or both to form a structure $A_x A'_{1-x}B_yB'_{1-y}O_{3-\delta}$. In addition to the ABO_3 perovskites, Nowick and Du [14] introduced two other perovskite type oxides, $A_2(B'B'')O_6$, where B′ and B″ are 3+ and 5+, and $A_3(B'B'')O_9$, where B′ and B″ are 2+ and 5+. Vacancies can also be produced by changing the B′/B″ ratio. These materials also showed very good mixed conductivity [15]. Facing so many possible alternatives, the proper selection of effective doping strategies is essential.

6.2.3 Properties of Materials

The conduction properties of mixed conducting materials are usually expressed by conductivity for both ions and electrons. The total conductivity of a mixed conducting material can be given as the sum of the ionic conductivity and the electronic conductivity:

$$\sigma_{\mathrm{total}} = \sigma_{\mathrm{ion}} + \sigma_{\mathrm{e}} \tag{6.2}$$

The total and ionic conductivities can be measured directly using the conventional four probe method and the electron blocking four probe method, respectively [16]. The electronic conductivity can, thus, be obtained by subtracting the ionic conductivity from the total conductivity. The useful concept of transport (or transference) number t_k is more frequently used to describe the properties of the mixed conducting materials, in practice. For any carrier k, t_k is defined by:

$$t_{\mathrm{k}} = \sigma_{\mathrm{k}}/\sigma_{\mathrm{total}} \tag{6.3}$$

In the case of mixed conduction:

$$\sigma_{total} = \sigma_{ion} + \sigma_e = \sigma_{total}(t_{ion} + t_e) \tag{6.4}$$

where t_{ion} and t_e represent the transport numbers of ionic and electronic defects respectively. t_{ion} plus t_e always equals unity to retain electron neutrality. t_{ion} and t_e should be as close as possible, at the same time, for a good mixed conducting material. For an ideal mixed conducting material, t_{ion} and t_e should both be equal to 0.5. Normally, t_e is much smaller than t_{ion} in most mixed conducting materials. When t_e is too low, an external circuit may be needed to assist electronic conduction so that ions transport can take place. An increase in both t_e and t_{ion} values can be obtained by doping with appropriate elements, which has been one of the frequently used strategies in the development of good mixed conducting materials.

Some mixed conducting oxides show interesting chemical properties including catalytic activity for oxidative reactions [8, 17], which is normally dependant on the nature of the B site cation. The A site cations strongly affect the mobility of the oxygen ion in the bulk phase [18]. Membrane reactors prepared from such materials can simultaneously be used as membranes for the separation of oxygen from air and catalysts for oxidation processes. As a result, additional catalysts are no longer required [19, 20]. However, mixed conducting materials without catalytic activity, but with high oxygen permeability, can also be used for fabricating membrane reactors, but additional catalysts are be required [21–26].

6.3 CURRENT STATUS OF OXYGEN PERMEABLE MEMBRANES

Oxygen permeable dense membranes with high ionic and electronic conductivities were first reported by Teraoka *et al.* [3, 27]. A perovskite-type oxide, $LaMO_3$, with cobalt and/or iron on the M site, was doped partially with Sr^{2+} on the lanthanum site, yielding a solid oxide solution with the general formula $La_{1-x}Sr_xCo_{1-y}Fe_yO_{3-\delta}$. This material exhibited an oxygen permeation flux two to four orders of magnitude higher than that of stabilized zirconia, and attracted considerable attention [27]. The highest oxygen permeation rate was observed as $0.023\,mol\,m^{-2}\,s^{-1}$ at 850°C and 1 atm oxygen pressure difference, using a 1.0 mm thick disk membrane prepared from $SrCo_{0.8}Fe_{0.2}O_{3-\delta}$ [3]. This permeation result is only slightly smaller than those found in mesoporous membranes (0.1–$0.3\,mol\,m^{-2}\,s^{-1}$) with $5\,\mu m$ thickness and 3 nm pores, but bigger than those observed for microporous membranes (0.01–$0.001\,mol\,m^{-2}\,s^{-1}$) with 50 nm thickness and 0.5–1.0 nm pores [5, 28]. It was noted by Steele [29] that an oxygen permeation flux higher than $1\,cm^3\,cm^{-2}\,min^{-1}$ ($0.0074\,mol\,m^{-2}\,s^{-1}$) is always required for a process to be economical, and usually a flux around $3.5\,cm^3\,cm^{-2}\,min^{-1}$ ($0.026\,mol\,m^{-2}\,s^{-1}$) is favourable for most technological applications. So far, many ionic and electronic oxygen conducting materials have been developed and studied for their transport properties, chemical and thermal stabilities, and performances in oxygen permeation by a number of researchers [30–46]. Most of the materials are of a perovskite structure including several important families of $Sr(Co,Fe)O_3$, $La(Co,Fe)O_3$ and $LaGaO_3$, etc., some of which have non-perovskite structures. Increasing ion conductivity and/or electron conductivity and improving the stability of the materials are the two major tasks in the current development of the materials and membranes [46–50]. Recent progress shows the following few types of materials are very promising.

6.3.1 Perovskite-type Oxides

6.3.1.1 Sr(Co,Fe)$O_{3-\delta}$(SCFO)

Typical perovskite oxides that have been extensively studied are $SrCoO_3$ or $SrFeO_3$ based materials. They are exceptional among other perovskite-type oxides due to their high oxygen permeation rate, the maximum value being up to 3.1 $cm^3 cm^{-2} min^{-1}$ (0.023 $mol\,m^{-2} s^{-1}$) at 850 °C [3]. It was found that the properties of these materials are very dependent on the chemical compositions and are able to be changed substantially when Sr or Co(Fe) is partially substituted with other metal ions with different valences using different doping strategies. The doping cation and its concentration remarkably affect the performance of the oxides including the total electrical conductivity, ionic transference number and phase stability.

Cation deficiency in sublattice B of SCFOs has a lower influence on oxygen permeability in comparison with that in sublattice A. Both oxygen permeability and ionic conductivity of oxides reduce while decreasing the mean radius of the cations in the sublattice A. For example, in solid solutions $Sr_{0.7}M_{0.3}CoO_{3-\delta}$ (M=La, Nd, Sm, Gd), the sequence by which they reduce is La>Nd>Sm>Gd [51]. Li *et al.* [52] studied the effect of the partial substitution of strontium with bismuth on the oxygen permeation flux through $Sr_{1-x}Bi_xFeO_3$ oxides (without cobalt). It was found that the permeability increases with increasing bismuth content.

The oxygen permeability of SCFOs depends not only on their ionic and electronic conductivity, but also on the rate of oxygen exchange between the oxides and the gas phase [53, 54]. Accordingly, some type of surface modification of the oxides such as acid treatment could greatly improve the oxygen permeability [55]. Miura *et al.* [56] attributed this result to the removal of some impurities, such as SrO, by the acid treatment. Since the surface exchange rate and oxygen ionic conductivity increase and decrease, respectively, with increase of the oxygen pressure in SCFOs, bulk oxygen permeation becomes the permeation determining factor at high oxygen pressures [34].

However, SCFO perovskites are usually regarded to be thermodynamically and structurally unstable at lower temperatures and low partial pressures of oxygen, especially in reducing environments, owing to the presence of alkaline earth metal cations. The stability of the materials must be paid considerable attention during the development of new materials and modification of currently available materials.

The oxygen conductivity of titanium-containing $SrFeO_3$ was studied by Kharton *et al.* [57, 58], because it was thought that the phase stability of the material could be improved by titanium. It was found that the ion transference numbers of the oxide in air vary from 0.005 to 0.08, increasing with temperature. Since oxygen permeation through the oxide membranes is limited by both bulk ionic conductivity and surface exchange rates, depositing porous layers of the same material or a mixture of platinum and praseodymium oxide onto the membrane feed side surface leads to a significant increase in the oxygen permeability [57, 58]. The oxide is stable at temperatures above 770 K with respect to CO_2, but unstable at lower temperatures with the surface decomposition of the perovskite, associated with the formation of strontium carbonate. The average thermal expansion coefficient of strontium ferrite-titanate is $(27.0 \pm 0.4) \times 10^{-6} K^{-1}$ at 780–1040 K [57].

The effect of partial substitution of cobalt(iron) with various metal cations such as chromium, copper, manganese, nickel, titanium etc. were studied by several researchers [31, 42, 58–60]. It was found that the introduction of moderate amounts of higher valence cations such as titanium or chromium into the cobalt or iron sublattice might decrease the electron conductivity and thus oxygen permeability, but improve the structural stability [61]. In addition, the stability of these perovskites in CO_2 containing atmospheres can be improved by creation of a moderate cation deficiency in the strontium sublattice, although the conductivity also decreases [60, 61].

Shao *et al.* [25, 42] studied the effect of barium in doping $SrCo_{0.8}Fe_{0.2}O_{3-\delta}$. The partial substitution of strontium with barium ions of a larger radius forms a perovskite oxide of the form $Ba_{1-x}Sr_xCo_{0.8}Fe_{0.2}O_{3-\delta}$. It was found that the phase stability is greatly improved by the introduction of barium into $SrCo_{0.8}Fe_{0.2}O_{3-\delta}$ while the conductivity is not decreased. The highest oxygen permeation rate is up to 1.19 $cm^3\,cm^{-2}\,min^{-1}$ for a 1.5 mm thickness membrane at 850 °C. Suitable barium doping content was in the range $x = 0.3$–0.5 and 0.3 was found to be the optimum value. A long term oxygen permeation study of more than 1000 h at 850 °C indicated that only a very slow decay of the performance for $Ba_{0.5}Sr_{0.5}Co_{0.8}Fe_{0.2}O_{3-\delta}$ occurred due to surface segregation and bulk phase decomposition [42]. Recently, Wang *et al.* [62] applied a $Ba_{0.5}Sr_{0.5}Co_{0.8}Fe_{0.2}O_{3-\delta}$ tubular membrane reactor packed with LiLaNiO/γ-Al_2O_3 catalyst to partial oxidation of methane to syngas. Their results proved very promising with a 94 % methane conversion and higher than 95 % CO selectivity achieved by operating steadily in pure methane for 500 h. However, more than 500 h operating in a syngas atmosphere was not performed. In order to improve the stability of $Ba_{0.5}Sr_{0.5}Co_{0.8}Fe_{0.2}O_{3-\delta}$, the strontium ions in the A site of SCFO can be completely substituted with the same valence state, but larger radius size barium ions and stable higher valence state titanium ions to form the oxide $BaTi_{0.2}Co_{0.5}Fe_{0.3}O_{3-\delta}$. Its oxygen permeability can also be kept at a relative high level, about 0.9 $cm^3\,cm^{-2}\,min^{-1}$ at 950 °C [45, 46]. They also studied novel zirconium doped materials, $BaCo_{0.4}Fe_{0.6-x}Zr_xO_{3-\delta}$ ($x = 0$–0.4) [45]. $BaCo_{0.4}Fe_{0.4}Zr_{0.2}O_{3-\delta}$ shows a high oxygen permeation rate of 0.9 $cm^3\,cm^{-2}\,min^{-1}$ and long term stable operation at 800 °C.

6.3.1.2 La(Co,Fe)$O_{3-\delta}$ (LCFO)

La(Co,Fe)$O_{3-\delta}$ based ceramics are another type of mixed conducting oxides, which exhibit significant oxygen ionic conductivity with prevailing electronic conduction [35, 59, 63–65]. Its total electrical conductivity may attain a value up to 1000 $S\,cm^{-1}$. Although the oxygen permeability is much less than that of Sr(Co, Fe)$O_{3-\delta}$ based oxides [66], some problems suffered in most perovskites derived from Sr(Co, Fe)$O_{3-\delta}$ are minimized in La(Co,Fe)$O_{3-\delta}$ based ceramics.

In La(Co, Fe)$O_{3-\delta}$ oxides, lanthanum can be partially substituted with M^{2+} metal cations (M = Sr, Ba, Ca). The metal cations substituted for cobalt in $LaCoO_{3-\delta}$ include gallium, chromium, iron, lead or nickel and so on. The partial substitution of B site cobalt with different metal cations was studied in detail by Kharton *et al.* [32, 33, 35, 67, 68] who found that the oxygen permeation fluxes, electrical conductivities and thermal expansions of $LaCo_{1-x}Cr_xO_3$ ($x = 0.1$–0.4) solid solutions all decrease with increasing chromium concentration. Surface exchange is the rate limiting step of oxygen transport. Their study [32] further showed that introduction of nickel into the cobalt sublattice of $LaCo_{1-x}Ni_xO_3$ was found to result in a sharp decrease in the oxygen ionic conductivity which was the flux limiting factor. However, the electronic conductivity of the solid solution increases with nickel content. The substitution of cobalt with either chromium or nickel led to a decrease in thermal expansion, while the substitution of cobalt(iron) with titanium in the oxides led to an increase in thermal expansion as well as oxygen ionic and electronic conductivities.

The cobalt may also be substituted simultaneously with both nickel and iron cations with the formation of a rhombohedrally distorted perovskite structure [32, 67]. In the $LaCo_{1-x-y}Fe_xNi_yO_3$ ($x = 0.1$–0.2; $y = 0.1$–0.3) oxide system, the introduction of nickel into the perovskite lattice leads to increasing oxygen nonstoichiometry and electrical conductivity and decreasing thermal expansion, but the effect of doping by iron is the opposite. Oxygen permeation through $LaFe_{1-x}Ni_xO_{3-\delta}$ membranes was found to increase with nickel content due to increasing oxygen vacancy concentration and bulk ionic conductivity, which are the permeation flux limiting factors [33]. For solid solutions with equal concentrations of nickel and iron, oxygen permeation is limited predominantly by the bulk ionic conductivity. Further substitution of cobalt with nickel results in a sharp increase in the permeation flux due to increasing oxygen nonstoichiometry, which is accompanied by surface exchange limitations of oxygen transport [32, 67].

For the substitution of A site lanthanum, strontium was found to be the best candidate. The oxygen permeation in $La_{0.3}Sr_{0.7}CoO_{3-\delta}$ is predominantly controlled by bulk diffusion across the membrane with a range of thickness 0.57–2.15 mm, and the characteristic thickness, which determines the transition from predominant control by diffusion to that by surface exchange, is extrapolated to be about 80 μm [64].

The partial substitution of lanthanum with lead was also studied by Kharton *et al.* [69] who found that the catalytic and electrochemical activities of the ceramic were enhanced remarkably. Furthermore, a higher degree of stability was observed in lead substituted perovskites in relation to poisoning by sulphur dioxide, compared with strontium substituted ones. In the oxide system $La_{1-x}Pb_xFeO_{3-\delta}$ ($x = 0.1$–0.3), electrical conductivity and oxygen permeation fluxes increase with increasing x, but both the oxygen permeability and conductivity values of the oxides are considerably less than those of $La_{1-x}Sr_xFeO_{3-\delta}$. Permeation fluxes through $La_{1-x}Pb_xFeO_{3-\delta}$ are limited by the bulk ionic conductivity.

It was found that the double site substitution perovskite oxides in the form of $La_{1-x}Sr_xCo_{1-y}Fe_yO_{3-\delta}$ (LSCF) give high oxygen permeability and also possess good stability under air atmosphere, thus they were recognized as a very promising materials for air separation [3, 70–74]. Armstrong *et al.* [73] also studied the properties of $LaSr_3Fe_{3-x}Co_xO_{10}$ oxides, the total conductivity increasing with increasing cobalt content. However, above a transition temperature $T > T_t \approx 400\,°C$, the conductivity decreases with increasing temperature due to the loss of oxygen vacancies [73]. The oxygen permeation flux across the oxide membranes at high temperatures $800\,°C \leq T \leq 950\,°C$ increases with increasing cobalt content and the flux values are approximately one order of magnitude lower than that of perovskite $SrFe_{0.2}Co_{0.8}O_{3-\delta}$, but better crystal structural stability in low oxygen partial pressures and at high temperatures is exhibited compared to the latter. Xu and Thomson [74] found that $La_{0.6}Sr_{0.4}Co_{0.2}Fe_{0.8}O_{3-\delta}$ is very stable under air/nitrogen gradients at temperatures up to 960 °C. Its phase transition temperature is around 825 °C [20]. In other work [37, 75], LSCF showed a slight decline in oxygen permeation after long term operation over 1100 h. Another composite of $La_{0.2}Sr_{0.8}Co_{0.2}Fe_{0.8}O_{3-\delta}$ perovskite oxide was studied by Li *et al.* [75].

The cubic perovskite $La_{0.2}Sr_{0.8}Cu_{0.4}Co_{0.6}O_{3-\delta}$ also has high electronic and ionic conductivities at high temperatures. The electronic conductivity increases with increasing temperature initially, but then decreases due to the loss of oxygen atoms from the structure. The oxygen permeability is comparable to the values reported for other cobalt containing perovskite oxides with high strontium contents. The temperature of phase transformation from the disordered cubic perovskite to an ordered structure related to brownmillerite, is 975 °C under an oxygen partial pressure of 0.01 atm [76].

6.3.1.3 $LaGaO_3$ (LGO)

$LaGaO_3$ based oxides exhibit lower relative electronic conductivity, but higher ionic conductivity [68] and have thus attracted much attention. The concept of substitution of cobalt by gallium is assumed to increase the ionic character of the B–O bonds in the ABO_3 perovskite lattice. This can increase the stability of $LaCoO_3$ in a reducing atmosphere but would decrease the electronic conductivity [30]. $LaGaO_3$ also shows a relatively low thermal expansion coefficient (TEC) and a large electrolyte domain [77]. Kharton *et al.* [68] investigated the effects of different cation dopants (titanium, chromium, iron, cobalt, nickel) and the dopant concentrations on the physicochemical and transport properties of $LaGaO_3$ based materials. Oxygen ion transference numbers vary from 2×10^{-6} to 0.98 throughout the series and p-type electronic conductivity increases with increasing transition metal content. Substitution of Ga with higher valence cations (titanium, chromium) decreases ionic conductivity whereas small amounts of iron or cobalt (~5 %) increase ionic conductivity. The $LaGa_{1-x}Ni_xO_{3-\delta}$ ($x = 0.2$–0.5) perovskites have significant oxygen permeation fluxes as well as thermal expansion coefficients [78].

The partial substitution of lanthanum with an alkaline earth cation in $LaGaO_3$, in general, decreases the electronic or hole conduction. Ishihara *et al.* [79] studied doping of $LaGaO_3$ with strontium for the lanthanum site and magnesium for the gallium site and found that the electrical conductivity increases with the level of strontium doping and reaches a maximum value at $x = 0.1$. The electrical conductivity of $La_{0.9}Sr_{0.1}GaO_3$ increases with partial substitution of magnesium, aluminium or indium for gallium, with magnesium (concentration 0.2 mol) producing the largest increase in the conductivity [79]. As a result, $La_{1-x}Sr_xGa_{1-y}Mg_yO_{3-\delta}$ (LSGM) is the optimum composition exhibiting highest ionic conductivity [80]. The high oxide ionic conductivity in LSGM results from the high mobility of oxygen vacancies in addition to the high concentration of oxygen vacancies. Meanwhile, the oxygen diffusion coefficient was found to be higher than the surface exchange coefficient in LSGM [40]. However, LSGM is a pure ionic conductor that barely possesses electronic conductivity. By further doping with other cations, it can become a mixed conductor. Ullmann *et al.* [81, 82] studied the perovskite-type oxides $La_{1-x}A_xM_{1-y}B_yO_{3-\delta}$ with $A = Sr^{2+}$, Ln^{3+}, Ce^{4+}, M = Fe, Co, Ga and B = Co, Fe, Mg in the concentration range $x = 0.1$–1 mol and $y = 0.1$–0.5 mol. The electrical conductivity of the LSGM could be increased by doping cobalt or iron. Increasing cobalt or iron doping concentrations leads to increased p-type conductivity. By applying higher cobalt or iron doping concentrations the oxide becomes a mixed conductor.

The characteristic values of above three types of perovskite and related oxides are summarized in Tables 6.1 and 6.2.

Apart from the advantages mentioned above, perovskite-type or related oxides still have a few disadvantages:

- Relatively low oxygen permeability. Oxygen permeation rate of most materials in this class is lower than $1\,cm^3\,cm^{-2}\,min^{-1}$.
- Peroviskite related phase (highest oxygen permeability in oxidizing conditions) thermodynamically and/or dimensionally unstable, under large oxygen chemical potential gradients such as air/methane.
- Perovskite-type oxides exhibit an ordering in the oxygen sublattice and a high reactivity with gas species such as CO_2 at temperatures below 1000–1070 K, leading to degradation in behaviour with time.

Table 6.1 Conductivity and oxygen permeability values of mixed conducting oxides

Composition	Conductivity (800–1000 °C) σ, S cm^{-1}	Transference number t_{ion}	Permeation flux (800–1000 °C) J, mol cm^{-2} s^{-1}	Ref
$SrCo_{0.33}Fe_{0.67}O_{3-\delta}$ 1.5 mm	~100	0.012	$(0.89–2.06) \times 10^{-6}$	[66]
$SrCo_{0.8}Fe_{0.2}O_{3-\delta}$ 1.5 mm	~100		$(4.3–9.3) \times 10^{-7}$	[53]
$Sr_{0.7}M_{0.3}CoO_{3-\delta}$ (M = La,Nd,Sm,Gd) 1.4 mm			$(2.6–0.41) \times 10^{-7}$ ↓ at 1153 K	[78]
$Sr_{1-x}La_xCoO_{3-\delta}$ (x = 0.3–0.35) 1.4 mm			$(2.8–1.7) \times 10^{-7}$ ↓	
$Sr_{0.97}Fe_{0.8}Ti_{0.2}O_{3-\delta}$ 1.0 mm	10–31.6	0.0044–0.012	$(0.66–3.98) \times 10^{-7}$	[57, 61]
$Sr_{0.97}Fe_{0.4}Ti_{0.6}O_{3-\delta}$ 1.0 mm	~1	0.009–0.041	$(0.25–1.12) \times 10^{-7}$	
$SrCo_{0.8}Ti_{0.2}O_{3-\delta}$	63.1–100			
$CaFe_{0.2}Ti_{0.8}O_{3-\delta}$ 1.14 mm	0.06–0.1	0.005–0.031	$(0.32–7.9) \times 10^{-9}$	[129]
$Ba_{0.5}Sr_{0.5}Co_{0.8}Fe_{0.2}O_{3-\delta}$ 1.5 mm			$(0.67–1.15) \times 10^{-7}$	[42]
$BaBi_xCo_{0.2}Fe_{0.8-x}O_{3-\delta}$ (x = 0.1–0.5) 1.5 mm			$(2.23–7.44) \times 10^{-7}$	[41]
$LaCo(M)O_{3-\delta}$ 1.0 mm	~800		1.86×10^{-7}	[35]
$LaCo_{0.9}Cr_{0.1}O_{3-\delta}$ 1.4 mm	400–500		$(0.4–0.8) \times 10^{-7}$	[32]
$LaCo_{0.8}Cr_{0.2}O_{3-\delta}$ 1.4 mm	~355			
$LaCo_{0.7}Cr_{0.3}O_{3-\delta}$ 1.4 mm	316–355			
$LaCo_{0.6}Cr_{0.4}O_{3-\delta}$ 1.4 mm	158–200			
$LaCo_{0.9}Ni_{0.1}O_{3-\delta}$	~1000			
$LaCo_{0.7}Ni_{0.3}O_{3-\delta}$ 1.0 mm	630–794		$(0.4–1.2) \times 10^{-7}$	
$LaFe_{1-x}Ni_xO_{3-\delta}$ (x = 0.2–0.5) 1.0 mm	50–316		$(0.43–8.6) \times 10^{-8}$	[33]
$LaCo_{0.8}Fe_{0.1}Ni_{0.1}O_{3-\delta}$ 1.0 mm	630–660		$(0.46–1.08) \times 10^{-9}$	[67]
$LaCo_{0.7}Fe_{0.1}Ni_{0.2}O_{3-\delta}$ 1.0 mm	760–800		$(1.68–3.20) \times 10^{-9}$	
$LaCo_{0.6}Fe_{0.2}Ni_{0.2}O_{3-\delta}$ 1.0 mm	530–580		$(0.27–1.17) \times 10^{-9}$	
$LaCo_{0.5}Fe_{0.2}Ni_{0.3}O_{3-\delta}$ 1.0 mm	470–500		$(1.57–3.11) \times 10^{-9}$	
$La_{0.3}Sr_{0.7}CoO_{3-\delta}$ 1.15 mm			$(0.32–1.0) \times 10^{-6}$	[64, 130]
$La_{0.2}Sr_{0.8}CoO_{3-\delta}$ 0.5 mm			6.44×10^{-7}	[20]
$La_{0.9}Pb_{0.1}FeO_{3-\delta}$ 1.0 mm	0.5–0.8		$(1.0–2.0) \times 10^{-8}$ at 1000 °C	[69]
$La_{0.8}Pb_{0.2}FeO_{3-\delta}$ 1.0 mm	2.5–4.0			
$La_{0.7}Pb_{0.3}FeO_{3-\delta}$ 1.0 mm	3.5–5.0			
$La_{0.6}Pb_{0.4}FeO_{3-\delta}$ 1.0 mm	~17			
$La_{0.6}Sr_{0.4}Co_{0.2}Fe_{0.8}O_{3-\delta}$	252	0.0009	$(0.3–3.0) \times 10^{-7}$	[36, 37, 75]
$La_{0.4}Sr_{0.6}Co_{0.2}Fe_{0.8}O_{3-\delta}$	219	0.0018		
$La_{0.2}Sr_{0.8}Co_{0.2}Fe_{0.8}O_{3-\delta}$	120	0.0052	$(1.1–8.9) \times 10^{-7}$	
$La_{0.2}Sr_{0.8}Co_{0.8}Fe_{0.2}O_{3-\delta}$	310	0.0028		
$La_{0.8}Sr_{0.2}Co_{0.8}Fe_{0.2}O_{3-\delta}$ 0.5 mm			$(2.5–8.3) \times 10^{-7}$	[77]
$La_{0.6}Sr_{0.4}Co_{0.8}Fe_{0.2}O_{3-\delta}$			4.46×10^{-9}	[70]
$La_{0.2}Sr_{0.8}Co_{0.6}Cu_{0.4}O_{3-\delta}$ 0.8 mm	250–600		$(1.2–6.0) \times 10^{-7}$	[76]
$LaGa_{1-x}Ni_xO_{3-\delta}$ (x = 0.2–0.6) 1.0 mm			$(1.9–6.6) \times 10^{-8}$	[33]
$LaGa_{0.8}Ni_{0.2}O_{3-\delta}$ (1.0 mm)	~2	0.0008–0.012	$(4.47–8.91) \times 10^{-8}$	[35, 131]

Table 6.1 *Continued*

Composition	Conductivity (800–1000°C) σ, S cm^{-1}	Transference number t_{ion}	Permeation flux (800–1000°C) J, mol cm^{-2} s^{-1}	Ref
$LaGa_{0.5}Ni_{0.5}O_{3-\delta}$	20–32	0.00012–0.0022		
$LaGa_{0.4}Co_{0.6}O_{3-\delta}$	12.6–63			
$LaGa_{0.6}Co_{0.4}O_{3-\delta}$	2–4			
$LaGa_{0.85}Mg_{0.15}O_{3-\delta}$	0.02–0.09	0.94–0.92		
$LaGa_{0.4}Mg_{0.2}Cr_{0.4}O_{3-\delta}$	0.32–0.33	0.0044–0.031		
$LaGa_{0.4}Mg_{0.2}Fe_{0.4}O_{3-\delta}$	~5	0.012–0.046		
$LaGa_{0.4}Mg_{0.2}Co_{0.4}O_{3-\delta}$ 1.0 mm	~18	0.0008–0.0017	$(5.0–6.31) \times 10^{-8}$	
$LaGa_{0.3}Mg_{0.1}Co_{0.6}O_{3-\delta}$ 1.0 mm	79–126	$(0.22–4.1) \times 10^{-4}$	$(0.04–1) \times 10^{-8}$	
$LaGa_{0.4}Mg_{0.25}Co_{0.35}O_{3-\delta}$ 1.0 mm	9–10		$(3.8–5.0) \times 10^{-8}$	
$La_{0.8}Sr_{0.2}Ga_{0.8}Fe_{0.2}O_{3-\delta}$ 0.5 mm			$(0.35–0.52) \times 10^{-6}$	[77]
$La_{0.8}Sr_{0.2}Ga_{0.7}Fe_{0.3}O_{3-\delta}$ 0.5 mm	0.33–0.79		$(0.41–1.03) \times 10^{-6}$	
$La_{0.8}Sr_{0.2}Ga_{0.6}Fe_{0.4}O_{3-\delta}$			$(0.43–0.83) \times 10^{-6}$	
$La_{0.8}Sr_{0.2}Ga_{0.6}Fe_{0.4}O_{3-\delta}$			$(0.37–0.73) \times 10^{-6}$	
$La_{0.8}Sr_{0.2}Ga_{0.7}Ni_{0.3}O_{3-\delta}$ 0.5 mm	~3.98		$(0.25–1.0) \times 10^{-6}$	
$La_{0.8}Sr_{0.2}Ga_{0.7}Co_{0.3}O_{3-\delta}$ 0.5 mm	2–10		$(0.23–1.1) \times 10^{-6}$	
$La_{0.9}Sr_{0.1}GaO_{3-\delta}$	0.025–0.076	0.97–0.98		
$La_{0.9}Sr_{0.1}Ga_{0.95}M_{0.05}O_{3-\delta}$ M = Cr, Fe, Co	0.015–0.126	0.77–0.95		
$La_{0.9}Sr_{0.1}Ga_{0.8}Fe_{0.2}O_{3-\delta}$	0.033–0.071	0.14–0.47		
$La_{0.86}Sr_{0.14}Ga_{0.96}Ti_{0.04}O_{3-\delta}$	0.019–0.060	0.91–0.89		

6.3.2 Non-perovskite-type Oxides

Apart from the perovskite oxides mentioned above, some non-perovskite-type oxides also exhibit mixed oxygen ionic and electronic conducting properties and have attracted considerable attention [47, 83–85]. One type of mixed phase oxides, $SrFeCo_{0.5}O_x$, discovered by Balachandran *et al.* [84], shows a comparable electronic and ionic transference number, making it unique among mixed conductors. The properties and the phase stability of this oxide were investigated [47, 83] and a tubular membrane was also fabricated as a reactor for methane conversion to syngas by Balachandran *et al.* [84] who found that the methane conversion efficiency was higher than 98 % and CO selectivity was up to 90 % at 850°C. An oxygen permeation rate of 2–4 cm^3 cm^{-2} min^{-1} was also achieved. Under such a reducing atmosphere, high structural stability was observed whilst operating for more than 1000 h [21]. This clearly suggests that the structural stability and mechanical integrity of the material in reducing atmospheres are better than many other cobalt containing perovskite oxides. In addition, studies carried out by Kim *et al.* [86, 87] concluded that the oxygen transport in

Table 6.2 TEC values in air and phase transformation temperature of mixed conducting oxides

Composition	$\alpha \times 10^6$, K^{-1} (800–1000 °C)	Ref
$SrCo_{0.8}Fe_{0.2}O_{3-\delta}$	29.4 ± 0.8	[57, 61]
$SrCo_{0.8}Ti_{0.2}O_{3-\delta}$	28.8 ± 0.9	
$Sr_{0.97}Co_{0.8}Ti_{0.2}O_{3-\delta}$	27.0 ± 0.4	
$Sr_{0.97}Fe_{0.4}Ti_{0.6}O_{3-\delta}$	16.6 ± 0.1	
$Sr_{0.70}La_{0.30}CoO_{3-\delta}$	28.8	[51, 77]
$Sr_{0.70}Nd_{0.30}CoO_{3-\delta}$	31.5	
$Sr_{0.70}Sm_{0.30}CoO_{3-\delta}$	25.9	
$Sr_{0.70}Gd_{0.30}CoO_{3-\delta}$	20.4	
$Sr_{1-x}La_xCoO_{3-\delta}$ ($x = 0.3$–0.4)	27.9–18↓	
$Sr_{0.65}La_{0.35}Co_{0.9}O_{3-\delta}$	16.9 ± 0.9	
$Sr_{0.89}Co_{0.1}Fe_{0.1}Cr_{0.01}O_{3-\delta}$	30.5 ± 0.9	
$SrCo(Ti)O_{3-\delta}$	25–29	[132]
$Ba_{0.5}Sr_{0.5}Co_{0.8}Fe_{0.2}O_{3-\delta}$		[42]
$LaCoO_{3-\delta}$	21.9 ± 0.3	[35]
$LaCo_{1-x}Cr_xO_{3-\delta}$ ($x = 0.1$–0.4)	23.7–20.3	[32, 67]
$LaCo_{1-x}Ni_xO_{3-\delta}$ ($x = 0.1, 0.3$)	19.6, 17.3	
$LaCo_{0.9-y}Fe_{0.1}Ni_yO_{3-\delta}$ ($y = 0.1, 0.2$)	20.8, 18.6	
$LaCo_{0.8-y}Fe_{0.2}Ni_{0.2}O_{3-\delta}$ ($y = 0.2, 0.3$)	19.2, 18.0	
$La_{0.8}Sr_{0.2}Co_{0.6}Fe_{0.4}O_{3-\delta}$		[20]
$La_{1-x}Pb_xFeO_{3-\delta}$ ($x = 0.1, 0.2, 0.3$)	10.0, 10.4, 10.1	[69]
$LaGa_{0.3}Co_{0.6}Mg_{0.1}O_{3-\delta}$	19.8 ± 0.3	[131]
$LaGa_{0.4}Co_{0.4}Mg_{0.2}O_{3-\delta}$	15.4 ± 0.3	
$LaGa_{0.4}Co_{0.35}Mg_{0.25}O_{3-\delta}$	12.4 ± 0.2	
$LaGa_{1-x}Ni_xO_{3-\delta}$ ($x = 0.2$–0.6)	10.8–11.6	[78]
$LaCo_{0.6}Ga_{0.3}Mg_{0.1}O_{3-\delta}$	20.0 ± 0.3	
$LaCo_{0.9}Mg_{0.1}O_{3-\delta}$	19.5 ± 0.3	
$LaFe_{1-x}Ni_xO_{3-\delta}$ (x = 0.2–0.5)	8.9–11.9↑	

this oxide is dependent primarily on the bulk diffusion rate. Recently, a further study suggested that a suitable preparation technique could improve the oxygen permeation properties of $SrFeCo_{0.5}O_x$ [88]. The relationships between structures and properties of more complex oxides, $Sr_4Fe_{6-x}Co_xO_{13+\delta}$ with $1 \leq x \leq 3$ were also extensively studied by Ma *et al.* [48] who found that the total conductivity and the oxygen flux increase with cobalt substitution and temperature. The conductivity of $Sr_4Fe_4Co_4O_{13+\delta}$ was found to be 20 and 6 S cm^{-1} at 900 °C in flowing air and argon environments, respectively. Its ionic transference number is about 0.4. The oxygen permeation rate was observed up to 10 cm^3 cm^{-2} min^{-1} for a 0.75 mm thick tubular membrane in methane conversion reaction conditions for more than 1000 h. Based on the above studies, it is concluded that this oxide exhibits better structural and chemical stabilities at high temperatures and in low oxygen partial pressure atmospheres compared to the perovskite $SrFe_{0.2}Co_{0.8}O_{3-\delta}$ although the oxygen permeability has been shown to be lower than that of the latter [85, 89]. The effect of preparation procedure of this material on the phase relationships was also investigated systematically [90].

Another type of non-perovskite-type mixed conductor is $La_2Cu_{1-x}Co_xO_{4+\delta}$ with $0 \leq x \leq 0.3$ investigated by Bochkov *et al.* [91]. $La_2Cu_{1-x}Co_xO_{4+\delta}$ ($0 \leq x \leq 0.3$) solid solutions with orthorhombic K_2NiF_4 type structures were found to be mixed conductors also, with electrical conductivity of up to $13\,S\,cm^{-1}$ [91]. Incorporating cobalt into the copper sublattice of $La_2CuO_{4+\delta}$ leads to increasing oxygen content, and the oxygen permeability increases with cobalt concentration at $x = 0.02–0.10$, but decreases with further dopant additions. Since no alkaline earth metal cations are contained in the oxide, the sintered ceramics of this material exhibit sufficient stability, maintaining a single phase for more than 12 months.

6.3.3 Summary of Ceramic Oxygen Permeable Materials

Due to the large number of oxygen permeable materials available, it is quite difficult to have an extensive comparison. Some general observations are summarized below:

(1) The $A^{3+}B^{3+}O_3$ type perovskites. (i.e., $Sr(Co,Fe)O_{3-\delta}$) have higher oxygen ionic conductivity than the $A^{2+}B^{4+}O_3$ type perovskites which are not good oxygen ionic conductors [2].

(2) The optimum tolerance factor is near unity from the viewpoint of lattice stability and equivalency in oxygen sites. A large specific free volume is desirable in view of the mobility of oxygen. The specific free volume linearly increases with the decrease of the tolerance factor for each series with a fixed A site cation. The optimum tolerance factor for $A^{III}B^{III}O_3$ type perovskites, to obtain the maximum electrical conductivity, is around 0.96 due to the balance between the specific free volume and the tolerance factor. The change in the tolerance factor due to doping should be as small as possible [2].

(3) The optimum oxygen deficiency δ to obtain the maximum electrical conductivity is around 0.2. When δ is considerably larger than 0.2, the electrical conductivity decreases due to short range interaction among vacancies [2].

(4) SrCoFeO based solid solutions with cubic perovskite structure exhibit the highest oxygen permeation fluxes compared to other mixed oxide membranes. However, high thermal expansion coefficients and interaction with gas species, such as carbon dioxide and water vapour, complicate the employment of SrCoFeO as an oxygen separation membrane.

(5) Perovskite-type oxides containing alkaline earth metal cations are reported to exhibit high reactivities with gas species such as CO_2, SO_2 and water vapour, whereas relatively low oxygen permeation fluxes are typical for perovskites without alkaline earth elements.

(6) In LaSrFeO and LaGaO oxides, lanthanum site doping with a cation of a lower oxidation state leads to a large number of oxide vacancies being formed and, therefore, to an increase in oxide ion conductivity.

(7) A site cations strongly affect the oxygen nonstoichiometry of the oxide and the mobility of oxygen ions in the bulk phase [92]. The creation of vacancies in both cation sublattices of ABO_3 perovskites results in a decrease in the oxygen permeation flux [51].

(8) Oxygen permeation in most perovskite-type oxide membranes is controlled by both bulk diffusion and surface exchange rates. The critical membrane thickness that is introduced to distinguish between the regions where bulk diffusion and surface exchange kinetics are rate limiting has been reported to be in the range of 20 to 3000 μm depending upon composition, temperature and oxygen partial pressure [66].

6.4 DUAL PHASE MEMBRANES

According to the Wagner theory, the oxygen permeation rate is proportional to $[(\sigma_i\sigma_e)/(\sigma_i + \sigma_e)]$ [93]. Therefore, a good mixed conducting material must have high ionic conductivity as well as considerable electronic conductivity. It is not easy to meet all the requirements in a single component material. Mazanec *et al.* [94] first introduced the dual phase membrane concept for partial oxidation of methane, which was composed of yttria stabilized zirconia (YSZ) for oxygen ionic conduction and a metal such as Ag or Au for improving the electronic conduction. In dual phase membranes, the ionic conductor is usually the continuous phase and the metal is uniformly dispersed, but networked into the whole membrane. By using a dual phase membrane, an enhanced oxygen permeation flux is thus achieved. The concept of the dual phase membranes gives rise to additional opportunities in the selection of materials and fabrication methods of the membranes, and thus it is interesting to touch on this subject briefly.

Figure 6.3 gives a schematic description of a dual phase membrane in which one continuous phase is for ion conduction and the other is dispersed for improving electron conduction. Following this idea, Chen and his coworkers [95–98] carried out a detailed research on dual phase membranes using a several dual phase composites such as yttria stabilized zircornia palladium [95, 99], erbia stabilized bismuth sliver or gold [97–99]. It was found that in order to form a networked electron conducting phase, about 40 vol% of a metal is needed to mix with the ceramic material. The oxygen transport kinetics are believed to be controlled by the bulk diffusion of oxygen ions in the oxide phase for membranes with a thickness down to about 1.0 mm and jointly controlled by bulk diffusion and surface exchange when the membrane thickness decreases further [97].

Kim and Lin [100] prepared a number of dual phase membranes by coating a porous YSZ layer on α-alumina support membranes and then impregnating a continuous Pd phase on the wall of the pores of the YSZ layer. The residual pores of the Pd modified YSZ layer were then plugged with YSZ using a chemical vapour deposition technique (CVD). The structure of a composite dual phase membrane is shown in Figure 6.3(a), while a general structure of the dual phase membrane for oxygen transport is schematically described in Figure 6.3(b).

Some more complex fluorite-type materials were also studied, such as $Bi_{1.5-x}Y_{0.5}Cu_xO_{3-\delta}$ in the fluorite phase (δ phase) with different CuO concentrations ($x = 0.005$–0.015) [101] and

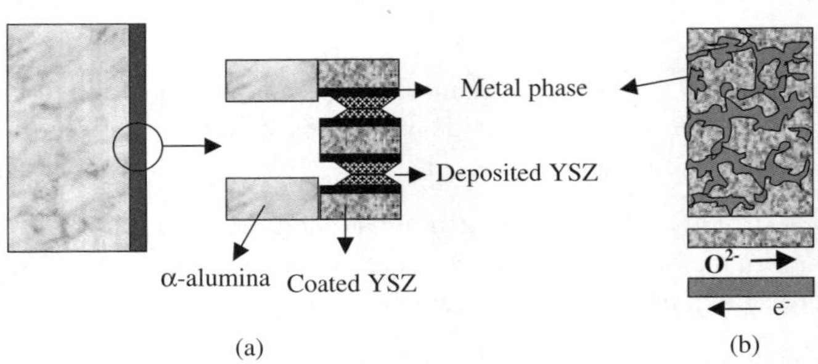

Figure 6.3 Schematic description of dual phase membranes: (a) composite dual phase membrane; (b) general transport of dual phase membrane [128]. Copyright (2006) Taylor & Francis, Catalysis Reviews–Science and Engineering, **48** (2), Liu, Y., Tan, X. and Li, K., *Mixed conducting ceramics for catalytic membrane processing*, p. 145–198

$Bi_{1.5}Y_{0.3}Sm_{0.2}O_3$ (BYS) [102]. The oxygen permeation fluxes through the $Bi_{1.5-x}Y_{0.5}Cu_xO_{3-\delta}$ membranes with a thickness of 2 mm is in the range of (5.0–9.2) $\times 10^{-9}$ mol cm^{-2} s^{-1} at 800 °C, and is controlled by bulk diffusion [101]. In other work [102], percolative and nonpercolative fluorite type $Bi_{1.5}Y_{0.3}Sm_{0.2}O_3$ (BYS) and silver dual phase membranes were prepared by mixing BYS and Ag powders. The percolative BYS/Ag 40 % and nonpercolative BYS/Ag 30 % membranes exhibit about a 50 fold and 10 fold increase, respectively, in oxygen permeation flux compared to BYS membranes. For a 1.3 mm thick BYS/Ag 40 % membrane, an oxygen permeation flux of 5.8×10^{-7} mol cm^{-2} s^{-1} was measured at 850 °C. For the nonpercolative BYS/Ag membrane, the presence of Ag enhances the surface reaction rate of oxygen permeation. The continuous silver phase in the percolative BYS/Ag membrane increases both the electronic conductivity and surface reaction rate, which results in a substantial improvement in oxygen permeance. Recently, $Bi_{1.5}Y_{0.3}Sm_{0.2}O_3$ (BYS) tubular membranes were prepared and tested in the oxidative coupling of methane and selective oxidation of ethane to ethylene [39].

Apart from dual phase membranes prepared from a ceramic and metal, dual phase membranes have also been prepared successfully using two ceramic materials, one with a large ionic conductivity and the other exhibiting good electronic conductivity [62, 103]. The oxygen permeability and structural properties of the dual phase membranes fabricated from gadolinia doped ceria electrolyte $Ce_{0.8}Gd_{0.2}O_{2-\delta}$ perovskite-type $La_{0.8}Sr_{0.2}Fe_{0.8}Co_{0.2}O_{3-\delta}$ [103] and $La_{0.15}Sr_{0.85}Ga_{0.3}Fe_{0.7}O_{3-\delta}$-$Ba_{0.5}Sr_{0.5}Fe_{0.2}Co_{0.8}O_{3-\delta}$ [62] were studied in oxygen separation and the performances of the membranes were examined.

6.5 OXYGEN TRANSPORT

6.5.1 Transport Mechanism

Lattice diffusion in oxides, often synonymously termed bulk diffusion, takes place because of the presence of point defects such as vacancies and interstitial atoms. It follows different mechanisms depending on the type of the defects [104]:

(1) Vacancy mechanism: when an atom or ion on a normal site jumps into an adjacent unoccupied lattice site, the vacancy moves into the site left by the atom.

(2) Interstitial mechanism: it occurs when an atom on an interstitial site moves to one of the neighbouring interstitial site. The jump of the interstitial atom involves a considerable distortion of the crystal lattice, hence this mechanism would only be likely when the diffusing atom is smaller than the atoms on the normal lattice positions. Diffusion of interstitially dissolved atoms (e.g. H, C, N, or O) in metals provides the best known examples of this mechanism.

(3) Interstitialcy mechanism: if the distortion of the regular lattice is too large for interstitial diffusion to be favorable, interstitial atoms may migrate via the interstitialcy mechanism. This mechanism involves the movement of an interstitial atom onto a normal lattice site, displacing the atom located on that site into another interstitial position.

(4) Other mechanisms: the crowdion is a variant of the interstitialcy mechanism. In this case it is assumed that an extra atom is crowded into a line of atoms, and that it thereby

displaces several atoms along the line from their equilibrium positions. The energy to move such a defect is small, but it can move only along the line or along an equivalent direction.

As is known, a large number of oxygen vacancies are generated in solid oxides formed by doping aliovalent cations. In the presence of oxygen, the oxygen vacancies tend to be filled with oxygen atoms with formation of two electron holes at high temperatures as a result of charge compensation:

$$\frac{1}{2}O_2 + V_O^{\bullet\bullet} \Leftrightarrow O_O^x + 2h^\bullet \tag{6.5}$$

where the charged defects are defined using Kröger–Vink notation. That is, O_O^x stands for lattice oxygen, $V_O^{\bullet\bullet}$ for oxygen vacancy and $h^\bullet$ for positive electron hole.

Obviously, the vacancy mechanism is the best one to describe oxygen transport in mixed conducting oxides. In addition to bulk diffusion, oxygen permeating through a mixed conducting oxide membrane also undergoes surface exchange reactions at both the oxygen rich and oxygen lean sides of the membrane, which involve many sub-steps, such as oxygen adsorption, dissociation, recombination and charge transfer [40, 105]. Therefore, the mechanism for oxygen permeation through a mixed ionic–electronic conducting membrane can be schematically described, as shown in Figure 6.4. It can be seen that the permeation process from the high oxygen partial pressure side to the low oxygen partial pressure side includes the following steps in series:

(1) Mass transfer of gaseous oxygen from the gas stream to the membrane surface (high pressure side).

(2) Reaction between the molecular oxygen and oxygen vacancies at the membrane surface (high pressure side).

(3) Oxygen vacancy bulk diffusion across the membrane.

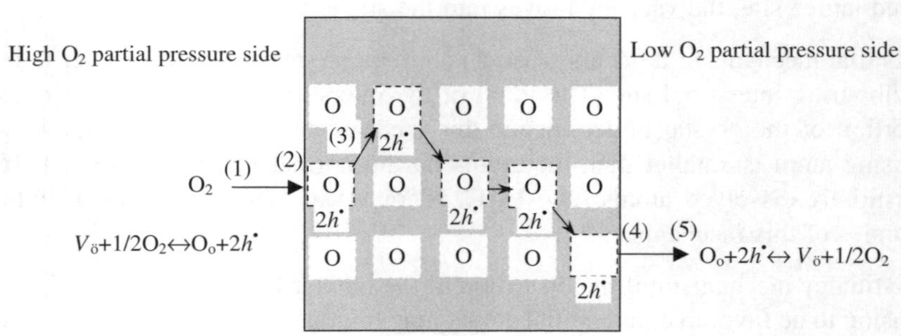

Figure 6.4 Schematic of oxygen migration in membranes [128]. Copyright (2006) Taylor & Francis, Catalysis Reviews–Science and Engineering, **48** (2), Liu, Y., Tan, X. and Li, K., *Mixed conducting ceramics for catalytic membrane processing*, p. 145–198

(4) Reaction between lattice oxygen and electron hole at the membrane surface (low pressure side).

(5) Mass transfer of oxygen from the membrane surface to the gas stream (low pressure side).

However, the resistances between the gas phase and membrane (steps 1 and 5) are usually small and negligible [36]; as a result, only membrane bulk diffusion and the surface reactions need to be taken into consideration in developing transport equations.

6.5.2 Transport Equations

The flux of charged species (oxygen ion, oxygen vacancy, proton, electron and electron hole) in a mixed conductor is usually described by the Nernst–Planck equation [106]:

$$J_i = -\frac{\sigma_i}{z_i^2 F^2} \nabla \mu_i + C_i \upsilon \tag{6.6}$$

where μ_i and υ are the electrochemical potential of charged species and the local velocity of the inert marker, respectively. The electrochemical potential for each charged species consists of a chemical potential or an activity term and a local electrostatic potential term, ϕ:

$$\mu_i = \mu_i^0 + RT \ln a_i + z_i F \phi \tag{6.7a}$$

where μ_i^0 a_i, and ϕ are the standard chemical potential, activity and the Galvanic (internal) potential, respectively.

In Equation (6.6), σ_i is the conductivity of charged species i, which may be correlated to the concentration and diffusivity by the Nernst–Einstein equation:

$$\sigma_i = \frac{z_i^2 F^2}{RT} C_i D_i \tag{6.7b}$$

where D_i is the diffusion coefficient of charged species i.

Under the following conditions: (1) the local velocity of the inert marker is negligible; (2) the ideal state is applied (activity coefficient is unit); (3) no external current is imposed and (4) one dimensional model is applied [107], the transport flux of charged defects at steady state expressed in terms of diffusivity and concentration can be obtained by substituting Equations (6.7a) and (6.7b) into Equation (6.6):

$$J_i = -D_i C_i \left[\frac{1-t_i}{C_i} \cdot \frac{dC_i}{dx} - \sum_{j \neq i} \frac{z_i}{z_j} \cdot \frac{t_j}{C_j} \cdot \frac{dC_j}{dx} \right] \tag{6.8}$$

where t is the transport number of defect i:

$$t_i = \frac{\sigma_i}{\sum \sigma_j} = \frac{z_i^2 D_i C_i}{\sum_j z_j^2 D_j C_j} \tag{6.8a}$$

For mixed ionic–electronic conducting membranes, oxygen vacancy, $V_O^{\bullet\bullet}$, and electron hole, $h^{\bullet}$, are the only mobile charges [36]. As a result, Equation (6.8) is then reduced to:

$$J_V = -\frac{(C_h + 4C_V)D_V D_h}{C_h D_h + 4C_V D_V} \cdot \frac{\mathrm{d}C_V}{\mathrm{d}x} \tag{6.9}$$

Furthermore, for perovskite membranes whose ionic transference number is close to zero ($C_h D_h >> C_V D_V$ and $C_h >> C_V$) [59, 108], Equation (6.9) is further simplified as:

$$J_V = -D_V \cdot \frac{\mathrm{d}C_V}{\mathrm{d}x} \tag{6.10}$$

Based on the stoichoimetric relations of oxygen and the oxygen vacancy, the oxygen flux can be written as:

$$J_{O_2} = -\frac{D_V}{2} \cdot \frac{\mathrm{d}C_V}{\mathrm{d}x} \tag{6.11}$$

or

$$J_{O_2} = \frac{D_V (C_V'' - C_V')}{2L} \tag{6.11a}$$

On the membrane surfaces, the following reversible reactions take place at the high and low pressure sides, respectively:

$$\frac{1}{2}O_2 + V_O^{\bullet\bullet} \xleftrightarrow{k_f/k_r} O_O + 2h^{\bullet} \tag{6.12a}$$

$$O_O + 2h^{\bullet} \xleftrightarrow{k_f/k_r} \frac{1}{2}O_2 + V_O \tag{6.12b}$$

where k_f and k_r are, respectively, the forward and reverse reaction rate constants of the above surface reactions.

Because of the high electronic conductivity, the electron hole concentrations are essentially constant on both membrane surfaces, and thus the reverse and forward reaction rates of the surface reactions of (6.12a) and (6.12b) are pseudo zero order at steady state under isothermal conditions [36]. Therefore, the local rate of oxygen consumed (step 2) or formed (step 4) on the membrane surfaces can be, respectively, expressed in the following two equations if they are assumed to be elementary:

$$J_{O_2} = k_f (p'_{O_2})^{0.5} C_V' - k_r \tag{6.13a}$$

$$J_{O_2} = k_r - k_f (p''_{O_2})^{0.5} C''_V \tag{6.13b}$$

where p'_{O_2} and p''_{O_2} are the partial pressures of oxygen in the upstream and downstream, respectively. Combination of Equations (6.11) and (6.13) gives the oxygen permeation flux in term of oxygen partial pressures as:

$$J_{O_2} = \frac{k_r D_V \left[(p'_{O_2})^{0.5} - (p''_{O_2})^{0.5}\right]}{2Lk_f (p'_{O_2} p''_{O_2})^{0.5} + D_V \left((p'_{O_2})^{0.5} + (p''_{O_2})^{0.5}\right)} \tag{6.14a}$$

It can be seen that the total permeation resistance is composed of three parts: exchange reaction at upstream membrane surface, bulk diffusion and exchange reaction at downstream membrane surface.

For tubular/hollow fibre membranes, the local oxygen permeation rate through a hollow fibre (or tubular) membrane can be given as [109]:

$$\frac{dN_{O_2}}{dl} = \frac{k_r \left[(p''_{O_2})^{0.5} - (p''_{O_2})^{0.5}\right]}{\dfrac{(p''_{O_2})^{0.5}}{2\pi R_o} + \dfrac{k_f \ln(R_o/R_{in})(p''_{O_2})^{0.5}(p''_{O_2})^{0.5}}{\pi D_V} + \dfrac{(p'_{O_2})^{0.5}}{2\pi R_{in}}} \tag{6-14b}$$

and the oxygen permeation flux with respect to log mean membrane area, $dA_m = 2\pi R_m dl$ in which $R_m = (R_o - R_{in})/\ln(R_o/R_{in})$ can be expressed by:

$$J_{O_2} = \frac{k_r D_V \left[(p'_{O_2})^{0.5} - (p''_{O_2})^{0.5}\right]}{2(R_o - R_{in}) \cdot k_f (p'_{O_2} p''_{O_2})^{0.5} + \dfrac{R_m}{R_{in}} \cdot D_V (p'_{O_2})^{0.5} + \dfrac{R_m}{R_o} \cdot D_V (p''_{O_2})^{0.5}} \tag{6.14c}$$

where R_o and R_{in} are the outer and inner radius of the tubular membrane. It can be seen the membrane thickness here is referred to the tube (hollow fibre) wall thickness.

6.5.3 Transport Analysis

Both experimental and theoretical studies on oxygen transfer through a mixed conducting membrane were carried out by Tan *et al.* [110] who developed a LSCF hollow fibre membrane module containing six hollow fibres of 27 cm in length. The experimental setup for the oxygen transfer in the LSCF hollow fibre membrane module is shown in Figure 6.5. It can be seen that the membrane module was placed in a CARBOLITE tubular furnace of length 15 cm. Therefore, the sealing joints were kept 6 cm away from the inlet of the furnace tube. Before experiments, the temperature profile of the tubular furnace was measured using a thermocouple and the results are shown in Figure 6.6. The temperature profile of the furnace is necessary for calculation of the oxygen permeation rate using Equation (6.14) with known kinetic parameters. The feed gas flow rates were controlled and measured using a mass flow controller. Argon was used as a sweep gas and was introduced either in the shell side (lumen feed mode) or in the fibre lumen (shell feed mode) depending on the operating modes. Oxygen concentrations in both the air stream and the argon stream effluents were measured online using an oxygen meter. The overall oxygen permeation rate

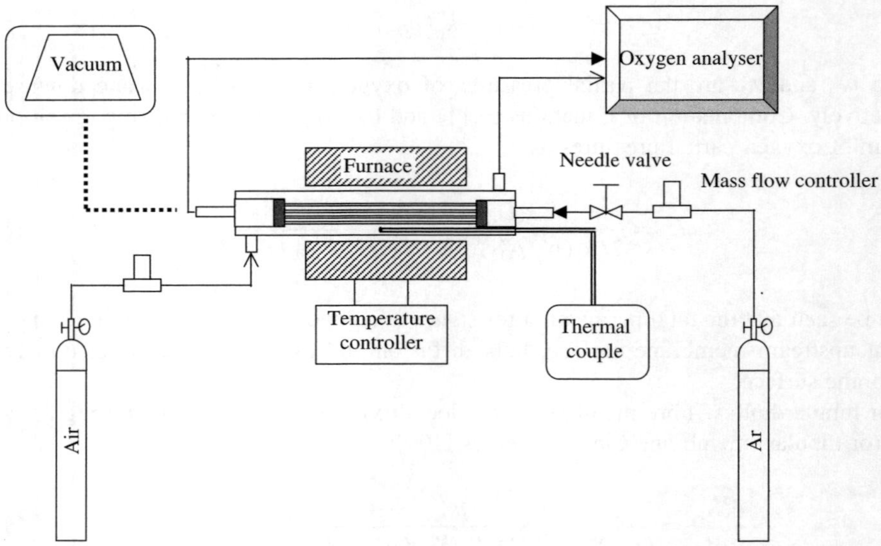

Figure 6.5 Experimental setup for oxygen permeation [110]. Copyright (2005) John Wiley & Sons, Ltd, American Institute of Chemical Engineers Journal, **51** (7), Tan, X., Liu, Y. and Li, K., *Mixed conducting ceramic hollow-fiber membranes for air separation*, p. 1991–2000

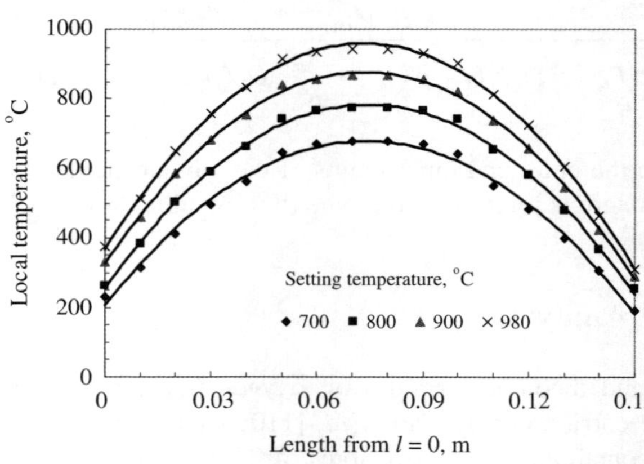

Figure 6.6 Temperature profile in the membrane module [110]. Copyright (2005) John Wiley & Sons, Ltd, American Institute of Chemical Engineers Journal, **51** (7), Tan, X., Liu, Y. and Li, K., *Mixed conducting ceramic hollow-fiber membranes for air separation*, p. 1991–2000

was calculated from the concentration changes of oxygen in the argon stream or in the air stream:

$$N_t = \frac{F_{Ar} y_e}{(1 - y_e)} \quad \text{from the argon stream} \tag{6.15a}$$

or

$$N_t = \frac{F_{Air}(0.21 - x_e)}{(1 - x_e)} \quad \text{from the air stream} \tag{6.15b}$$

where F is the feed gas flow rate, in mol s^{-1}, x_e and y_e are, respectively, oxygen concentrations in the air or in the argon effluents.

The experimental and modeling results obtained by Tan *et al.* [110] for the shell feed operation is illustrated in Figure 6.7, where the oxygen permeation rate is plotted against the argon or air flow rates at different operating temperatures. The modeling results plotted in solid lines are calculated using literature kinetic parameters obtained from a disc shaped symmetric membrane [36] and are compared with the experimental data. As can be seen, both experimental and modeling results show the same trend. However, the experimental results are better than the modeling ones except for at an operating temperature of 900 °C. This difference results from the different membrane surface structure, as the LSCF hollow fibre membranes prepared by Tan *et al.* [110] with a phase inversion technique possess an asymmetric structure, where the inner surface layer of the membranes is porous and thus the

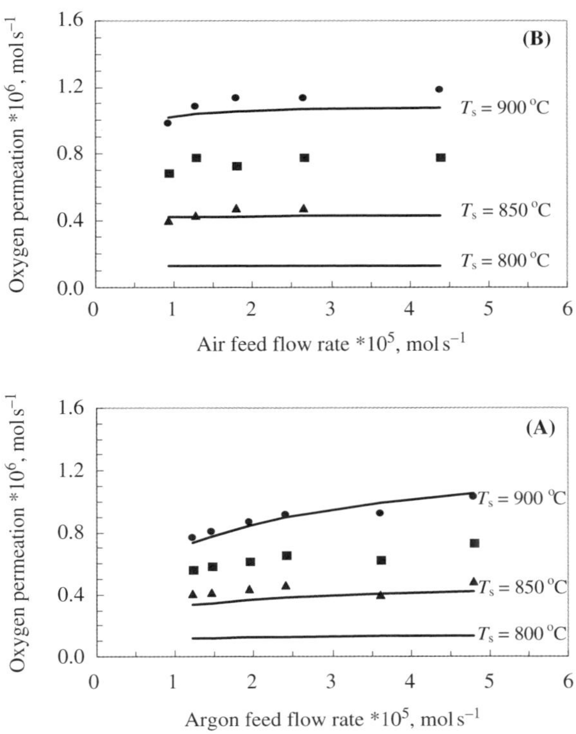

Figure 6.7 Oxygen permeation rate as a function of the gas flow rate for the shell feed and cocurrent operating mode: (A) air feed flow rate is constant; (B) argon feed flow rate is constant (experimental data: ▲ 800 °C, ■ 850 °C, • 900 °C) [110]. Copyright (2005) John Wiley & Sons, Ltd, American Institute of Chemical Engineers Journal, **51** (7), Tan, X., Liu, Y. and Li, K., *Mixed conducting ceramic hollow-fiber membranes for air separation*, p. 1991–2000

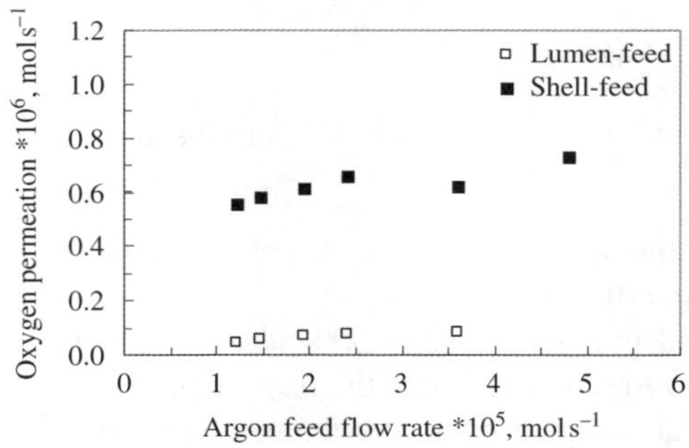

Figure 6.8 Comparison between lumen feed and shell feed operations ($T_s = 850\,°C$) [110]. Copyright (2005) John Wiley & Sons, Ltd, American Institute of Chemical Engineers Journal, **51** (7), Tan, X., Liu, Y. and Li, K., *Mixed conducting ceramic hollow-fiber membranes for air separation*, p. 1991–2000

membrane area for the surface exchange reaction is higher than that of symmetric membrane [36]. Therefore, the downstream surface exchange resistance is greatly decreased due to the higher surface area. As a result, the experimental results are better than modeling values at lower temperatures where the downstream surface exchange is, in general, the controlling step. As the temperature increases, bulk diffusion plays an increasingly important role in oxygen permeation and gradually becomes the controlling step. Therefore, the enhancement of oxygen permeation due to the better membrane surface structure diminishes as the operating temperature increases; hence the experimental data is close to the modeling results (900 °C).

Figure 6.8 compares the results between the lumen feed operation and the shell feed operation at 850 °C. As can be seen, the experimental data for the lumen feed operation is much inferior to that obtained from the shell feed operation, further confirming that the downstream surface exchange kinetics play an important role in oxygen permeation at low operating temperatures.

Figure 6.9 illustrates the theoretical importance of each resistance (upstream surface exchange, bulk diffusion and downstream surface exchange) for oxygen permeation at 800 and 1000 °C, respectively. It can be seen that at low operating temperatures, i.e., 800 °C, the upstream surface exchange resistance shows a negligible effect, as the simulation results obtained without considering upstream surface exchange resistance (dashed line) are overlapped with the solid line calculated for the actual case, i.e. considering all three resistances. As the downstream surface exchange resistance is assumed to be negligible, the oxygen permeation rate (dotted line) is considerably higher, suggesting that the downstream surface exchange is the controlling step for oxygen permeation. When the temperature is elevated to 1000 °C, the difference is dramatically reduced, which indicates that the contribution of downstream surface exchange to the overall resistance becomes small at high temperatures, and bulk diffusion gradually becomes the controlling step.

6.6 AIR SEPARATION

Separation of oxygen from air or other oxygen containing gas mixtures has become of increasing interest in a number of applications which range from small scale oxygen production to

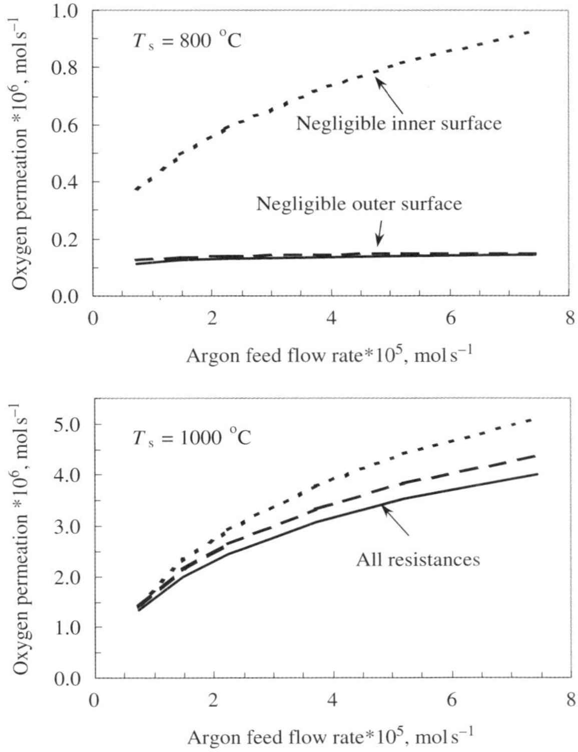

Figure 6.9 Modeling results for shell feed, cocurrent flow operation at different temperatures [110]. Copyright (2005) John Wiley & Sons, Ltd, American Institute of Chemical Engineers Journal, **51** (7), Tan, X., Liu, Y. and Li, K., *Mixed conducting ceramic hollow-fiber membranes for air separation*, p. 1991–2000

large scale combustion processes or partial oxidation of light hydrocarbons [111]. Among the current technologies for oxygen separation, membranes can be considered as a clean, efficient and economic means of producing oxygen, and ionic conducting or ionic and electronic conducting ceramic dense membranes are of particular interest because:

- These membranes are 100 % selective for oxygen. This results in production of high purity gas products, which cannot be achieved with even the best organic membranes.
- Additionally, the flux of these membranes is 3–4 orders of magnitude higher than that offered by organic membranes.

Most materials of interest for producing such ceramic membranes are based on fluorite (ZrO_2 or CeO_2) or perovskite (ABO_3) related structures. Perovskite materials such as $La_xSr_{1-x}Co_y Fe_{1-y}O_{3-\delta}$, are able to provide higher oxygen permeation flux without using an external electric circuit. Air Products has been developing ion or mixed conducting membrane technology for many years; more than 47 US patents were issued at the end of the last century [1, 112]. Recently, this technology has entered the commercial stage with three different processes for oxygen production or supply [1]. The first one is the electrically driven SEOS™ Oxygen

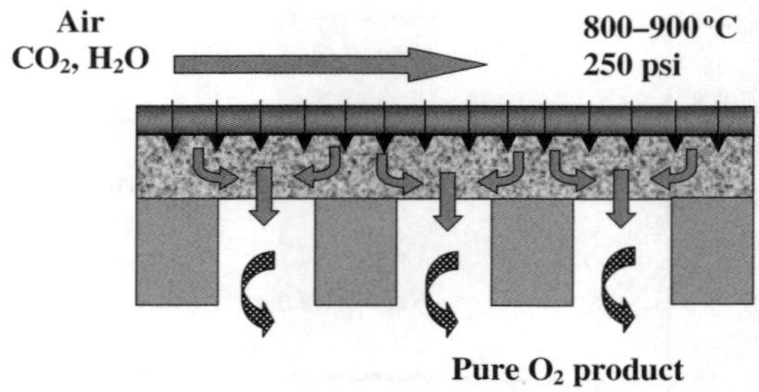

Figure 6.10 Thin film support membrane structure. This pressure resistant structure may be fabricated in a flat plate or tubular configuration [1]. Reprinted from Solid State Ionics, **134**, Dyer *et al.*, *Ion transport membrane . . .* , p. 21–33. Copyright (2000), with permission from Elsevier

Table 6.3 Comparison of cryogenic versus ion transport membrane (ITM) air separation unit (ASU) in integrated gasification combined cycle (IGCC) power applications [112]. Reprinted from Membrane Technology, **110** Stiegel, G., *Mixed conducting ceramic membranes for gas separation and reaction*, p. 5–7. Copyright (1999), with permission from Elsevier

	Cryogenic based case	ITM based case	Delta	% change
Capital investment, US$m	641	610	−31	−4.8
Capital investment, US$kw^{-1}	1567	1453	−114	−7.3
ASU capital investment, US$m	60.2	41.6	−18.6	−31
Power generation, MW	409	420	11	2.7
Thermal efficiency, % HHV	45.2	46.5	1.3	2.9
Cost of electricity, mills kwh^{-1}	55.5	51.9	−3.6	−6.5

Generator using membranes fabricated from an oxygen ion conductor with low electronic conductivity materials. With the aid of an external electric circuit, oxygen can be separated and compressed by this equipment from the low pressure side to the high pressure side. The second one is the pressure driven ITM Oxygen using membranes fabricated from mixed conducting materials. A multilayer membrane structure is designed as shown in Figure 6.10. An oxygen partial pressure differential imposed across the membrane at high temperature drives oxygen ions from the high partial pressure side to the low partial pressure side. The feed side pressure can be very high. To minimize the mechanical load imposed by the driving force, the process operating conditions constitute a medium pressure air feed stream, typically 100–300 psi and a low pressure oxygen permeate stream, typically at a fraction of an atmosphere. The cost and performance of the cryogenic based case and the ITM oxygen integrated IGCC facility are compared in Table 6.3.

Potential cost reductions in capital investment and electricity consumption can possibly be achieved with additional integration and increase of capacity in the factory. Better environmental performance would be another benefit of using this technology.

The last process is the chemical potential driven ITM Syngas technology using membranes. It is also fabricated from mixed conducting materials. The driving force for oxygen transport in the membrane is developed by depleting the oxygen partial pressure on one side of the membrane through a chemical reaction. A relatively low pressure air feed is allowed and oxygen can be transported to a higher pressure reaction product stream. This technology can be used in the syngas process [1].

6.6.1 Design Equations

Models for air separation in a ceramic membrane module consisting of n tubes (or hollow fibres) can be developed. They are dependent on the practical operating modes as schematically illustrated in Figure 6.11. In the formulation of mathematical models, the following general assumptions have to be made in order to reduce the complexity of the models:

(1) The membrane module is operated at steady state under isothermal conditions.

(2) The transport fluxes of charged defects are only in the radial direction of the hollow fibre membrane and are negligible in the axial direction because of the long hollow fibres used.

(3) The gas phase mass transfer resistances are negligible. Therefore, oxygen partial pressures on the membrane surfaces are, respectively, equal to those in the module shell or lumen.

(4) The ideal gas law is used to describe the gas behaviour of single components and the gas mixture.

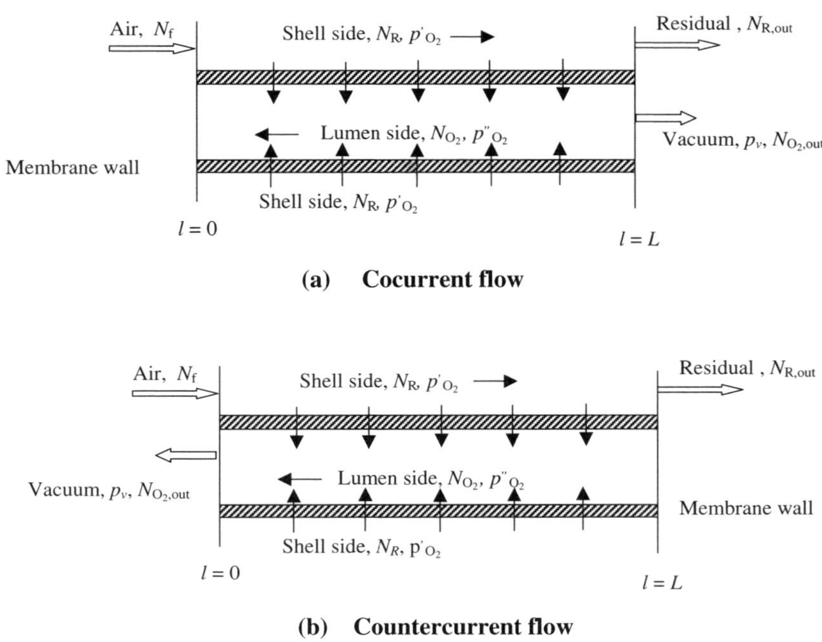

Figure 6.11 Operating modes of the membrane module for air separation [109]. Copyright (2002) John Wiley & Sons, Ltd, American Institute of Chemical Engineers Journal, **48**, Tan, X. and Li, K., *Modeling of air separation in a LSCF hollow-fiber membrane module*, p. 1469–1477

(5) The plug flow conditions are attained in both gas streams and the axial dispersion is negligible.

Based on the above assumptions, the oxygen permeation equation derived in the preceding section, i.e. Equation (6.14) can be readily employed, while the overall and component material balances can be written based on operating flow patterns as shown in Figure 6.11 and are given as follows.

6.6.1.1 Cocurrent Flow

$$\text{Overall:} \quad N_f = N_R + N_{O_2} \tag{6.16a}$$

$$O_2: \quad 0.21N_f = N_R \frac{p'_{O_2}}{p} + N_{O_2} \tag{6.16b}$$

In addition, the Hagen–Poiseuille equation is employed to describe the pressure profile in the lumen:

$$\frac{dp''_{O_2}}{dl} = -\frac{8\mu RT}{n\pi R_{in}^4 p''_{O_2}} \cdot N_{O_2} \tag{6.17}$$

Boundary conditions for this case are:

$$l = 0 \quad N_R = N_f \quad N_{O_2} = 0 \quad p'_{O_2} = 0.21p$$

$$l = L \quad p''_{O_2} = p_v$$

6.6.1.2 Countercurrent Flow

$$\text{Overall:} \quad N_R = N_{R,out} + N_{O_2} \tag{6.18a}$$

$$O_2: \quad N_R \frac{p'_{O_2}}{p} = N_{R,out} \frac{p'_{O_2,out}}{p} + N_{O_2} \tag{6.18b}$$

Again, the Hagen–Poiseuille equation is employed to describe the pressure profile in the lumen:

$$\frac{dp''_{O_2}}{dl} = \frac{8\mu RT}{n\pi R_{in}^4 p''_{O_2}} \cdot N_{O_2} \tag{6.19}$$

Boundary conditions for this case are:

$$l = 0 \quad N_R = N_f \quad p'_{O_2} = 0.21p \quad p''_{O_2} = p_v$$

$$l = L \quad N_{O_2} = 0$$

The above design equations derived for both the cocurrent and countercurrent flow patterns are a group of ordinary differential equations. Equations (6.16) and (6.17), describing the cocurrent flow, and Equations (6.18) and (6.19) describing the countercurrent flow, together with the transport equation of Equation (6.14b), can be solved numerically using the Runge–Kutta method.

6.6.2 Performance Analysis

Performance analysis of a $La_{0.6}Sr_{0.4}Co_{0.2}F_{0.8}O_{3-\delta}$ (LSCF) hollow fibre membrane module for air separation was carried out both theoretically and experimentally by Tan and Li [109, 113] whose calculations were based on the operating and the LSCF membrane parameters given in Table 6.4 unless otherwise specified. Effects of process operating conditions and characteristics of the membrane employed on oxygen permeation flux or productivity, α (defined as ratio of oxygen fed into the module to that produced ($\alpha = N_{O_2,out}/0.21N_f$) were analyzed and presented below.

6.6.2.1 Effect of Operating Pressures and Temperatures

Calculated values for these effects are illustrated in Figures 6.12 to 6.14. In Figure 6.12, the oxygen permeation flux is plotted against the vacuum pressure applied in the hollow fibre lumen for different operating temperatures, while keeping the shell side pressure at 1 atm. It can be seen from the figure that as the vacuum level is increased, the oxygen flux is improved, but only up to a certain value of the vacuum pressure, which may be termed the effective vacuum pressure. Any further increase in the vacuum level (decrease in the vacuum pressure) after this effective vacuum pressure does not result in better oxygen permeation performance, as shown in the figure. This behaviour is true for different operating temperatures and flow patterns, and it is interesting to note that the effective vacuum pressure, in this case about

Table 6.4 Values of parameters used in simulation of LSCF hollow fibre membrane module

Membrane area[a]	$A_t = 200$, cm^2
OD of the hollow fibre	$d_o = 0.1$, cm
ID of the hollow fibre	$d_{in} = 0.05$, cm
Length of the reactor (hollow fibre)	$L = 20$, cm
Operating pressure in shell side	$p = 1$, atm
Vacuum pressure in fibre lumen	$p_v = 0.01$, atm
Operating temperature	$T = 1173$, K
Feed flow rate	$N_f = 5$, cm^3 s^{-1}
Diffusion coefficient of oxygen vacancy, cm^2 s^{-1}	$D_V = 1.58 \times 10^{-2} \exp\left(-\frac{8852.5}{T}\right)$, [36]
Forward reaction rate constant of (6a), cm atm$^{-0.5}$ s^{-1}	$k_f = 5.90 \times 10^{6} \exp\left(-\frac{27291}{T}\right)$, [36]
Reverse reaction rate constant of (6b), mol cm^{-2} s^{-1}	$k_r = 2.070 \times 10^{4} \exp\left(-\frac{29023}{T}\right)$, [36]

[a] Defined as: $A_t = 2n\pi L \cdot \dfrac{R_o - R_{in}}{\ln(R_o/R_{in})}$

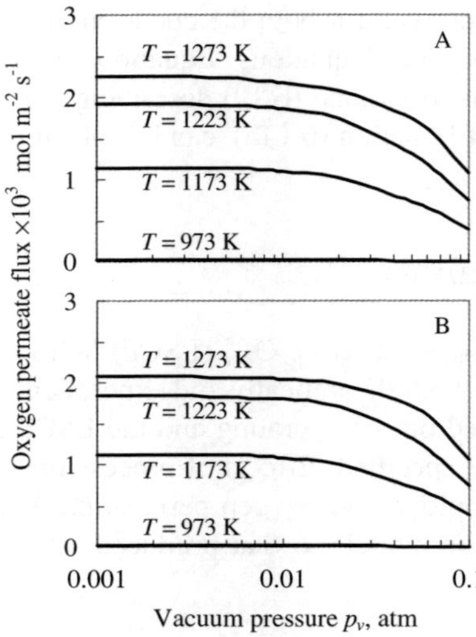

Figure 6.12 Effect of vacuum pressure on oxygen permeation flux ($p = 1$ atm): A, cocurrent; B, countercurrent [109]. Copyright (2002) John Wiley & Sons, Ltd, American Institute of Chemical Engineers Journal, **48**, Tan, X. and Li, K., *Modeling of air separation in a LSCF hollow-fiber membrane module*, p. 1469–1477

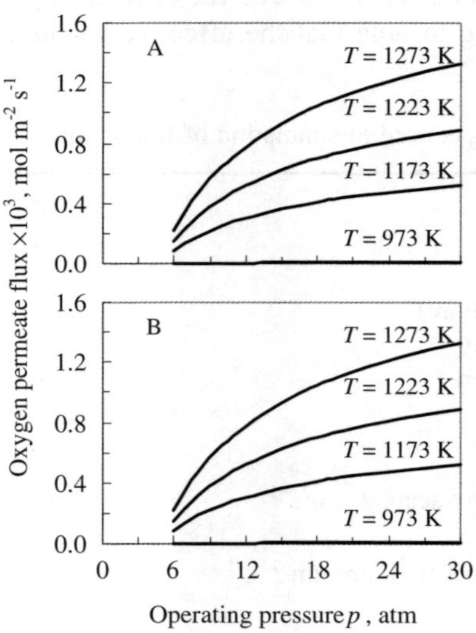

Figure 6.13 Effect of operating pressure on oxygen permeation flux ($p_v = 1$ atm): A, cocurrent; B, countercurrent [109]. Copyright (2002) John Wiley & Sons, Ltd, American Institute of Chemical Engineers Journal, **48**, Tan, X. and Li, K., *Modeling of air separation in a LSCF hollow-fiber membrane module*, p. 1469–1477

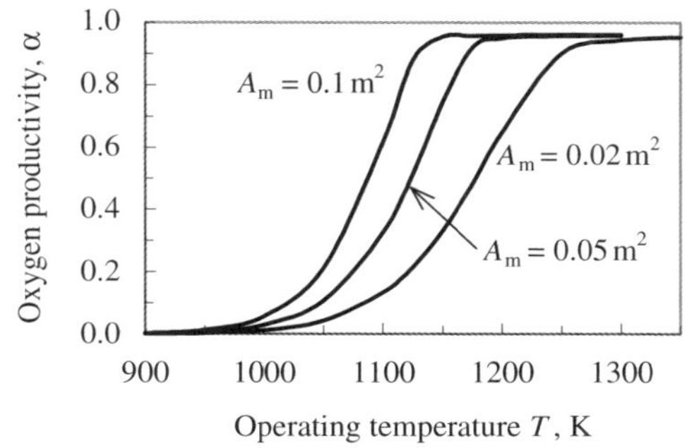

Figure 6.14 Effect of operating temperature on oxygen productivity [109]. Copyright (2002) John Wiley & Sons, Ltd, American Institute of Chemical Engineers Journal, **48**, Tan, X. and Li, K., *Modeling of air separation in a LSCF hollow-fiber membrane module*, p. 1469–1477

0.01 atm, remains almost identical and does not seem to be dependent on the operating temperature and flow pattern.

In Figure 6.13, the effect of shell side pressure on the oxygen permeation flux is shown for different operating temperatures, keeping the lumen side pressure at 1 atm. As expected, elevation of the operating pressure in the shell side generally increases the oxygen permeation flux. However, the trend of such an increase is clearly nonlinear indicating that the mass transfer resistance of the membrane is a function of the operating pressure applied. Further inspection of Figures 6.12 and 6.13 indicates that the LSCF hollow fibre module operated at an elevated pressure may not be a better choice as compared to that operated under vacuum conditions. As can be seen from the calculated data, at an operating temperature of 1273 K, an oxygen permeation flux of about $2.3 \times 10^{-3}\,mol\,m^{-2}\,s^{-1}$ could be easily achieved when the vacuum pressure applied at the lumen side of the module is reduced to 0.01 atm (Figure 6.12). Such an oxygen permeation flux would not be possible to achieve at elevated pressure operation, even when the shell side pressure is increased to 30 atm, as shown in Figure 6.13. It thus follows that elevated pressure operation commonly employed in conventional polymeric hollow fibre membrane modules for gas separation is not suitable for LSCF hollow fibre modules. Instead, vacuum operation at the lumen side of the membrane module is preferable, as it yields much higher oxygen production rates.

The effect of operating temperature on the oxygen permeation is shown in Figure 6.14, where the oxygen productivity is plotted against temperature. It can be seen that the oxygen productivity is less than 0.1 at an operating temperature of 1000 K or lower. The oxygen productivity increases drastically after 1000 K, and then approaches a maximum value at 1300 K for $A_t = 0.02\,m^2$, at 1180 K for $A_t = 0.05\,m^2$ and at 1130 K for $A_t = 0.1\,m^2$, respectively. The results indicate that the membrane area plays an important role in the production of oxygen. For a given membrane area, module optimum production of oxygen could be achieved with careful selection of operating pressure and temperatures as illustrated in the above figures.

6.6.2.2 Effect of Flow Patterns

Predicted values for this effect are given in Figures 6.15 and 6.16. In Figure 6.15, the vacuum level in the hollow fibre lumen is plotted against oxygen productivity, α, for both the

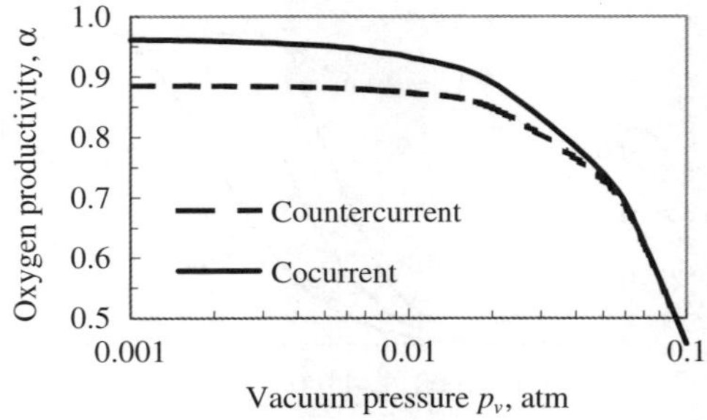

Figure 6.15 Comparison between cocurrent and countercurrent flow patterns ($T = 1273$ K) [109]. Copyright (2002) John Wiley & Sons, Ltd, American Institute of Chemical Engineers Journal, **48**, Tan, X. and Li, K., *Modeling of air separation in a LSCF hollow-fiber membrane module*, p. 1469–1477

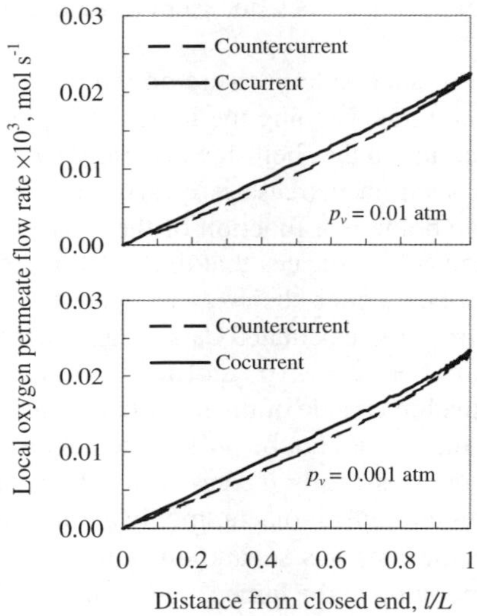

Figure 6.16 Plot of local oxygen permeate flow rate for cocurrent and countercurrent flow patterns [109]. Copyright (2002) John Wiley & Sons, Ltd, American Institute of Chemical Engineers Journal, **48**, Tan, X. and Li, K., *Modeling of air separation in a LSCF hollow-fiber membrane module*, p. 1469–1477

cucurrent and countercurrent flow patterns. It can be seen that cocurrent flow exhibits higher oxygen productivity compared to the countercurrent flow pattern when the vacuum pressure is less than 0.05 atm. When the vacuum pressure is higher than 0.05 atm, the difference between the two flow patterns becomes small. This is clearly in contrast to the results for polymeric hollow fibre membrane modules where the countercurrent flow pattern always leads to better separation. Such contrary behaviour is due to the high local oxygen permeation available in the module for the cocurrent flow pattern. As illustrated in Figure 6.16, the profile of the local permeation flow of oxygen for cocurrent flow pattern is always greater than that for countercurrent flow pattern throughout the entire hollow fibre length, thus resulting in

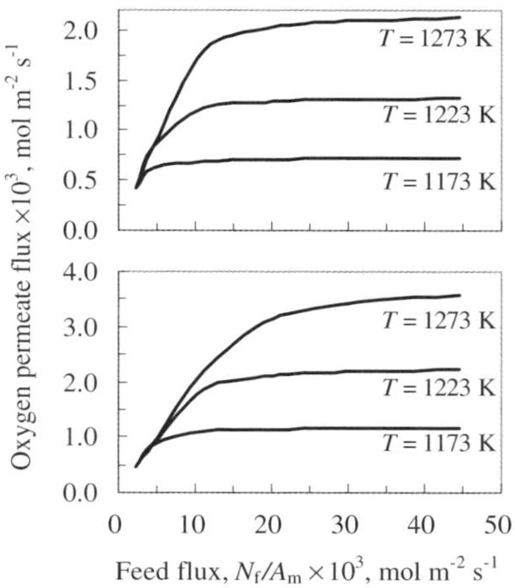

Figure 6.17 Effect of feed flow rate on oxygen permeation flux: (a) $p = 20\,\text{atm}$, $p_v = 1\,\text{atm}$; (b) $p = 1\,\text{atm}$, $p_v = 0.01\,\text{atm}$ [109]. Copyright (2002) John Wiley & Sons, Ltd, American Institute of Chemical Engineers Journal, **48**, Tan, X. and Li, K., *Modeling of air separation in a LSCF hollow-fiber membrane module*, p. 1469–1477

high oxygen productivity. Therefore, the cocurrent flow pattern has been used for all the subsequent analysis.

6.6.2.3 Effect of Feed Flow Rate

The effect of feed flow rate for a given membrane area on oxygen permeation flux is illustrated in Figure 6.17. As can be seen, the oxygen permeation flux increases as the feed flow rate is increased, but only up to $12 \times 10^{-3}\,\text{mol}\,\text{m}^{-2}\,\text{s}^{-1}$, beyond which the oxygen permeation flux remains relatively unchanged as shown in Figure 6.17(a). As the hydrodynamics of the feed flow provide negligible effect on the oxygen permeation because of the dominance of the membrane resistance, the dependence of the oxygen flux at lower feed flow rate is due to the change in oxygen driving force. At a feed flow rate of $20 \times 10^{-3}\,\text{mol}\,\text{m}^{-2}\,\text{s}^{-1}$ or greater, the partial pressure of oxygen in the shell side remains constant (close to 0.21) throughout the module, and hence the maximum value of the oxygen permeation flux is attained. It should be noted that for vacuum pressure operation, such a maximum oxygen permeation flux can not be reached unless the feed flow rate is further increased to 25×10^{-3}–$40 \times 10^{-3}\,\text{mol}\,\text{m}^{-2}\,\text{s}^{-1}$ as shown in Figure 6.17(b). The reason for this difference is probably due to the high oxygen permeation flux achievable under vacuum pressure operation. With the increased permeation characteristics, the maximum oxygen permeation flux can only be achieved by further increase in the feed flow rate.

6.6.2.4 Effect of Membrane Area

As illustrated above, the operating pressure, p, in the shell side of the module and the vacuum pressure, p_v, in fibre lumen determine the oxygen permeation flux for a given membrane

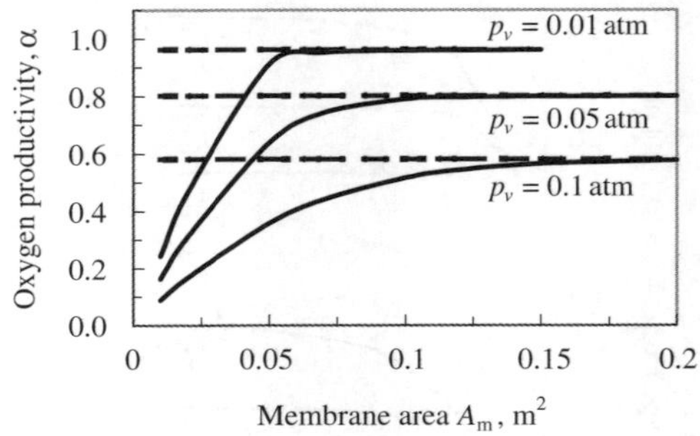

Figure 6.18 Effect of membrane area on the oxygen productivity [109]. Copyright (2002) John Wiley & Sons, Ltd, American Institute of Chemical Engineers Journal, **48**, Tan, X. and Li, K., *Modeling of air separation in a LSCF hollow-fiber membrane module*, p. 1469–1477

module. If the pressure buildup in the fibre lumen is negligible, the maximum oxygen productivity, α_{max} can be derived from Equation 6.16 with the condition that partial pressure oxygen in shell side of module is equal to that in the lumen, i.e. $p'_{O_2} = p_v$ as:

$$\alpha_{max} = \frac{p - p_v/0.21}{p - p_v} \tag{6.20}$$

Figure 6.18 illustrates the effect of membrane area on the oxygen productivity at different vacuum pressures, keeping the shell side pressure and operating temperature at 1 atm and 1173 K, respectively. The dashed lines in Figure 6.18 are the maximum oxygen productivities obtained from Equation (6.20) for the different vacuum pressures. It can be seen that for a given feed flow rate, the oxygen productivity increases as the membrane area is increased. However, further increase in the membrane area results in no effect once the maximum value of the oxygen productivity is reached. It is interesting to note that the higher the vacuum level, the less the membrane area required to achieve high oxygen productivity. It therefore suggests that economically there is a trade off between the vacuum level and membrane area required to obtain optimum pure oxygen production.

The membrane employed in this study has absolute selectivity for oxygen, hence a pure oxygen product is achievable on the permeate side of the module. However, purified nitrogen on the residual side is not simultaneously achieved, as the oxygen in air would not be entirely permeated unless an absolute vacuum was applied in the fibre lumen as shown in Figure 6.19. In addition, the desired membrane area must also be provided for a given feed flow rate. As a result, a high vacuum level as well as enough membrane area is essential to produce pure oxygen and nitrogen simultaneously.

6.6.2.5 Comparison with Experimental Data

Although so far there have been many studies on oxygen permeable ceramic materials, very limited experimental work on membrane modules for air separation has been conducted. Li *et al.* [75] performed some experiments on oxygen permeation through a tubular LSCF membrane, the dimensions of which were 8 mm in outer diameter, 15 cm in length and 1.5 mm in wall thickness. Air was introduced into the shell side of the module. Helium used as a

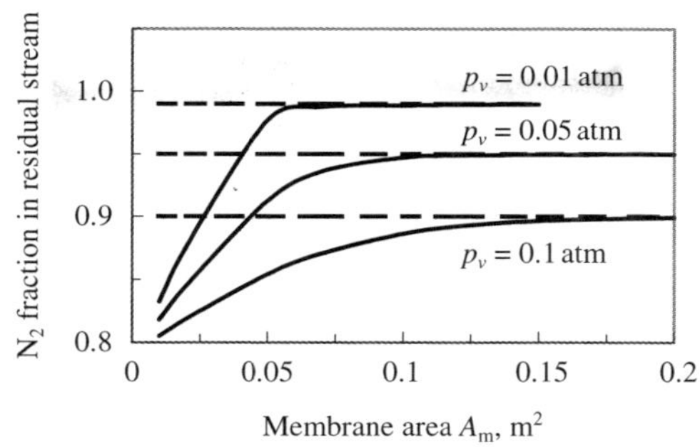

Figure 6.19 Effect of membrane area on the nitrogen purity in the residual stream [109]. Copyright (2002) John Wiley & Sons, Ltd, American Institute of Chemical Engineers Journal, **48**, Tan, X. and Li, K., *Modeling of air separation in a LSCF hollow-fiber membrane module*, p. 1469–1477

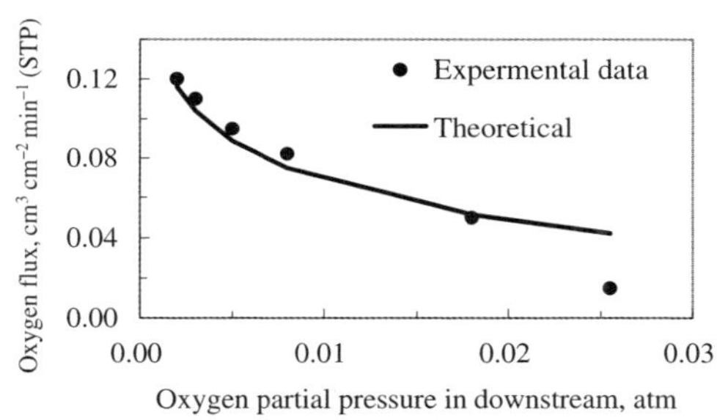

Figure 6.20 Permeation flux of oxygen at various downstream oxygen partial pressures at a temperature of 1123 K ($p = 1$ atm) [109]. Copyright (2002) John Wiley & Sons, Ltd, American Institute of Chemical Engineers Journal, **48**, Tan, X. and Li, K., *Modeling of air separation in a LSCF hollow-fibre membrane module*, p. 1469–1477

sweep gas was fed to the bore of the tube. Both upstream and downstream were maintained at atmospheric pressures. The partial pressure of oxygen in the bore side of the tube was maintained by adjusting the flow rate of helium. Figures 6.20 and 6.21 compares the experimental results with the theoretical values calculated based on equations outlined above. In Figure 6.20, the oxygen permeation flux is plotted against the oxygen partial pressures in the bore side at the temperature of 1123 K while the oxygen partial pressure in the shell side was kept at a constant value, i.e. 0.21 atm. As can be seen, all the oxygen permeation fluxes except one point at the highest oxygen partial pressure are in good agreement with the calculation results predicted using kinetic parameters given by Xu and Thomson [36].

Figure 6.21 plots the oxygen permeation through the membrane at different operating temperatures while keeping the partial pressures of oxygen at the shell and bore sides at 0.21 atm and 0.001 atm, respectively. The increasing trend of oxygen permeation with temperature is also in good agreement with the modeling results. However, the theoretical results show some deviations at temperatures above 1073 K. Nevertheless, the comparisons given in Figures 6.20 and 6.21 indicate the theoretical equations developed together with the kinetic parameters given by Xu and Thomson [36] can be used reasonably well in predicting the performances of hollow fibre LSCF membranes for oxygen separation from air.

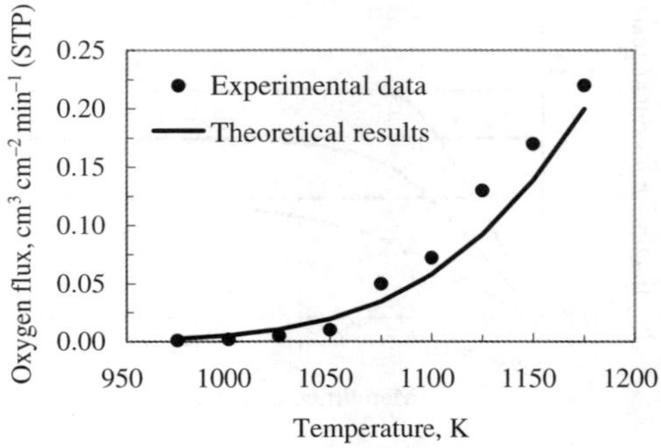

Figure 6.21 Temperature dependence of the oxygen permeation flux ($p'_{O_2} = 1$ atm, $p''_{O_2} = 0.001$ atm) [109]. Copyright (2002) John Wiley & Sons, Ltd, American Institute of Chemical Engineers Journal, **48**, Tan, X. and Li, K., *Modeling of air separation in a LSCF hollow-fiber membrane module*, p. 1469–1477

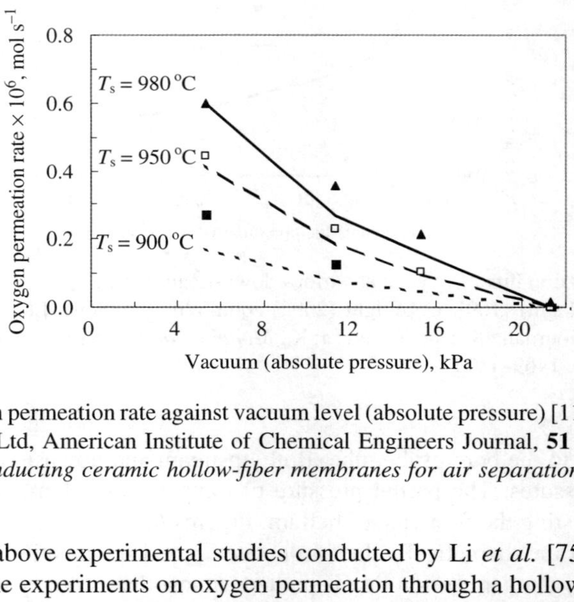

Figure 6.22 Oxygen permeation rate against vacuum level (absolute pressure) [110]. Copyright (2005) John Wiley & Sons, Ltd, American Institute of Chemical Engineers Journal, **51** (7), Tan, X., Liu, Y. and Li, K., *Mixed conducting ceramic hollow-fiber membranes for air separation*, p. 1991–2000

Apart from the above experimental studies conducted by Li *et al.* [75], Tan *et al.* [110] also performed some experiments on oxygen permeation through a hollow fibre LSCF membrane module containing six hollow fibres of 27 cm in length. In their experiments, air flowed in the shell side of the module while the vacuum was applied to the fibre lumen at the same end of the module as the air effluent outlet. The effect of the vacuum level on the performance of the membrane module is given in Figure 6.22, where the oxygen permeation rate is plotted against the absolute pressures at different operating temperatures. As is expected, the oxygen permeation rate increases as the applied vacuum level is increased (i.e. the absolute pressure is decreased). The theoretical results are also plotted in the figure with the solid curves. It can be seen that the experimental data is in good agreement with theoretical results and is, in general, greater than the corresponding calculated values. This may be due to the fact that the oxygen molecules in the inner porous layer (downstream) are able to be released easily from the membrane surface under vacuum operation. In conclusion, vacuum operation favors oxygen permeation kinetically in LSCF hollow fibre membrane modules [110].

6.6.2.6 Production of Oxygen Using Hollow Fibre Modules

Tan and Li [113] developed two LSCF hollow fibre membrane modules for oxygen production. Table 6.5 summarizes the properties of the membrane modules prepared. Since the modules are developed for vacuum operation, hollow fibres used are all 'test tube' type with the open end side glued onto a glass holder for suction. All the hollow fibres were individually tested to be gas tight before the module fabrication using a gas permeation apparatus developed by Tan *et al.* [110].

The system for oxygen production is illustrated in Figure 6.23 where a self cleaning dry vacuum system was connected to the fibre lumen of the module at the open end. A large part of the hollow fibres from the closed end was located in a CARBOLITE furnace tube. The temperature profile within the furnace tube was measured in advance of experiments. The flow rate of the oxygen pulled out of the membrane module by the vacuum pump was measured using a soap bubble meter. Due to slight leakage of the vacuum pump, the product gas from the vacuum pump is not pure oxygen, but over 60 %. Therefore, an oxygen analyzer was used to measure the oxygen concentration in the stream in order to check the mass balance of permeated oxygen.

Table 6.5 Properties of the membrane modules for oxygen production

	Module A	Module B
Operating mode	Vacuum	Vacuum
OD of the hollow fibre	0.146, cm	0.146, cm
ID of the hollow fibre	0.114, cm	0.114, cm
Number of the hollow fibres	10	22
Length of the hollow fibre	24 cm	24
Fibre length at constant temperature zone of the furnace	6 cm	6 cm
Effective membrane area	24.37 cm^2	53.61 cm^2
CARBOLITE furnace type	MTF 12/38/250	MTF 12/38/250
Maximum temperature	1200 °C	1200 °C

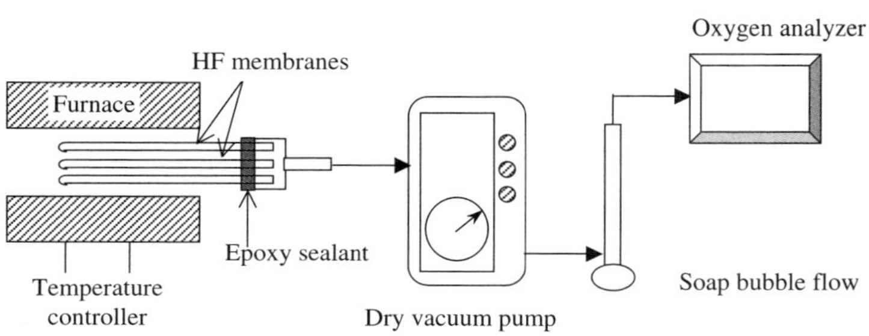

Figure 6.23 Oxygen production apparatus, vacuum operation [113]. Copyright (2007) John Wiley & Sons, Ltd, American Institute of Chemical Engineers Journal, Tan, X. and Li, K., *Oxygen production using dense ceramic hollow fibre membrane modules with different operating modes*, in press

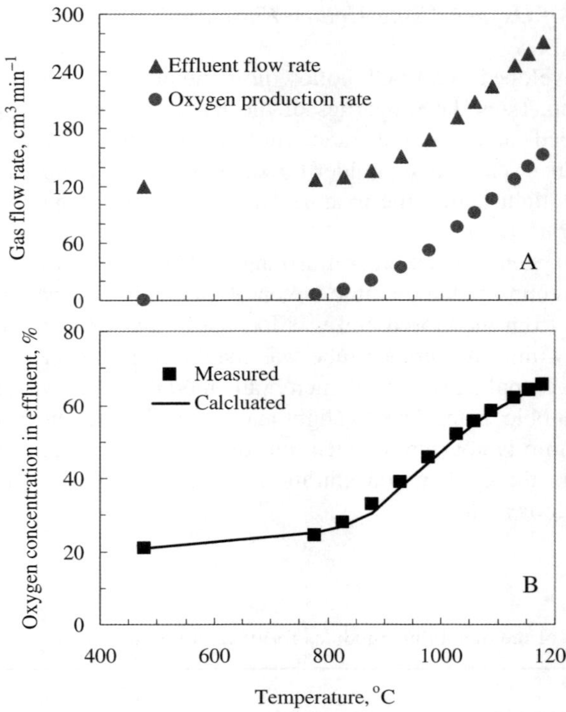

Figure 6.24 Oxygen production using the vacuum operation (Module A, vacuum level = 100.6 kPa). [113]. Copyright (2007) John Wiley & Sons, Ltd, American Institute of Chemical Engineers Journal, Tan, X. and Li, K., *Oxygen production using dense ceramic hollow fiber membrane modules with different operating modes*, in press

Figure 6.24A shows the experimental results of oxygen production with Module A by applying a vacuum level of 100.6 kPa. As can be seen, at temperatures of lower than 500 °C, no oxygen permeation takes place, as the data shows that the oxygen concentration in the product stream is 21 % and flow rate of the stream is 119 $cm^3 min^{-1}$ due to the leakage of the vacuum pump. After the temperature is increased to above 780 °C, oxygen permeation takes place. Both the oxygen concentration and the total product flow rate increase as temperature is increased because the oxygen permeation rate increases with increase in operating temperature. At an operating temperature of 1180 °C, the flow rate in the product stream reaches 270.5 $cm^3 min^{-1}$ giving a corresponding pure oxygen production rate of 270.5 − 119 = 151.5 $cm^3 min^{-1}$ if the membrane is gas tight. Figure 6.24B compares the measured oxygen concentrations in the product stream to the calculated ones. It can be seen that they are in excellent agreement, further confirming that the leakage is only due to the vacuum pump and the membrane used for the experiments is gas tight.

Comparison between the experimental data and theoretical results is given in Figure 6.25, where the oxygen permeation fluxes obtained from different modules at different vacuum levels are plotted against operating temperatures. It can be seen that the experimental data is in good agreement with the theoretical curves with the exception that, at higher temperatures, the experimental data is slightly lower than the theoretical results. This is probably due to the lower oxygen concentration in the tubular furnace (oxygen concentration in feed side).

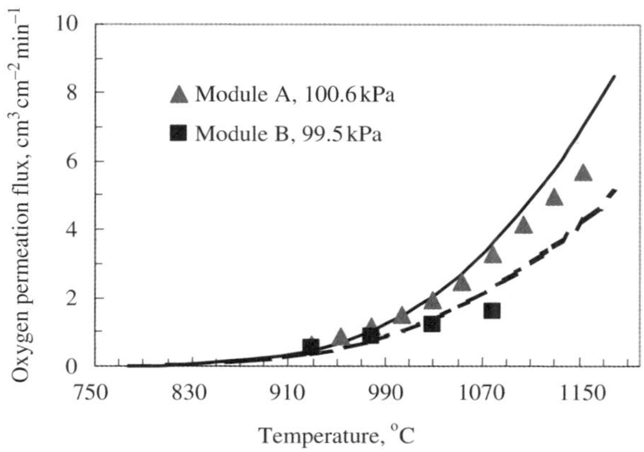

Figure 6.25 Comparison between experimental data and theoretical results (vacuum operation) [113]. Copyright (2007) John Wiley & Sons, Ltd, American Institute of Chemical Engineers Journal, Tan, X. and Li, K., *Oxygen production using dense ceramic hollow fiber membrane modules with different operating modes*, in press

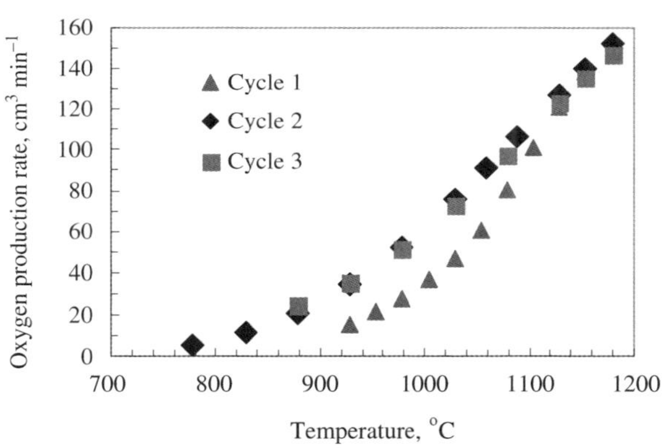

Figure 6.26 Results for the different operating cycles (Module A, vacuum level = 100.6 kPa) [113]. Copyright (2007) John Wiley & Sons, Ltd, American Institute of Chemical Engineers Journal, Tan, X. and Li, K., *Oxygen production using dense ceramic hollow fiber membrane modules with different operating modes*, in press

Since the hollow fibre membranes are located horizontally in the centre part of the furnace tube, the hot air in the furnace is the only feed source. At high temperatures, because of the high oxygen permeation flux, the permeated oxygen from the furnace cannot be compensated in time by air outside the furnace, leading to a lower oxygen concentration in the air of the furnace, hence giving the slight difference between the experimental and theoretical results.

Figure 6.26 shows the experimental results of different operating cycles of the Module A with the vacuum level of 100.6 kPa. One operating cycle means a whole operating process, i.e. heating up the module from room temperature, collecting experimental data at high temperature and cooling the module back to the ambient temperature. It can be seen that the oxygen production rate in cycle 1 is lower than those in cycle 2 and cycle 3, especially at lower temperatures. However, as the temperature is increased to 1078 °C, the difference in oxygen production rates between cycle 1 and the other cycles is diminished. Afterwards, the oxygen production rates in the cycle 2 and cycle 3 are almost identical, suggesting the membrane is relatively stable between the cycles.

6.7 FURTHER DEVELOPMENT – CHALLENGES AND PROSPECTS

Although significant advances have been made in the last decade, there are still many challenges that have to be faced so as to allow the utilization of ceramic mixed conducting membranes in the practical applications mentioned above.

First of all, the permeation rate of oxygen through the membranes must be high enough to satisfy the requirements of industrial applications. Currently, most of the related research work uses small disc membranes, as they are relatively easy to fabricate and have been used to test the properties of the materials during the stages of material development and screening. Such disk membranes only supply very limited membrane area and are also too thick. Some researchers use tubular membranes because the tubular membranes provide a relatively large active area compared to disk membranes. However, transport resistance in the tubular membranes is very high due to their symmetric structure. The rate of oxygen permeation depends on the thickness of the membrane, usually being inversely related to the membrane thickness. Therefore, it is necessary to develop a technique for fabricating thin membranes to maximize the permeation flux. Many attempts are being made to reduce the thickness of the membranes [54, 114–117]. Most of these efforts have been based on the concept of composite membranes where a dense thin film is coated on a porous substrate by various methods such as spin coating or dip coating, chemical vapour deposition and some other methods. The film thickness was well controlled. An approximately 2 μm thick membrane was obtained by spin coating of the $SrCe_{0.95}Yb_{0.05}O_{3-\delta}$ suspension on a porous $SrZr_{0.95}Y_{0.05}O_{3-\delta}$ substrate [117]. When the top layer material is different from substrate material, the top layer is usually unstable especially at elevated temperature. Jin *et al.* [116] successfully prepared a crack free asymmetric membrane using the same materials for both the top layer and substrate by simply coating a slurry containing $La_{0.6}Sr_{0.4}Co_{0.2}Fe_{0.8}O_{3-\delta}$ powders directly onto the surface of a green support, followed by a sintering process. This asymmetric membrane had a 200 μm thick dense top layer and exhibited an oxygen flux about three to four times higher than that of a 2 mm dense disk. The same idea was also taken and applied by Air Products to produce a large membrane consisting of a dense thin top layer and a thick porous support. A multiple planar stack module containing such membranes has been used in a pilot plant for air separation [1, 112]. Although a multiple planar stack can be employed to enlarge the membrane area to that required at plant scale [1], many problems, such as sealing, connecting and pressure resistance, have to be faced. Also the packing densities of both planar stack and tubular membrane modules are still too low and lead to a high fabrication costs.

Hollow fibre membranes are, in this context, a good alternative. So far, ceramic hollow fibre membranes are not commercially available, mainly due to the lack of techniques available to produce the ceramic membrane with a thin and dense separation layer in a hollow fibre form. Hollow fibre membranes are frequently produced from polymeric materials and can provide the largest membrane area per unit packing volume; as high as 8000–10 000 $m^2 m^{-3}$ packing density has been achieved. However, this configuration is seldom used in ceramic membranes because of the poor mechanical strength. An early effort was made for gas separation by the development of glass hollow fibre membranes [118], but the membranes were too fragile to be used in industrial processes. Many other attempts have been made to develop fabrication techniques for making hollow fibre ceramic membranes [38, 119–125]. Brinkman *et al.* [120] developed aluminium oxide and silica nitride hollow fibre membranes using an extrusion technique, but these membranes are porous and symmetric. One step multilayer inorganic hollow fibres were also explored using a triple orifice spinneret [126]. Xu *et al.* [125] reported a composite hollow fibre membrane by synthesis of NaA zeolite on a porous ceramic hollow fibre. Unfortunately, the fabrication of such a thin membrane presents considerable difficulties, such as mismatch of the ceramic material with the porous substrate. Luyten *et al.* [38] synthesised dense hollow fibre $LaSrCoFeO_{3-x}$ membranes, but no performance was reported. Several ceramic hollow fibre membranes prepared from different materials were developed [110, 121–124, 127] using a combined phase inversion and sintering method where a typical asymmetric structure is obtained because of the phase inversion. The fabrication process by phase inversion is fast and simple. The characteristics of the hollow fibres produced via phase inversion are that the membrane structure is asymmetric and possesses a very thin top layer integrated on to a porous support of the same material. The phase inversion technique was applied using a polymer solution containing suspended ceramic powders. A dry jet wet spinning process was carried out to spin hollow fibre precursors, which were then sintered at high temperature to obtain performance and strength. Effects of sintering temperature and ratio of polymer to ceramic powder on the gas tight property and strength were investigated.

Secondly, the membrane materials must possess good thermal and chemical stability whilst also maintaining high ionic and electronic conductivities. Also, for use in chemical reactions, a high catalytic activity is also desirable. Therefore, development of membrane materials is always an important challenge for the future.

Finally, design and engineering problems, such as connecting and sealing of the membrane systems at high temperatures, have to be overcome before extensive commercial applications can be realized.

NOTATION

a	Activity of charged defect
A_t	Membrane area, $A_t = \dfrac{2\pi(R_o - R_{in})L}{\ln(R_o/R_{in})}$, m^2
C	Defect concentration, $mol\,m^{-3}$ or $mol\,cm^{-3}$
D	Effective diffusivity of defect, $m^2\,s^{-1}$ or $cm^2\,s^{-1}$
D_V	Effective diffusivity of oxygen vacancy, $m^2\,s^{-1}$
F	Faraday constant
F	Feed gas flow rate, $mol\,s^{-1}$
$h^{\bullet}$	Electron hole
J	Permeation flux of defect, $mol\,m^{-2}\,s^{-1}$
k_r	Reverse surface reaction rate constant of (6.12), $mol\,m^{-2}\,s^{-1}$
k_f	Forward surface reaction rate constant of (6.12), $m\,Pa^{-0.5}\,s^{-1}$
l	Variable length of hollow fibre membrane, m
L	Thickness of membrane, m
L	Length of hollow fibre membrane, m

n	Number of hollow fibre membranes
N	Molar flow rate, mol s^{-1}
N_t	Overall oxygen permeation molar flow rate, mol s^{-1}
O_O^x	Lattice oxygen
p	Operating pressure in the shell side or gas partial pressure, Pa
p_v	Operating vacuum level, Pa
r	Ionic radius
R	Ideal gas constant
R_{in}	Inner radius of tubular membrane, cm
R_m	Log-mean radius of tubular membrane, $R_m = (R_o - R_{in})/\ln(R_o/R_{in})$
R_o	Outer radius of hollow fibre membrane, m
T	Operating temperature, K
T_s	Setting value of the furnace temperature, °C
t	Transference number of defect
$V_O^{\bullet\bullet}$	Oxygen vacancy
x	Location coordinate, m
x_e	Oxygen mole fraction in air effluent
y_e	Oxygen mole fraction in argon effluent
z	Charge number of defect

Greek Letters

α	Oxygen productivity, $a = N_{O_2}/0.21N_f$
α_{max}	Maximum oxygen productivity
σ	Conductivity
μ	Chemical potential
μ	Oxygen viscosity, Pa s
μ_i^0	Standard chemical potential
ϕ	Electric potential
υ	Local velocity of inert defect

Superscripts

$'$	Upstream (shell)
$''$	Downstream (lumen)

Subscripts

A	A site cation
B	B site cation
e	Electron
f	Feed
h	Electron hole
i, j	Hole, proton or vacancy
k	Carrier

ion	Ion
V	Vacancy
out	Outlet
O_2	Oxygen
total	Total
R	Retentate

REFERENCES

1. Dyer, P. N., Richards, R. E., Russek, S. L. and Taylor, D. M., *Ion transport membrane technology for oxygen separation and syngas production.* Solid State Ionics, **134**: 21–33 (2000).
2. Knauth, P. and Tuller, H. L., *Solid state ionics: roots, status, and future prospects.* Journal of the American Ceramic Society, **85** (7): 1654–1680 (2002).
3. Teraoka, Y., Zhang, H. M., Furukawa, S. and Yamazoe, N., *Oxygen permeation through perovskite-type oxides.* Chemistry Letters, (11): 1743–1746 (1985).
4. Tuller, H. L., Schoonman, J. and Riess, I., *Oxygen Ion and Mixed Conductors and Their Technological Applications.* (2000) Dordrecht, the Netherlands: Kluwer Academic.
5. Bouwmeester, H. J. M. and Burggraaf, A. J., *Dense ceramic membranes for oxygen separation,* in *Fundamentals of Inorganic Membrane Science and Technology*, Burggraaf, A. J. and Cot, L., Editors. (1996) Amsterdam: Elsevier Science B. V., p. 435–528.
6. Hayashi, H., Inaba, H., Matsuyama, M., Lan, N. G., Dokiya, M. and Tagawa, H., *Structural consideration on the ionic conductivity of perovskite-type oxides.* Solid State Ionics, **122**: 1–15 (1999).
7. Yamamura, Y., Ihara, C., Kawasaki, S., Sakai, H., Suzuki, K., Takami, S., Kubo, M. and Miyamoto, A., *Material design of perovskite-based oxygen ion conductor by molecular dynamics method.* Solid State Ionics, **160**: 93–101 (2003).
8. Gellings, P. J. and Bouwmeester, H. J. M., *Ion and mixed-conducting oxides as catalysts.* Catalysis Today, **12**: 1–101 (1992).
9. Thursfield, A. and Metcalfe, I. S., *The use of dense mixed ionic and electronic conducting membranes for chemical production.* Journal of Materials Chemistry, **14**: 2475–2485 (2004).
10. Kroger F. A. and Vink H. J., *Solid state physics advances in Research and Applications*, Seitz, F. and Turnbull, T., Editors. Vol 3 (1957), New York: Academic, p. 307.
11. Galasso, F. S., *Structure, Properties and Preparation of Perovskite-Type Compounds.* (1969) London: Pergamon Press.
12. Etsell, T. H. and Flengas, S. N., *The electrical properties of solid oxide electrolytes.* Chemical Reviews, **70**: 339–376 (1970).
13. Gur, T. M., Belzner, A. and Huggins, R. A., *A new class of oxygen selective chemically driven nonporous ceramic membranes. Part I. A-site doped perovskites.* Journal of Membrane Science, **75**: 151–162 (1992).
14. Nowick, A. S. and Du, Y., *High-temperature protonic conductors with perovskite-related structures.* Solid State Ionics, **77**: 137–146 (1995).
15. Nowick, A. S. and Vaysleyb, A. V., *Isotope effect and proton hopping in high-temperature protonic conductors.* Solid State Ionics, **97**: 17–26 (1997).
16. Ma, B., Balachandran, U. and Park, J.-H., *Electrical transport properties and defect structure of* $SrFeCo_{0.5}O_x$. Journal of the Electrochemical Society, **143** (5): 1736–1744 (1996).
17. McCarty, J. G. and Wise, H., *Perovskite catalysts for methane combustion.* Catalysis Today, **8** (2): 231–248 (1990).
18. Lin, Y. S. and Zheng, Y., *Catalytic properties of oxygen semipermeable perovskite-type ceramic membrane materials for oxidative coupling of methane.* Journal of Catalysis, **164**: 220–231 (1996).

19. Lu, Y., Dixon, A. G., Moser, W. R., Ma, Y. and Balachandran, U., *Oxygen-permeable dense membrane reactor for the oxidative coupling of methane.* Journal of Membrane Science, **170**: 27–34 (2000).
20. Zeng, Y., Lin, Y. S. and Swartz, S. L., *Perovskite-type ceramic membrane: synthesis, oxygen permeation and membrane reactor performance for oxidative coupling of methane.* Journal of Membrane Science, **150**: 87–98 (1998).
21. Balachandran, U., Dusek, J. T., Maiya, P. S., Ma, B., Mieville, R. L., Kleefisch, M. S. and Udovich, C. A., *Ceramic membrane reactor for converting methane to syngas.* Catalysis Today, **36**: 265–272 (1997).
22. Jin, W., Gu, X., Li, S., Huang, P., Xu, N. and Shi, J., *Experimental and simulation study on a catalyst packed tubular dense membrane reactor for partial oxidation of methane to syngas.* Chemical Engineering Science, **55**: 2617–2625 (2000).
23. Jin, W., Li, S., Huang, P., Xu, N., Shi, J. and Lin, Y. S., *Tubular lanthanum cobaltite perovskite-type membrane reactors for partial oxidation of methane to syngas.* Journal of Membrane Science, **166**: 13–22 (2000).
24. Maiya, P. S., Anderson, T. J., Mieville, R. L., Dusek, J. T., Picciolo, J. J. and Balanchandran, U., *Maximizing H_2 production by combined partial oxidation of CH_4 and water gas shift reaction.* Applied Catalysis A: General, **196**: 65–72 (2000).
25. Shao, Z., Xiong, G., Dong, H., Yang, W. and Lin, Y. S., *Synthesis, oxygen permeation study and membrane performance of a $Ba_{0.5}Sr_{0.5}Co_{0.8}Fe_{0.2}O_{3-\delta}$ oxygen-permeable dense ceramic reactor for partial oxidation of methane to syngas.* Separation and Purification Technology, **25**: 97–116 (2001).
26. Shao, Z., Dong, H., Xiong, G., Cong, Y. and Yang, W., *Performance of a mixed-conducting ceramic membrane reactor with high hydrogen permeability for methane conversion.* Journal of Membrane Science, **183**: 181–192 (2001).
27. Teraoka, Y., Zhang, H. M., Okamoto, K. and Yamazoe, N., *Mixed ionic-electronic conductivity of $La_{1-x}Sr_xCo_{1-y}FeyO_{3-\delta}$ perovskite-type oxides.* Materials Research Bulletin, **23** (1): 51–58 (1988).
28. Lin, Y. S., *Microporous and dense inorganic membranes: current status and prospective.* Separation and Purification Technology, **25**: 39–55 (2001).
29. Steele, B., *Ceramic ion conucting membranes and their technological applications.* Comptes rendus de L'Academic des Science, Paris, **1** (2): 533–543 (1998).
30. Kharton, V. V., Vispup, A. P., Naumovich, E. N. and Lapchuk, N. M., *Mixed electronic and ionic conductivity of $LaCo(M)O_3$ (M = Ga, Cr, Fe or Ni): I. Oxygen transport in perovskite $LaCoO_3$-$LaGaO_3$.* Solid State Ionics, **104**: 67–78 (1997).
31. Kharton, V. V., Shuangbao, L., Kovalevsky, A. V. and Naumovich, E. N., *Oxygen permeability of perovskites in the system $SrCoO_3$-$SrTiO_3$.* Solid State Ionics, **96**: 141–151 (1997).
32. Kharton, V. V., Kovalevsky, A. V., Tikhonovich, V. N., Naumovich, E. N. and Viskup, A. P., *Mixed electronic and ionic conductivity of $LaCo(M)O_3$ (M = Ga, Cr, Fe or Ni) II. Oxygen permeation through Cr- and Ni-substituted $LaCoO_3$.* Solid State Ionics, **110**: 53–60 (1998).
33. Kharton, V. V., Viskup, A. P., Naumovich, E. N. and Tikhonovich, V. N., *Oxygen permeability of $LaFe_{1-x}Ni_xO_{3-\delta}$ solid solutions.* Materials Research Bulletin, **34** (8): 1311–1317 (1999).
34. Kharton, V. V., Viskup, A. P., Yaremchenko, A. A., Kerko, P. F., Naumovich, E. N. and Kovalevsky, A. V., *Ionic transport in $SrCo_{0.85}Ti_{0.15}O_{3-\delta}$ ceramics at high oxygen pressures.* Materials Research Bulletin, **34** (12/13): 1921–1928 (1999).
35. Kharton, V. V., Naumovich, E. N., Kovalevsky, A. V., Viskup, A. P., Figueiredo, F. M., Bashmakov, I. A. and Marques, F. M. B., *Mixed electronic and ionic conductivity of $LaCo(M)O_3$ (M = Ga, Cr, Fe or Ni) IV. Effect of preparation method on oxygen transport in $LaCoO_{3-x}$.* Solid State Ionics, **138**: 135–148 (2000).
36. Xu, Sherman J. and Thomson, William J., *Oxygen permeation rates through ion-conducting perovskite membranes.* Chemical Engineering Science, **54** (17): 3839–3850 (1999).
37. Li, S., Jin, W., Xu, N. and Shi, J., *Synthesis and oxygen permeation properties of $La_{0.2}Sr_{0.8}Co_{0.2}Fe_{0.8}O_{3-\delta}$ membranes.* Solid State Ionics, **124**: 161–170 (1999).

38. Luyten, J., Buekenhoudt, A., Adriansens, W., Cooymans, J., Weyten, H., Servaes, F. and Leysen, R., *Preparation of LaSrCoFeO*$_{3-x}$ *membranes.* Solid State Ionics, **135**: 637–642 (2000).
39. Akin, F. T. and Lin, Y. S. J., *Oxygen permeation through oxygen ionic or mixed-conducting ceramic membranes with chemical reactions.* Journal of Membrane Science, **231**: 133–146 (2004).
40. Ishihara, T., Kilner, J. A., Honda, M., Sakai, N., Yokokawa, H. and Takita, Y., *Oxygen surface exchange and diffusion in* $LaGaO_3$ *based perovskite type oxides.* Solid State Ionics, **113–115**: 593–600 (1998).
41. Shao, Z., Xiong, G., Cong, Y. and Yang, W., *Synthesis and oxygen permeation study of novel perovskite-type* $BaBixCo_{0.2}Fe_{0.8-x}O_{3-\delta}$ *ceramic membranes.* Journal of Membrane Science, **164**: 167–176 (2000).
42. Shao, Z., Yang, W., Cong, Y., Dong, H., Tong, J. and Xiong, G., *Investigation of the permeation behavior and stability of a* $Ba_{0.5}Sr_{0.5}Co_{0.8}Fe_{0.2}O_{3-\delta}$ *oxygen membrane.* Journal of Membrane Science, **172**: 177–188 (2000).
43. Shao, Z., Xiong, G., Tong, J., Dong, H. and Yang, W., *Ba Effect in doping* $Sr(Co_{0.8}Fe_{0.2})O_{3-\delta}$ *on the phase structure and oxygen permeation properties of the dense ceramic membranes.* Separation and Purification Technology, **25**: 419–429 (2001).
44. Tong, J., Yang, W., Zai, R. C. B. and Lin, L., *Titanium-based perovskite-type mixed conducting ceramic membranes for oxygen permeation.* Materials Letters, **56**: 958–962 (2002).
45. Tong, J., Yang, W., Zhu, B. and Cai, R., *Investigation of ideal zirconium-doped perovskite-type ceramic membrane materials for oxygen separation.* Journal of Membrane Science, **203**: 175–189 (2002).
46. Tong, J., Yang, W., Cai, R., Zhu, B., Xiong, G. and Lin, L., *Investigation on the structure stability and oxygen permeability of Titanium-doped perovskite-type oxides of* $Ba_{0.8}Ti_{0.2}Co_xFe_{1-x}O_{3-\delta}$ $(x = 0.2–0.6)$. Separation and Purification Technology, **32**: 289–299 (2003).
47. Ma, B. and Balachandran, U., *Phase stability of* $SrFeCo_{0.5}O_x$ *in reducing environments.* Materials Research Bulletin, **33** (2): 223–236 (1998).
48. Ma, B., Hodges, J. P., Jorgensen, J. D., Miller, D. J., Richardson Jr., J. W. and Balachandran, U., *Structure and property relationships in mixed-conducting* $Sr_4(Fe_{1-x}Co_x)_6O_{13\pm\delta}$ *materials.* Journal of Solid State Chemistry, **141**: 576–586 (1998).
49. Mitchell, B. J., Rogan, R. C., Richardson Jr., J. W., Ma, B. and Balachandran, U., *Stability of the cubic perovskite* $SrFe_{0.8}Co_{0.2}O_{3-\delta}$. Solid State Ionics, **146**: 311–321 (2002).
50. Wachsman, E. D. and Clites, T. L., *Stable mixed-conducting bilayer membranes for direct conversion of methane to syngas.* Journal of the Electrochemical Society, **149**: A242–A246 (2002).
51. Kovalevsky, A. V., Kharton, V. V., Tikhonovich, V. N., Naumovich, E. N., Tonoyan, A. A., Reut, O. P. and Boginsky, L. S., *Oxygen permeation through* $Sr(Ln)CoO_{3-\delta}$ *(Ln = La, Nd, Sm, Gd) ceramic membranes.* Materials Science and Engineering, **B52**: 105–116 (1998).
52. Li, S., Cong, Y., Fang, L., Yang, W., Lin, L., Meng, J. and Ren, Y., *Oxygen permeation properties of the mixed conducting membranes without cobalt.* Materials Research Bulletin, **33** (2): 183–188 (1998).
53. Lee, T. H., Yang, Y. L., Jacobson, A. J., Abeles, B. and Zhou, M., *Oxygen permeation in dense* $SrCo_{0.8}Fe_{0.2}O_{3-\delta}$ *membranes: surface exchange kinetics versus bulk diffusion.* Solid State Ionics, **100**: 77–85 (1997).
54. Lee, T. H., Yang, Y. L., Jacobson, A. J., Abeles, B. and Milner, S., *Oxygen permeation in dense* $SrCo_{0.8}Fe_{0.2}O_{3-\delta}$ *membranes with porous electrodes.* Solid State Ionics, **100**: 87–94 (1997).
55. Liu, W., Zhang, G. G., Xie, S., Chen, C. S., Meng, G. Y. and Peng, D. K., *Electrical conduction and oxygen transport in* $SrFeCo_{0.5}O_x$ *oxide membranes.* Solid State Ionics, **135** (1–4): 727–730 (2000).
56. Miura, N., Okamoto, Y., Tamaki, J., Morinaga, K. and Yamazoe, N., *Oxygen semipermeability of mixed conductive oxide thick-film prepared by slip casting.* Solid State Ionics, **79**: 195–200 (1995).
57. Kharton, V. V., Kovalevsky, A. V., Viskup, A. P., Figueiredo, F. M., Frade, J. R., Yaremchenko, A. A. and Naumovich, E. N., *Faradaic efficiency and oxygen permeability of* $Sr_{0.97}Ti_{0.6}Fe_{0.4}O_{3-\delta}$ *perovskite.* Solid State Ionics, **128**: 117–130 (2000).

58. Kharton, V. V., Viskup, A. P., Kovalevsky, A. V., Jurado, J. R., Naumovich, E. N., Vecher, A. A. and Frade, J. R., *Oxygen ionic conductivity of Ti-containing strontium ferrite.* Solid State Ionics, **133**: 75–65 (2000).
59. Kharton, V. V., Naumovich, E. N. and Nikolaev, A. V., *Materials of high-temperature electrochemical oxygen membranes.* Journal of Membrane Science, **111**: 149–157 (1996).
60. Jurado, J. R., Figueiredo, F. M. and Frade, J. R., *Overpotential terms on the electrochemical permeability of $Sr_{0.97}(Ti,Fe)O_{3-y}$ materials.* Solid State Ionics, **122**: 197–204 (1999).
61. Kharton, V. V., Viskup, A. P., Kovalevsky, A. V., Jurado, J. R., Naumovich, E. N., Vecher, A. A. and Frade, J. R., *Oxygen ionic conductivity of Ti-containing strontium ferrite.* Solid State Ionics, **133**: 57–65 (2000).
62. Wang, H., Cong, Y. and Yang, W., *Investigation on the partial oxidation of methane to syngas in a tubular $Ba_{0.5}Sr_{0.5}Co_{0.8}Fe_{0.2}O_{3-\delta}$ membrane reactor.* Catalysis Today, **82**: 157–166 (2003).
63. Kharton, V. V., Naumovich, E. N., Vecher, A. A. and Nikolaev, A. V., *Oxide ion conduction in solid solutions $Ln_{1-x}Sr_xCoO_{3-\delta}$ (Ln = La, Pr, Nd).* Journal of Solid State Chemistry, **120–136**: 128–136 (1995).
64. Chen, C. H., Bouwmeester, H. J. M., van Doorn, R. H. E., Kruidhof, H. and Burggraaf, A. J., *Oxygen permeation of $La_{0.3}Sr_{0.7}CoO_{3-a}$.* Solid State Ionics, **98**: 7–13 (1997).
65. Stevenson, J. W., Armstrong, T. R., Carneim, R. D., Pederson, L. R. and Weber, W. L., *Electrochemical properties of mixed conducting perovskites $La_{1-x}M_xCo_{1-y}Fe_yO_{3-\delta}$ (M = Sr, Ba, Ca).* Journal of the Electrochemical Society, **143** (9): 2722–2729 (1996).
66. Aasland, S., Tangen, I. L., Wiik, K. and Odegard, R., *Oxygen permeation of $SrFe_{0.67}Co_{0.33}O_{3-\delta}$.* Solid State Ionics, **135** (1–4): 713–717 (2000).
67. Kharton, V. V., Viskup, A. P., Bochkov, D. M., Naumovich, E. N. and Reut, O. P., *Mixed electronic and ionic conductivity of $LaCo(M)O_3$ (M = Ga, Cr, Fe or Ni) III. Diffusion of oxygen through $LaCo_{1-x-y}Fe_xNi_yO_{3-\delta}$ ceramics.* Solid State Ionics, **110**: 61–68 (1998).
68. Kharton, V. V., Viskup, A. P., Yaremchenko, A. A., Baker, R. T., Gharbage, B., Mather, G. C., Figueiredo, F. M., Naumovich, E. N. and Marques, F. M. B., *Ionic conductivity of La(Sr)Ga(Mg,M) $O_{3-\delta}$ (M = Ti, Cr, Fe, Co, Ni): effects of transition metal dopants.* Solid State Ionics, **132**: 119–130 (2000).
69. Kharton, V. V., Viskup, A. P., Naumovich, E. N., Tonoyan, A. A. and Reut, O. P., *Oxygen ionic transport in A-site-deficient perovskite $La(Pb)FeO_3$.* Materials Research Bulletin, **33** (7): 1087–1093 (1998).
70. ten Elshof, J. E., Bouwmeester, H. J. M. and Verweij, H., *Oxidative coupling of methane in a mixed-conducting perovskite membrane reactor.* Applied Catalysis A: General, **130**: 195–212 (1995).
71. Tai, L. W., Nasrallah, M. M. anderson, H. U., Sparlin, D. M. and Sehlin, S. R., *Structure and electrical properties of $La_{1-x}SrxCo_{1-y}Fe_yO_{3-d}$ Part I. The system $La_{0.8}Sr_{0.2}Co_{1-y}FeyO_{3-d}$.* Solid State Ionics, **76**: 259–271 (1995).
72. Bouwmeester, H. J. M., Kruidhof, H. and Burggraaf, A. J., *Importance of the surface exchange kinetics as rate limiting step in oxygen permeation through mixed-conducting oxides.* Solid State Ionics, **72**: 185–194 (1994).
73. Armstrong, T., Prado, F. and Manthiram, A., *Synthesis, crystal chemistry, and oxygen permeation properties of $LaSr_3Fe_{3-x}Co_xO_{10}$ (0×1.5).* Solid State Ionics, **140**: 89–96 (2001).
74. Xu, S. J. and Thomson, W. J., *Stability of $La_{0.6}Sr_{0.4}Co_{0.2}Fe_{0.8}O_{3-\delta}$ perovskite membrane in reducing and nonreducing environments.* Industrial and Engineering Chemistry Research, **37**: 1290–1299 (1998).
75. Li, S., Jin, W., Huang, P., Xu, N., Shi, J. and Lin, Y. S., *Tubular lanthanum cobalitite perovskite type membrane for oxygen permeation.* Journal of Membrane Science, **166**: 51–61 (2000).
76. Kim, S., Yang, Y. L., Jacobson, A. J., Christoffersen, R. and Abeles, B., *Oxygen permeation, electrical conductivity and stability of perovskite oxide $La_{0.2}Sr_{0.8}Cu_{0.4}Co_{0.6}O_{3-x}$.* Solid State Ionics, **104**: 57–65 (1997).

77. Ishihara, T., Yamada, T., Arikawa, H., Nishiguchi, H. and Takita, Y., *Mixed electronic-oxide ionic conductivity and oxygen permeating property of Fe-, Co- or Ni-doped $LaGaO_3$ perovskite oxide.* Solid State Ionics, **135**: 631–636 (2000).
78. Kharton, V. V., Yaremchenko, A. A., Kovalevsky, A. V., Viskup, A. P., Naumovich, E. N. and Kerko, P. F., *Perovskite-type oxides for high-temperature oxygen separation membranes.* Journal of Membrane Science, **163**: 307–317 (1999).
79. Ishihara, T., Matsuda, H. and Takita, Y., *Doped $LaGaO_3$ perovskite type oxide as a new oxide ionic conductor.* Journal of the American Chemical Society, **116**: 3801–3803 (1994).
80. Huang, K., Tichy, R. S. and Goodenough, J. B., *Superior perovskite oxide-ion conductor; strontium- and magnesium-doped $LaGaO_3$: I, Phase relationships and electrical properties.* Journal of the American Ceramic Society, **81** (10): 2565–2575 (1998).
81. Ullmann, H. and Trofimenko, N., *Composition, structure and transport properties of perovskite-type oxides.* Solid State Ionics, **119**: 1–8 (1999).
82. Ullmann, H., Trofimenko, N., Naoumidis, A. and Stover, D., *Ionic/electronic mixed conduction relations in perovskite-type oxides by defect structure.* Journal of the European Ceramic Society, **19**: 791–796 (1999).
83. Balachandran, U., Ma, B., Maiya, P. S., Mieville, R. L., Dusek, J. T., Picciolo, J. J., Guan, J., Dorris, S. E. and Liu, M., *Development of mixed-conducting oxides for gas separation.* Solid State Ionics, **108**: 363–370 (1998).
84. Balachandran, U., Dusek, J. T., Meieville, R. L., Poeppel, R. B., Kleefisch, M. S., Pei, S., Kobylinski, T. P., Udovich, C. A. and Bose, A. C., *Dense ceramic membranes for partiasl oxidation of methane to syngas.* Applied Catalysis A: General, **133**: 19–29 (1995).
85. Armstrong, T., Prado, F., Xia, Y. and Manthiram, A., *Role of perovskite phase on the oxygen permeation properties of the $Sr_4Fe_{6-x}Co_xO_{13+}$ system.* Journal of the Electrochemical Society, **147**: 435–438 (2000).
86. Kim, S., Yang, Y. L., Jacobson, A. J. and Abeles, B., *Diffusion and surface exchange coefficients in mixed ionic electronic conducting oxides from the pressure dependence of oxygen permeation.* Solid State Ionics, **106**: 189–195 (1998).
87. Kim, S., Yang, Y. L., Christoffersen, R. and Jacobson, A. J., *Determination of oxygen permeation kinetics in a ceramic membrane based on the composition $SrFeCo_{0.5}O_{3.25-d}$.* Solid State Ionics, **109**: 187–196 (1998).
88. Ikeguchi, M., Yoshino, Y., Kanie, K., Nomura, M., Kikuchi, E. and Matsukata, M., *Effect of preparation method on oxygen permeation properties of $SrFeCo_{0.5}O_x$ membrane.* Separation and Purification Technology, **32**: 313–318 (2003).
89. Ma, B., Balachandran, U., Jorgensen, J. P., Miller, D. J. and Richardson Jr., J. W., *Phase stability of the mixed-conducting Sr-Fe-Co-O system.* Materials Research Society Symposium Proceedings, **496**: 167–172 (1998).
90. Xia, Y., Armstrong, T., Prado, F. and Manthiram, A., *Sol-gel synthesis, phase relationships, and oxygen permeation properties of $Sr_4Fe_{6-x}Co_xO_{13+d}$.* Solid State Ionics, **130**: 81–90 (2000).
91. Bochkov, D. M., Kharton, V. V., Kovalevsky, A. V., Viskup, A. P. and Naumovich, E. N., *Oxygen permeability of $La_2Cu(Co)O_{4+\delta}$ solid solutions.* Solid State Ionics, **120**: 281–288 (1999).
92. Mizusaki, Y., Mima, S., Yamauchi, K., Fukei, H. and Tagawa, *Nonstoichiometry of the perovskite-type oxides $La_{1-x}Sr_xCoO_{3-\delta}$.* Journal of Solid State Chemistry, **80**: 102–111 (1989).
93. Lin, Y. S., Wang, W. J. and Han, J. H., *Oxygen permeation through thin mixed-conducting solid state mmembranes.* American Institute of Chemical Engineers Journal, **40**: 786–798 (1994).
94. Mazanec, T. J., Cable, T. L. and Frye, Jr. J. G., *Electrocatalytic cells for chemical reaction.* Solid State Ionics, **53–56** (1): 111–118 (1992).
95. Chen, C. S., Boukamp, B. A., Bouwmeester, H. J. M., Cao, G. Z., Kruidhof, H., Winnubst, A. J. A. and Burggraaf, A. J., *Microstructural development, electrical properties and oxygen permeation of zirconia-palladium composites.* Solid State Ionics, **76**: 23–28 (1995).
96. Chen, P.-L. and Chen, I.-W., *Sintering of fine oxide powders: I, microstructural evolution.* Journal of the American Ceramic Society, **79** (12): 3129–3141 (1996).

97. Chen, C. S., Kruidhof, H., Bouwmeestera, H. J. M., Verweij, H. and Burggraaf, A. J., *Thickness dependence of oxygen permeation through stabilized bismuth oxide-silver composites.* Solid State Ionics, **99**: 215–219 (1997).
98. Chen, C. S. and Burggraaf, A. J., *Stabilized bismuth oxide ± noble metal mixed conducting composites as high temperature oxygen separation membranes.* Journal of Applied Electrochemistry, **29**: 355–360 (1999).
99. Chen, C. S., Kruidhof, H., Bouwmeestera, H. J. M., Verweij, H. and Burggraaf, A. J., *Oxygen permeation through oxygen ion oxide–noble metal dualphase composites.* Solid State Ionics, **86–88**: 569–572 (1996).
100. Kim, J. and Lin, Y. S., *Synthesis and oxygen-permeation properties of thin YSZ/Pd composite membranes.* American Institute of Chemical Engineers Journal, **46** (8): 1521–1529 (2000).
101. Han, J., Zeng, Y. and Lin, Y. S., *Oxygen permeation through fluorite-type bismuth-yttrium-copper oxide membranes.* Journal of Membrane Science, **132**: 235–243 (1997).
102. Kim, J. and Lin, Y. S., *Synthesis and oxygen permeation properties of ceramic-metal dual-phase membranes.* Journal of Membrane Science, **167** (1): 123–133 (2000).
103. Kharton, V. V., Kovalevsky, A. V., Viskup, A. P., Shaula, A. P., Figueiredo, F. M., Naumovich, E. N. and Marques, F. M. B., *Oxygen transport in $Ce_{0.8}Gd_{0.2}O_{2-\delta}$-based composite membranes.* Solid State Ionics, **160**: 247–258 (2003).
104. Shaw, C. K., *Mass transport in mixed conducting perovskite related oxides.* (2001) PhD thesis, Imperial College, London.
105. Lane, J. A. and Kilner, J. A., *Oxygen surface exchange on gadolinia doped ceria.* Solid State Ionics, **136–137**: 927–932 (2000).
106. Goto, K. S., *Solid state electrochemistry and its applications to sensors and electronic devices.* (1988) Amsterdam-Oxford-New York-Tokyo: Elsevier.
107. Tan, X., Liu, S., Li, K. and Hughes, R., *Theoretical analysis of ion permeation through mixed conducting membranes and its application to dehydrogenation reactions.* Solid State Ionics, **138** (1–2): 149–159 (2000).
108. Qiu, L., Lee, T. H., Liu, M. L., Yang, Y. L. and Jacobson, A. J., *Oxygen permeation studies of $SrCo_{0.8}Fe_{0.2}O_{3-\delta}$.* Solid State Ionics, **76**: 321–329 (1995).
109. Tan, X. and Li, K., *Modeling of air separation in a LSCF hollow-fibre membrane module.* American Institute of Chemical Engineers Journal, **48**: 1469–1477 (2002).
110. Tan, X., Liu, Y. and Li, K., *Mixed conducting ceramic hollow-fibre membranes for air separation.* American Institute of Chemical Engineers Journal, **51** (7): 1991–2000 (2005).
111. Guizard, C. G. and Julbe, A. C., *Nanophase ceramic ion transport membranes for oxygen separation and gas stream enrichment*, in *Recent Advances in Gas Separation by Microporous Ceramic Membranes*, Kanellopoulos, N. K. Editor. (1999) p. 435–469.
112. Stiegel, G., *Mixed conducting ceramic membranes for gas separation and reaction.* Membrane Technology, **110**: 5–7 (1999).
113. Tan, X. and Li, K., *Oxygen production using dense ceramic hollow fiber membrane modules with different operating modes.* American Institute of Chemical Engineers Journal: in press, (2007).
114. Itoh, H., Asano, H., Fukuroi, K., Nagata, M. and Iwahara, H., *Spin coating of a $Ca(Ti,Fe)O_3$ dense film on a porous substrate for electrochemical permeation of oxygen.* Journal of the American Ceramic Society, **80** (6): 1359–1365 (1997).
115. Kosacki, I. and Anderson, H. U., *The structure and electrical properties of $SrCe_{0.95}Yb_{0.05}O_3$ thin film protonic conductors.* Solid State Ionics, **97**: 429–436 (1997).
116. Jin, W., Li, S., Huang, P., Xu, N. and Shi, J., *Preparation of an asymmetric perovskite-type membrane and its oxygen permeability.* Journal of Membrane Science, **185**: 237–243 (2001).
117. Hamakawa, S., Li, L., Li, A. and Iglesia, E., *Synthesis and hydrogen permeation properties of membranes based on dense $SrCe_{0.95}Yb_{0.05}O_{3-\alpha}$ thin film.* Solid State Ionics, **148**: 71–81 (2002).
118. Way, J. D. and Roberts, D. L., *Hollow fibre inorganic membranes for gas separations.* Separation Science and Technology, **27** (1): 29–41 (1992).

119. Smid, J., Avci, C. G., Gunay, V., Terstra, R. A. and van Eijk, J. P. G. M., *Preparation and characterization of microporous ceramic hollow fibre membranes.* Journal of Membrane Science, **112**: 85–90 (1996).
120. Brinkman, H. W., Van Eijk, J. P. G. M., Meinema, H. A. and Terpstra, R. A., *Processing and characteristics of TNO's hollow fibre ceramic membranes.* International Ceramics, **1**: 45–47 (2000).
121. Tan, X., Liu, S. and Li, K., *Preparation and characterization of inorganic hollow fibre membranes.* Journal of Membrane Science, **188**: 87–95 (2001).
122. Liu, S., Tan, X., Li, K. and Hughes, R., *Preparation and characterisation of* $SrCe_{0.95}Yb_{0.05}O_{2.975}$ *hollow fibre membranes.* Journal of Membrane Science, **193** (2): 249–260 (2001).
123. Liu, S., Li, K. and Hughes, R., *Preparation of porous aluminium oxide* (Al_2O_3) *hollow fibre membranes by a combined phase-inversion and sintering method.* Ceramics International, **29**: 875–881 (2003).
124. Liu, S. and Li, K., *Preparation of* TiO_2/Al_2O_3 *composite hollow fibre membranes.* Journal of Membrane Science, **218**: 269–277 (2003).
125. Xu, X., Yang, W., Lui, J., Lin, L., Stron, N. and Brunner, H., *Synthesis of NaA zeolite membrane on a ceramic hollow fibre.* Journal of Membrane Science, **229**: 81–85 (2004).
126. de Jong, J., Benes, N. E., Koops, G. H. and Wessling, M., *Toward single step production of mutilayer inorganic hollow fibres.* Journal of Membrane Science, **239**: 2656–2669 (2004).
127. Liu, Y. and Li, K., *Preparation of* $SrCe_{0.95}Yb_{0.05}O_{3-\delta}$ *hollow fibre membranes: study on sintering processes.* Journal of Membrane Science, **259** (1–2): 47–54 (2005).
128. Liu, Y., Tan, X. and Li, K., *Mixed conducting ceramics for catalytic membrane processing.* Catalysis Reviews – Science and Engineering, **48** (2): 145–198 (2006).
129. Xie, S., Liu, W., Wu, K., Yang, P. H., Meng, G. Y. and Chen, C. S., *Mixed oxygen ionic and electronic conduction in* $CaFe_{0.2}Ti_{0.8}O_{3-\delta}$ *a combined oxygen permeation and electrical conductivity study.* Solid State Ionics, **118**: 23–28 (1999).
130. van Doorn, R. H. E., Bouwmeester, H. J. M. and Burggraaf, A. J., *Kinetic decomposition of* $La_{0.3}Sr_{0.7}CoO_{3-\delta}$ *perovskite membranes during oxygen permeation.* Solid State Ionics, **111**: 263–272 (1998).
131. Yaremchenko, A. A., Kharton, V. V., Viskup, A. P., Naumovich, E. N., Tikhonovich, V. N. and Lapchuk, N. M., *Mixed electronic and ionic conductivity of* $LaCo(M)O_3$ *(M = Ga, Cr, Fe or Ni). V. Oxygen permeability of Mg-doped* $La(Ga, Co)O_{3-\delta}$ *perovskites.* Solid State Ionics, **120**: 65–74 (1999).
132. Kharton, V. V., Li, S., Kovalevsky, A. V., Viskup, A. P., Naumovich, E. N. and Tonoyan, A. A., *Oxygen permeability and thermal expansion of* $Sr(Co)TiO_{3-\delta}$ *perovskites.* Materials Chemistry and Physics, **53**: 6–12 (1998).

7

Mixed Conducting Ceramic Membranes for Hydrogen Permeation

7.1 INTRODUCTION

Mixed ionic–electronic conducting solid oxides have been widely studied for their potential applications [1, 2]. As mentioned in the previous chapter, one of the applications that has attracted particular attention is mixed oxygen ionic–electronic conducting ceramic membranes for production of oxygen by separation from air or other oxygen containing gas mixtures [3]. Mixed proton–electron (hole) conducting materials are also available, such as $SrCe_{0.95}Yb_{0.05}O_{3-\delta}$ (SCYb) and $BaCe_{0.8}Y_{0.2}\,O_{3-\delta}$ (BCY). However, they have attracted much less attention as compared to the mixed oxygen ionic–electronic materials [4]. In principle, these mixed conducting materials also have potential applications in hydrogen separation and particularly in dehydrogenation reactions as proposed by Hamakawa *et al.* [5] and White *et al.* [6]. One of the advantages of using such a mixed conducting membrane is its structural simplicity, as no electrodes and external electric loadings are required as compared to pure proton conducting membranes.

Mixed proton–electronic (or hole) conducting behaviour has been observed for some perovskite oxide materials (ABO_3) and particularly $SrCe_{095}Yb_{0.05}O_{3-\delta}$ (SCYb) and $BaCe_{0.8}Y_{0.2}O_{3-\delta}$ (BCY). The high temperature proton conducting properties of several doped perovskite oxides (ABO_3) are a well known and intensively investigated phenomenon [7–11]. SCYb is one example of this group of materials. In a hydrogen free environment, SCYb shows p-type electric conduction and the oxygen vacancy is the predominant charged carrier, whereas it possesses pure protonic conduction in a hydrogen containing environment. However, it becomes a mixed proton–hole conductor when one side of the SCYb ceramic membrane is exposed to a reducing atmosphere, (e.g. hydrogen, methane, etc.) and the other side to an oxidizing atmosphere, such as oxygen. Under such an asymmetric conditions, with a necessary hydrogen partial pressure difference, hydrogen (or more appropriately protons) permeates through the membrane electrochemically without an external electric source [4, 5, 12].

Although investigations of the proton conducting properties of these perovskite oxides have been carried out for many years [13], little effort has been expended on investigations

Ceramic Membranes for Separation and Reaction K. Li

of the mixed proton–hole conducting behaviour apart from some experimental work carried by Hamakawa *et al.* [4, 5] who illustrated the possibility of hydrogen removal and non-oxidative methane coupling reactions using the mixed proton–hole SCYb conducting membranes.

Transport models of mixed conductors have been investigated and formulated initially by Wagner [14] and subsequently by many other researchers [3, 15, 16]. Although their models have provided important insight into the understanding of the proton–hole conducting behaviour of SCYb, they cannot be directly applied due to some restrictive assumptions. For example, most existing models consider only two types of mobile defects, i.e., oxygen vacancies and electrons/holes, so they cannot explain the most fascinating phenomenon in some mixed proton–hole conducting materials (i.e., $BaCe_{095}Y_{0.05}O_{3-\delta}$): a gradual transition from proton electronic conduction to oxygen ionic and electronic conduction with increasing temperature [17]. It is generally recognized that the conductivities of oxides in an asymmetric environment consist of three contributions: oxygen vacancy or oxygen ion, electron hole or electron, and proton [18]. Therefore, for the proton–hole conducting materials, more than two types of defects must be considered. A deeper theoretical analysis about the defect diffusion behaviour of perovskite ($SrCe_{095}Yb_{0.05}O_{3-\delta}$) was provided by Condon and Schober [19]. However, their numerical modelling work was focused on the material properties proton diffusion behaviour under different conditions and no engineering information was presented.

In this chapter, mixed conducting materials and membranes which show mixed proton–electronic (hole) conducting properties will be introduced, followed by a discussion of mechanisms of hydrogen (proton) permeation through the mixed conducting membranes in an asymmetric environment. Finally, bulk transport will be considered to quantify the net flux of hydrogen transferred through the mixed proton–hole conducting membranes and possible applications of these membranes in dehydrogenation reactions will be elaborated.

7.2 PROTON AND ELECTRON (HOLE) CONDUCTING MATERIALS AND MEMBRANES

The high temperature protonic conducting material, $SrCe_{0.95}Yb_{0.05}O_{3-\delta}$, was first described by Iwahara [8]. This material is a substituted solid solution based on the perovskite-type oxide $SrCeO_3$, in which 5 % of the cerium is replaced by ytterbium and which exhibits protonic conduction under a hydrogen containing atmosphere at elevated temperatures [8]. However, the undoped oxide $SrCeO_3$ does not show the proton conducting properties. Oxygen vacancies generated by doping play an important role in the formation of proton conducting properties including the formation and transport of protons. The interest in proton conducting materials is due to the following possible applications:

- As an electrolyte in hydrogen sensors [20]. Commercialized hydrogen analysers have been developed using $CaZr_{0.9}In_{0.1}O_{3-\delta}$ as a solid electrolyte and porous platinum as an electrode material. This represents an important milestone for the practical application of protonic conduction materials [21].
- As an electrolyte in hydrogen fuel cell, as they can work stably at 800–1000 °C [22].
- As a solid electrolyte in a steam electrolyzer [20].
- As a hydrogen pump for hydrogen extraction from various gas mixtures [8, 12].

- As a chemical reactor for hydrogenation or dehydrogenation of organic compounds [4, 23].

One particular area which is becoming of more and more interest to researchers is hydrogen. This includes the possibility of hydrogen production from hydrogen containing gases and continuous hydrogen removal from those thermodynamically unfavourable reactions, for example, methane coupling reactions using dense proton conducting membranes [24, 25]. Although, platinum or palladium alloy membranes have been studied for many years, they suffer from some fatal limitations [26]. Ceramic proton conducting membranes provide an interesting alternative [17]. Details of these applications in the form of an electrolyte have been described in articles and reviews [8, 20, 27–29].

7.2.1 Perovskite-type Oxides

The mixed proton and electron (hole) conducting materials have many aspects in common with mixed oxygen ionic and electronic conducting materials, but unlike oxygen permeable membranes, which work under either oxidizing versus reducing conditions (e.g. air versus methane) or oxidizing conditions on both sides of the membrane, the hydrogen permeable membranes mostly work under reducing conditions on both sides. Some special consideration of the stability of the membranes would be needed in the development strategies of hydrogen permeable materials [30].

Compared with a large number of oxygen conducting materials, the choice of proton conducting material is, so far, limited. Several perovskite type of oxides have been identified as having good capabilities in protonic conduction including doped $BaCeO_3$ [31], $BaZrO_3$ [30, 32, 33], $BPrO_3$ [30, 34], $SrZrO_3$, $CaZrO_3$ [35–37] and $SrTiO_3$ [38], since the first protonic conductor, $SrCe_{0.95}Yb_{0.05}O_{3-\delta}$, was reported. The general formula is written as $AB_{1-x}M_xO_{3-\delta}$, where the A element is taken from the group consisting of calcium, strontium or barium; the B element is taken from the group consisting of cerium, terbium, zirconium or thallium; the M element is taken from the group consisting of titanium, thulium, chromium, manganese, cobalt, nickel, copper, aluminium, gallium, yttrium, ytterbium, indium or tin, and x is less than the upper limit of solid solution formation range (normally less than 0.2) and δ is the oxygen deficiency per unit cell [13, 39, 40]. Their protonic conductivities in a hydrogen atmosphere are of the order of 10^{-3}–10^{-2} S cm^{-1} at 600–1000 °C. $BaCeO_3$ based oxides show the highest conductivity, but oxygen ions contribute to the conduction as the temperature is increased, and result in the transport number of protons decreasing with increase in temperature [41]. Although the conductivities of $SrCeO_3$ based oxides are lower, the transport number of protons is higher than that of $BaCeO_3$ based ones. The zirconate based oxides such as $SrZrO_3$ or $CaZrO_3$ show good chemical stability and good mechanical strength and they are more stable against carbon dioxide gas which reacts with cerate materials below 800 °C [37, 42], but the conductivity of zirconates is lower than those of cerates [43]. The major properties of these materials are summarized in Table 7.1.

In order to increase the electronic conductivity in the above mentioned perovskites, different types of doping have been carried out recently. Matsumoto *et al.* [44, 45] showed that $SrZrO_3$ and $BaCeO_3$ co-doped with ruthenium and yttrium exhibit enhanced mixed proton–electron conductivity. Based on X-ray absorption spectroscopy, the electronic conductivity was found to be p-type under reducing conditions. For a 0.5 mm thick membrane of $BaCe_{0.8}Y_{0.1}Ru_{0.1}O_{3-\alpha}$, the flux of hydrogen was measured to be 0.07 μmol cm^{-2} s^{-1} at 800 °C

Table 7.1 Properties of typical proton conducting materials

Composition	T (°C)	Properties	Ref
$BaCe_{0.95}Y_{0.05}O_{3-\delta}$	500–600	$\sigma_H > \sigma_O$	[17]
	700–800	$\sigma_H < \sigma_O$	
	800	$\sigma_H = 1.93 \times 10^{-3}\,\Omega^{-1}\,cm^{-1}$	
		$\sigma_O = 8.2 \times 10^{-3}\,\Omega^{-1}\,cm^{-1}$	
$BaCe_{0.95}Nd_{0.05}O_{3-\delta}$		$E = 0.54$ (lowest)	[91]
		$\sigma_H = 8.5 \times 10^{-3}\,\Omega^{-1}\,cm^{-1}$	
$SrCe_{0.95}Yb_{0.05}O_{3-\delta}$	600–800	$\sigma_H = 3.47 \times 10^{-3}–1.27 \times 10^{-2}\,\Omega^{-1}\,cm^{-1}$	[17]
		$\sigma_O = 1.2 \times 10^{-4}–2.1 \times 10^{-3}\,\Omega^{-1}\,cm^{-1}$	
		$\sigma_e = 7 \times 10^{-4}–3.2 \times 10^{-3}\,\Omega^{-1}\,cm^{-1}$	
	700	$\sigma_H = 7.2 \times 10^{-3}\,\Omega^{-1}\,cm^{-1}$	
		$\sigma_e = 2.1 \times 10^{-3}\,\Omega^{-1}\,cm^{-1}$	
		$E = 0.62$	
	800	$\sigma_e = 8.5 \times 10^{-3}\,\Omega^{-1}\,cm^{-1}$	[5]
	900	$J_H = 1.7 \times 10^{-8}\,mol\,cm^{-2}\,s^{-1}$	
$SrCe_{0.95}Tm_{0.05}O_{3-\delta}$	800	$\sigma_e = 1.29 \times 10^{-2}\,\Omega^{-1}\,cm^{-1}$	[47]
	900	$J_H = 2.9 \times 10^{-8}\,mol\,cm^{-2}\,s^{-1}$	
$SrZr_{0.9}Sc_{0.1}O_{3-\delta}$		$E = 0.59\,eV$, $\delta_H = 2.8 \times 10^{-2}\,\Omega^{-1}\,cm^{-1}$	[11]
$SrZr_{0.95}Yb_{0.05}O_{3-\delta}$			
$SrYb_{0.05}(Ce_{1-x}Zr_x)_{0.95}O_{3-\delta}$	<500	$E = 0.74–0.79\,eV$ in O_2	[43]
		$E = 0.56–0.62\,eV$ in H_2/H_2O	

with a hydrogen pressure ratio of approximately 400 between the feed and permeate sides of the membrane as shown in Figure 7.1, which is a factor of two higher than that of $SrZrO_3$ materials. Compared with other literature data on similar perovskites, the presence of ruthenium appears to have increased the permeability by one to two orders of magnitude.

Apart from doped $SrZrO_3$ and $BaCeO_3$, doped $SrCeO_3$ has also been extensively studied in terms of the actual measurement of hydrogen fluxes as well as the estimation of hydrogen fluxes based on models and conductivity data and the modeling of defect properties and dopant behaviour [46–50]. Hydrogen fluxes have recently been measured through asymmetric membranes made from $SrCeO_3$-based perovskites [49, 51]. Figures 7.2 and 7.3 illustrate H_2 flux variations with different upstream pressures of hydrogen for different membrane thicknesses obtained by Hamakawa *et al.* and Cheng *et al.*, respectively. As can be seen, the H_2 flux measured by Hamakawa *et al.* [49] for a 2 μm film made from $SrCe_{0.95}Yb_{0.05}O_{3-\alpha}$, operated at 40 kPa of H_2 on the feed side and He on the permeate side at 950 K, is around 70 μmol cm^{-2} min^{-1}, while the H_2 flux measured by Cheng *et al.* for a 0.15 mm film prepared from $SrCe_{0.95}Tm_{0.05}O_{3-\alpha}$, operated at 0.1 atm of H_2 on the feed side and 20 % O_2/He on the permeate side at 900 °C, is around 9×10^{-8} mol s^{-1} cm^{-2}. The effect of membrane thickness on the hydrogen flux was also studied by Hamakawa *et al.* and Cheng *et al.* [49, 51]. As shown in Figure 7.4, the H_2 flux measured by Hamakawa *et al.* is proportional to the inverse thickness of the membrane, suggesting that the surface kinetics do not affect the hydrogen permeation and the H_2 permeation is mainly controlled by bulk diffusion.

Based on the temperature dependence data on the H_2 flux in $SrCe_{0.95}Tm_{0.05}O_{3-\alpha}$ given by Cheng *et al.* (as shown in Figure 7.5) [51], the electronic conductivity is believed to be limiting for H_2 permeation. This has been indirectly confirmed by Liu *et al.* [52], as the experimental data obtained by them showed hydrogen permeation dependence on the oxygen partial

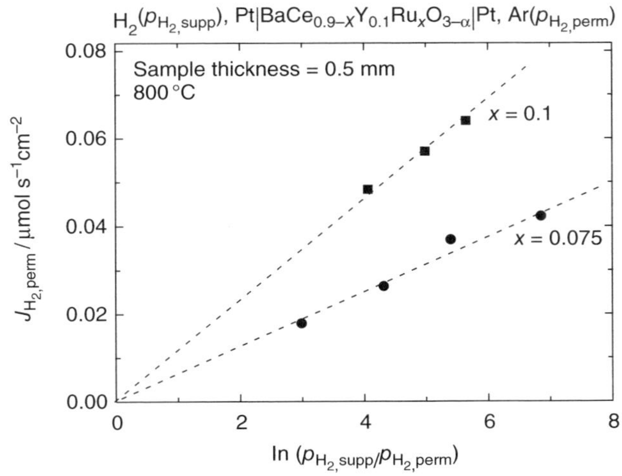

Figure 7.1 Hydrogen permeation fluxes in $BaCe_{0.825}Y_{0.1}Ru_{0.075}O_{3-\alpha}$ and $BaCe_{0.8}Y_{0.1}Ru_{0.1}O_{3-\alpha}$ at 800 °C as a function of the logarithm of the hydrogen pressure ratio (hydrogen supply side to permeated side). The sample thickness is 0.5 mm [44] Reprinted from Journal of the Electrochemical Society, **152** (3), Matsumoto, H. *et al.*, *Protonic-electronic mixed conduction and hydrogen permeation in* $BaCe_{0.9-x}Y_{0.1}Ru_xO_{3-\alpha}$, p. A488–A492. Copyright (2005) The Electrochemical Society

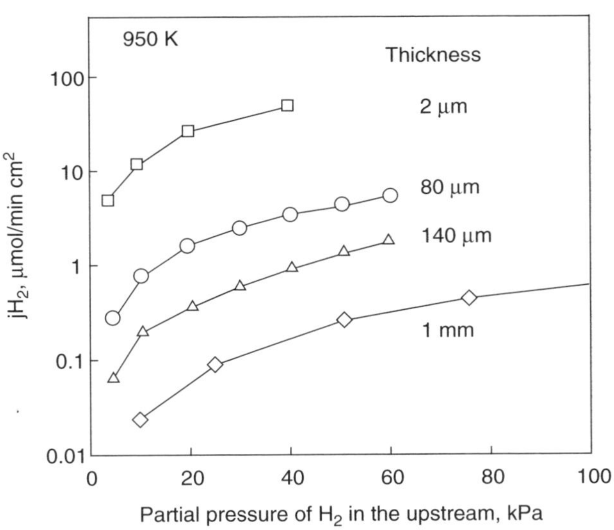

Figure 7.2 Hydrogen fluxes in $SrCe_{0.95}Yb_{0.05}O_{3-\alpha}$ membranes with different thickness as a function of the partial pressure of hydrogen in the feed gas [49]. Reprinted from Solid State Ionics, **148** (1–2), Hamakawa *et al., Synthesis and hydrogen . . .* p. 71–81. Copyright (2002), with permission from Elsevier

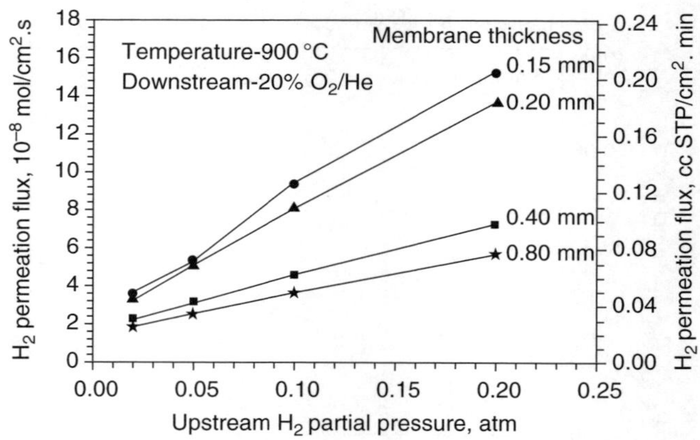

Figure 7.3 Hydrogen fluxes for the membranes of 5 % thulium doped $SrCeO_3$ as a function of partial pressure of hydrogen in the feed gas [51]. Reprinted from Solid State Ionics, **176** (35–36), Cheng *et al.*, *Synthesis and hydrogen* . . . p. 2653–2662. Copyright (2005), with permission from Elsevier

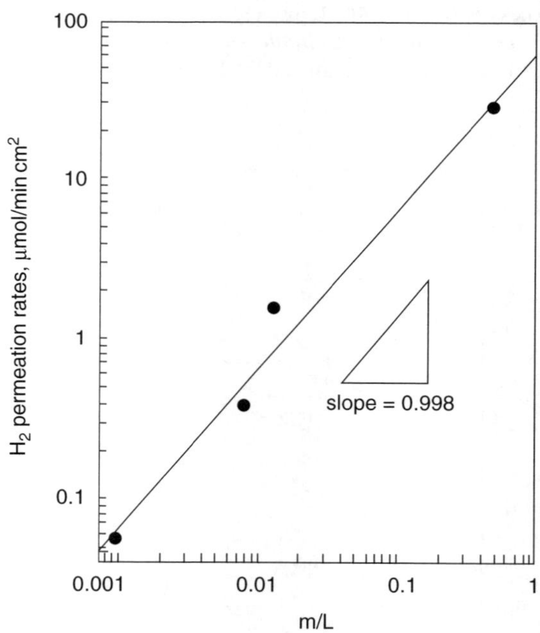

Figure 7.4 Hydrogen fluxes in the $SrCe_{0.95}Yb_{0.05}O_{3-\alpha}$ membranes as a function of the inverse thickness (in micrometers) of the membrane (950 K) [49]. Reprinted from Solid State Ionics, **148** (1–2), Hamakawa *et al., Synthesis and hydrogen* . . . , p. 71–81. Copyright (2002), with permission from Elsevier

pressure in the permeate stream, as shown in Figure 7.6. The H_2 flux increase with downstream oxygen partial pressure further indicates that the $SrCeO_3$ based perovskites behave as proton–hole mixed conductors.

Although $SrCeO_3$ based perovskites show the highest hydrogen fluxes, long term use of this material in industrial applications is still questionable mainly due to its stability in a CO_2

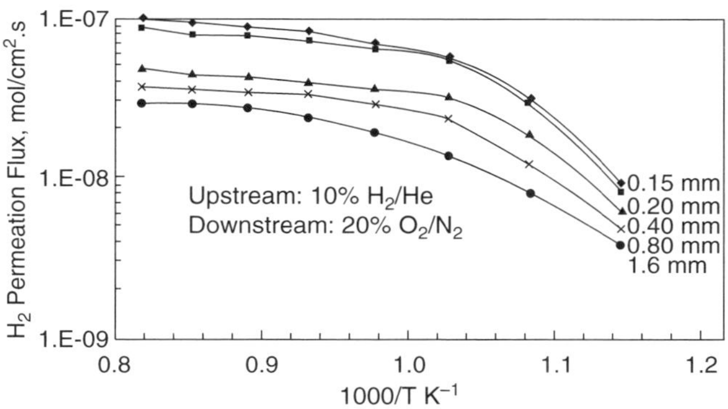

Figure 7.5 H_2 permeation flux versus $1000/T$ of 5 % thulium-doped $SrCeO_3$ membranes of different thicknesses [51]. Reprinted from Solid State Ionics, **176** (35–36), Cheng *et al., Synthesis and hydrogen . . .* p. 2653–2662. Copyright (2005), with permission from Elsevier

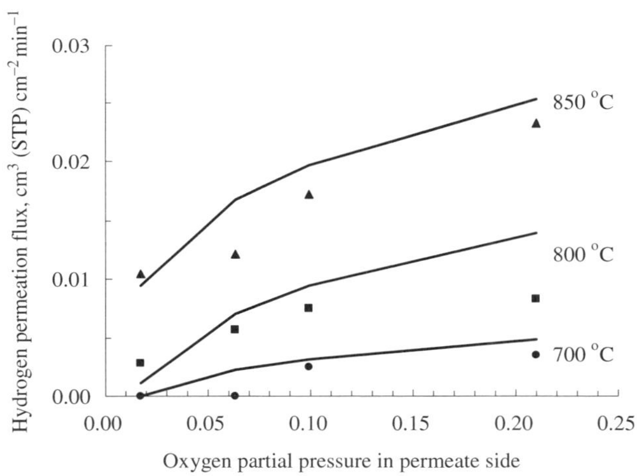

Figure 7.6 Hydrogen fluxes as a function of oxygen partial pressures in the permeate stream for the $SrCe_{0.95}Yb_{0.05}O_{3-\alpha}$ membrane [52]. Copyright (2006) John Wiley & Sons, Ltd, American Institute of Chemical Engineers Journal, **52** (4), Liu Y., Tan, X. and Li, K., *$SrCe_{0.95}Yb_{0.05}O_{3-\alpha}$ hollow-fiber membrane and its property in proton conduction*, p. 1577–1585

environment [53]. It should also be noted that $SrCeO_3$ becomes unstable under highly reduced environments. Therefore, zirconate based oxides such as $SrZrO_3$ or $CaZrO_3$, which show good chemical stability and good mechanical strength, may be an extractive alternative if proton and electron transport of these oxides can be substantially improved.

A new series of protonic materials with more complex perovskite structures have also been reported, where their conductivities are as high as that of $BaCeO_3$ based oxides at high temperature [40]. Their general formulas are $A_2(B'B'')O_6$ and $A_3(B'B''_2)O_9$, where A ions always have charge 2+ and B′ and B″ ions have charges of 3+ and 5+ for $A_2(B'B'')O_6$ and 2+ and 5+ for $A_3(B'B''_2)O_9$. To become protonic conductors, they must be taken off stoichiometry to create negative charges for which protons (or oxygen vacancies) can compensate [11, 54].

Table 7.2 Conductivity and activated energy of some complex perovskite type materials

Composition	E (eV)	σ	Ref.
$A_2(B'B'')O_6$			
$SrScNbO_6$	0.58 ± 0.09		[11]
$Sr_2Ca_{1.1}Nb_{0.9}O_{6-\delta}$	0.81	$1.8 \times 10^{-2}\,\Omega^{-1}\,cm^{-1}$	
$Sr_2Ca_{1.1}Ta_{0.9}O_{6-\delta}$	0.89	$3.4 \times 10^{-2}\,\Omega^{-1}\,cm^{-1}$	
$Sr_2(Nd_{1.1}Nb_{0.9})O_{6-\delta}$	1.04	$3.6 \times 10^{-2}\,\Omega^{-1}\,cm^{-1}$	
$A_3(B'B'')O_9$			
$Sr_3(Ca_{1.18}Nb_{1.82})O_{9-\delta}$	0.66	$5.5 \times 10^{-2}\,\Omega^{-1}\,cm^{-1}$	[40]
$Sr_3(Sr_{1.18}Nb_{1.82})O_{9-\delta}$	0.97	$2.3 \times 10^{-2}\,\Omega^{-1}\,cm^{-1}$	[54]
$Ba_3(Ca_{1.18}Nb_{1.82})O_{9-\delta}$	0.54	$2.1 \times 10^{-2}\,\Omega^{-1}\,cm^{-1}$	

Table 7.3 Properties of $Ce_{0.8}Yb_{0.2}O_{1.9}$ in H_2O/H_2 atmosphere

$Ce_{0.8}Yb_{0.8}O_{1.9}$	T (K)	E (eV)	σ_H (S cm^{-1})	J_H (mol cm^{-2} h^{-1})
	1000–1200	0.58 ± 0.05	1×10^{-6}–2×10^{-6}	1.7×10^{-7}–8×10^{-7}
	1300–1500	1.10 ± 0.06	8×10^{-6}–2×10^{-5}	2×10^{-6}–7.5×10^{-6}
	1500–1800	0.92 ± 0.06	2×10^{-5}–1×10^{-4}	7.5×10^{-6}–1.5×10^{-5}

Their formulas are written $A_2(B'_{1+x}B''_{1-x})O_{6-\alpha}$ and $A_3(B'_{1+x}B''_{2-x})O_{9-\alpha}$. In particular, these systems, such as $A_3(Ca_{1+x}Nb_{2-x})O_{9-\delta}$, are extremely sensitive to cation nonstoichiometry, x, with major effects on structural order, protonic mobility and mechanical properties. The protonic conduction in these complex perovskite related structural oxides have been systematically studied [55, 56]. These structures can accommodate many dopant cations on the A, B′ and B″ sites, which provides more opportunities to improve the ion and/or electron conductivities and other favourable properties. The properties of this new type of proton conducting materials are summarized in Table 7.2.

7.2.2 Non-perovskite-type Oxides

Doped rare earth sesquioxides are one of the classes of non-perovskite structured oxides observed to exhibit proton conductivity [57]. Terbium sesquioxide, Tb_2O_3, is the only compound examined so far that exhibits both electronic (p-type) and protonic conductivity in hydrogen atmospheres at high temperatures [57]. Hydrogen permeation is governed by the proton conductivity and is approximately $5 \times 10^{-5}\,S\,cm^{-1}$ (5 % of the total conductivity) at 900 °C.

Some fluorite type of oxides such as Y_2O_3 and CeO_2 have been found to have protonic conductivity in an H_2/H_2O atmosphere and at high temperatures [58, 59]. Although the hydrogen permeability of those materials is very low, their chemical and thermodynamic stabilities are very good. With a proper dopant, their hydrogen permeability can be greatly improved [60]; for example, $Ce_{0.8}Yb_{0.2}O_{1.9}$ was found to have a hydrogen permeation 100 times higher than that of un-doped CeO_2 [61]. Some characteristics of $Ce_{0.8}Yb_{0.2}O_{1.9}$ are listed in Table 7.3. A few other non-perovskite types of oxides have been known to exhibit protonic conductivity at high temperature, for example, $Ln_2Zr_2O_7$ and $LaPO_4$ systems [62, 63]. However, their conductivity is rather low.

Serra *et al.* [64] recently studied hydrogen permeation through commercial Al_2O_3 tubes and interpreted their data in terms of the flux of neutral hydrogen atoms. A typical flux for a 2.75 mm thick tube and 1.0 atm feed H_2 with the downstream being vacuum, was 4×10^{-8} mol $H_2\,s^{-1}\,m^{-2}$ at 1400 °C.

It was thought that the total conductivity derived from both the protonic and electronic conductivities must usually be at least 0.1 S cm^{-1}, depending on membrane thickness and other factors [65]. The technological challenges in the applications of protonic conducting materials, from the viewpoint of the materials, are their relatively low protonic conductivity possibly caused by the low electronic conductivity and poor chemical and thermodynamic stabilities of some high conductivity materials due to the reducing environment in which the proton conducting membranes work. The mechanical strength of these protonic materials may be another concern.

7.3 DUAL PHASE MEMBRANES

Dual phase membranes for hydrogen permeation consist of a ceramic phase as well as a metallic phase. The former is a pure proton conductor, while the latter is highly electron conducting. Combining these two phases together may provide high hydrogen permeation because both electronic and ionic conduction in the dual phase membrane become high, resulting in high hydrogen permeation.

Although most dual phase membranes studied are for the enhancement of oxygen permeation, attempts made by Argonne National Laboratory [66, 67] to improve hydrogen permeation by preparing a dual phase membrane have also been carried out recently. The $BaCe_{1-x}Y_xO_3$ ceramic membranes mixed with various concentrations of nickel were prepared. It was found that the surface properties of the membranes are significantly affected by the concentration of nickel [67] and the interfacial resistance of Ni and $BaCe_{1-x}Y_xO_3$ is much higher than the bulk resistance. A suggestion for further improvement of the hydrogen permeation by reducing the interfacial resistance to a reasonable level has been proposed [66].

Since metals or oxides alone may also transfer hydrogen, different combinations with respect to functional properties of both phases can be realized, for example: (1) a metal with low hydrogen conductivity in combination with a highly proton conductive oxide; (2) a metal or an alloy with high hydrogen permeability (e.g., Pt, Pd, Pd/Ag, Pd/Cu, Ni, Nb, Ta, Nb, V) together with a ceramic of low hydrogen permeability and (3) a combination where both the metallic and the ceramic phases conduct hydrogen (as protons for the ceramic and hydrogen for the metallic part). Balachandran *et al.* [68] investigated the properties of these three combinations with respect to the selectivity, permeability and mechanical stability of the dual phase membrane. It was suggested that when a pure proton conducting oxide is used as the oxide phase, the content of the metal used in the dual phase must be high enough to ensure sufficient electronic transport in the metallic phase so that the hydrogen flux can be maintained at its maximum. This type of membrane (1) has been tested with BCY as the ceramic matrix. It was observed that the hydrogen flux across the composite increases significantly as compared to a fully ceramic BCY membrane. For membranes where both phases exhibit H_2 permeability (3), the flux increases even more, reaching 20 cm^3(STP) $min^{-1}\,cm^{-2}$ at 900 °C [69].

Recently, a dual phase Pd/YSZ membrane for hydrogen separation from gas mixtures was developed by Balachandran *et al.* [69] who observed that the hydrogen flux through the

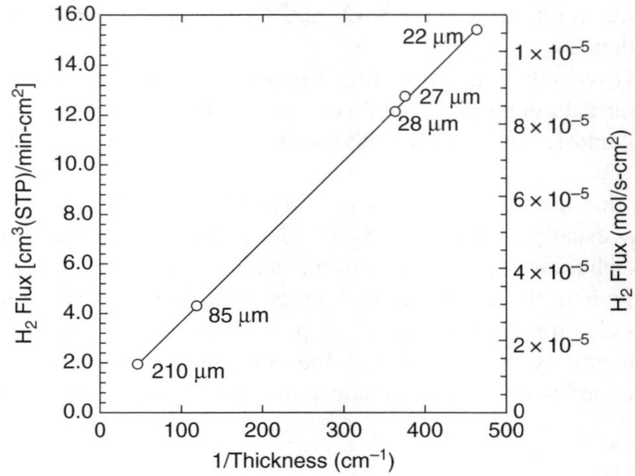

Figure 7.7 Hydrogen flux versus inverse of Pd/YSZ membrane thickness at 900 °C using 80 % H_2/balance He as the feed gas [69]. Reprinted from Fuel, **85** (2), Balachandran *et al.*, *Hydrogen separation by . . .* p. 150–155. Copyright (2006), with permission from Elsevier

membrane increases linearly with $\Delta pH_2^{1/2}$ between 600 and 900 °C and inversely with membrane thickness for thicknesses of ~22–210 μm at 900 °C, as shown in Figure 7.7. Their observations indicate that the flux is limited by the diffusion of hydrogen through the metal phase at that thickness range and surface exchange kinetics is obviously not a limiting step. In their study, the highest hydrogen flux was measured to be ~20.0 cm^3(STP) min^{-1} cm^{-2} for a ~22 μm thick membrane at 900 °C using 100 % H_2 as the feed gas.

In addition to the above promising results, stability of the membrane towards feed gas impurities was also studied. For instance, the hydrogen flux of Pd/YSZ membranes showed no degradation in a simulated synthesis gas containing up to ~400 ppm H_2S and was stable for at least 270 h under such conditions.

7.4 PROTON TRANSPORT

7.4.1 Transport Mechanism

Proton conducting oxides have no host constituent that liberates conducting protons, but take protons from water vapour or hydrogen molecules in ambient gases. The protons bond to oxygen ions, forming substitutional hydroxyl defects, $OH_O^{\bullet}$ [70]; which may be supported by IR studies [71]:

$$H_2O + V_O^{\bullet\bullet} + O_O^{\times} \xleftrightarrow{K_2} 2OH_O^{\bullet} \tag{7.1}$$

$$H_2 + 2O_O^{\times} + 2h^{\bullet} \xleftrightarrow{K_3} 2OH_O^{\bullet} \tag{7.2}$$

For simplicity, the oxygen ion in $OH_O^{\bullet}$ may be considered a part of the perfect lattice and thus denoted as an interstitial proton, $H_i^{\bullet}$.

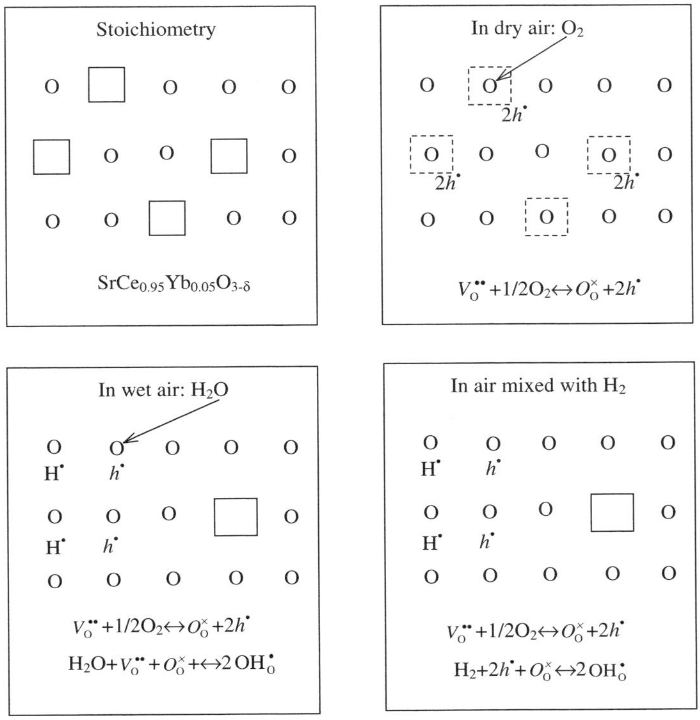

Figure 7.8 Proton formation in mixed conducting oxides [92]. Copyright (2006) Taylor & Francis, Catalysis Reviews – Science and Engineering, **48** (2), Liu, Y., Tan, X. and Li, K., *Mixed conducting ceramics for catalytic membrane processing*, p. 145–198

Figure 7.8 illustrates formation of the protons in $SrCe_{0.95}Yb_{0.05}O_3$ (SCYb). It suggests the oxygen vacancies formed by doping aliovalent cations play an important role in the formation of proton conduction.

Two mechanisms are usually applied to describe the proton transport oxides: (1) the free proton transport mechanism (Grotthuss type) and (2) the vehicle proton transport mechanism [72]. In free proton transport, protons jump between stationary oxygen ions. Each jump is followed by a rotation around the oxygen ion to get in position for the next jump. The jump is normally considered the rate limiting step and the rotation is easier. In the vehicle mechanism, the proton moves as a passenger on a larger species like an oxygen ion. The vehicle mechanism can only account for dominant long range transport processes in oxides when proceeding on interstitial lattice sites. Therefore, hydroxyl ion transport cannot provide a dominant conductivity, but accomplishes minority oxygen transport in a proton conductor. Figure 7.9 shows the migration of protons in mixed conducting oxides, following the free proton transport mechanism by changing the OH bonds one after another [55]. This mechanism is further supported by physical phenomena observed in experiments on quasielastic photo scattering and in a hole burning experiment [73]. The mobility of proton in the oxide is limited by proton transfer between fixed oxygen sites, but is facilitated by the thermal fluctuation of the oxygen ion separation [74]. In fact, the free proton transport is strongly related to the oxygen sublattice vibrations and thus has the activation energies similar to (and generally somewhat lower than) those for oxygen vacancy mobility.

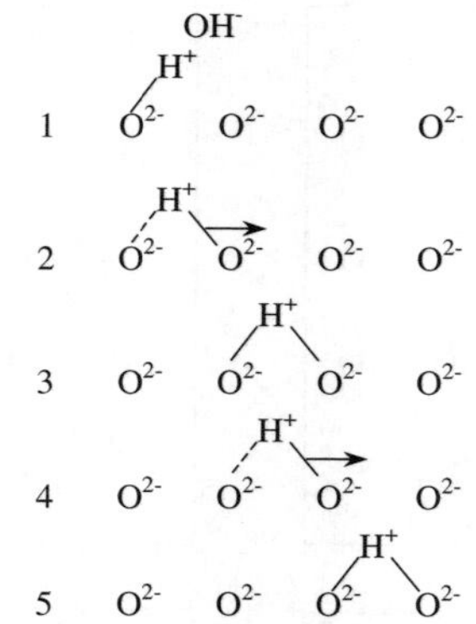

Figure 7.9 Proton migration in the hopping form in $SrCe_{0.95}Yb_{0.05}O_{2.975}$ oxide [92]. Copyright (2006) Taylor & Francis, Catalysis Reviews – Science and Engineering, **48** (2), Liu, Y., Tan, X. and Li, K., *Mixed conducting ceramics for catalytic membrane processing*, p. 145–198

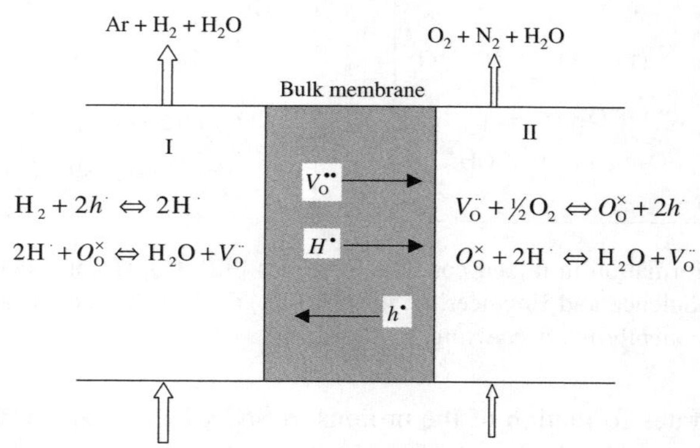

Figure 7.10 Transport diagram of ion permeation in the SCYb membrane [92]. Copyright (2006) Taylor & Francis, Catalysis Reviews – Science and Engineering, **48** (2), Liu, Y., Tan, X. and Li, K., *Mixed conducting ceramics for catalytic membrane processing*, p. 145–200

7.4.2 Transport Equations for Mixed Proton–Hole Conducting Membranes

Transport equations can be derived depending on the nature of the mixed conducting materials. As illustrated in the previous section, considerable hydrogen permeable ceramics show proton–hole conduction, Therefore, in the following section, transport equations are derived for this class of hydrogen permeable membranes only.

When a mixed proton–hole conducting membrane is exposed to gaseous hydrogen on one side and to oxygen on the other side, at high temperatures (i.e. asymmetric gas environment), the charged defects move in certain directions under the electrochemical gradient, as shown in Figure 7.10. As described in Chapter 6, the transport flux of charged defects in mixed conductors at steady state can be described by Equation (6.8), which is derived from Nernst–

Planck and activity equations. Along with the local electric neutrality condition, where SCYb is taken as an example and oxygen vacancy, proton and hole are included:

$$[Yb'_{Ce}] = [H_i^\bullet] + [h^\bullet] + 2[V_O^{\bullet\bullet}] \tag{7.3}$$

The transport equations can be given in matrix form as [75]:

$$\bar{A} \cdot \frac{d\bar{C}}{dx} = \bar{B} \tag{7.4}$$

where,

$$\bar{A} = \begin{bmatrix} 1 - t_V & -2t_p \dfrac{C_V}{C_p} & -2t_h \dfrac{C_V}{C_h} \\ -\dfrac{1}{2} t_V \dfrac{C_p}{C_V} & 1 - t_p & -t_h \dfrac{C_p}{C_h} \\ 2 & 1 & 1 \end{bmatrix} \bar{B} = \begin{bmatrix} -\dfrac{J_V}{D_V} \\ -\dfrac{J_p}{D_p} \\ 0 \end{bmatrix} \bar{C} = \begin{bmatrix} C_V \\ C_p \\ C_h \end{bmatrix} \tag{7.4a}$$

with the boundary conditions:

$$x = 0 \quad \bar{C}_i = C_i(I) \qquad x = L \quad \bar{C}_i = C_i(II) \tag{7.5}$$

where *I* and *II* represent the hydrogen side and oxygen side, respectively. Obviously, the form of the boundary condition equations depends on the surface reaction mechanisms. It has been shown in the previous section that the surface exchange kinetics was not important for hydrogen permeation and the transfer is controlled by bulk diffusion. Therefore, equilibrium models are used at the interface [75], in which all the surface reactions, namely Equations (6.5), (7.1) and (7.2), are assumed to be in equilibrium. As a result, the charged defect concentrations on the membrane surfaces can be given by:

$$C_p(I) = \frac{\sqrt{\left(1 + 1/\sqrt{K_3 p_{H_2}}\right)^2 + 8[Yb'_{Ce}]/K_2 p_{H_2O}} - \left(1 + 1/\sqrt{K_3 p_{H_2}}\right)}{4/K_2 p_{H_2O}} \tag{7.6a}$$

$$C_V(I) = C_p(I)^2 / K_2 p_{H_2O} \tag{7.6b}$$

$$C_h(I) = C_p(I) / \sqrt{K_3 p_{H_2}} \tag{7.6c}$$

and

$$C_V(II) = \frac{1}{16}\left[\sqrt{\left(\sqrt{K_1 p_{O_2}^{1/2}} + \sqrt{K_2 p_{H_2O}}\right)^2 + 8[Yb'_{Ce}]} - \left(\sqrt{K_1 p_{O_2}^{1/2}} + \sqrt{K_2 p_{H_2O}}\right)\right]^2 \tag{7.7a}$$

$$C_h(II) = \sqrt{K_1 p_{O_2}^{1/2} C_V(II)} \tag{7.7b}$$

$$C_p(II) = \sqrt{K_2 p_{H_2O} C_V(II)} \tag{7.7c}$$

where the equilibrium constants are defined as:

$$K_1 = \frac{C_h^2}{C_V p_{O_2}^{1/2}} \qquad K_2 = \frac{C_p^2}{C_V p_{H_2O}} \qquad K_3 = \frac{C_p^2}{C_h^2 p_{H_2}} \tag{7.8}$$

which were studied and measured by Uchida *et al.* [76] and Schober *et al.* [77]. Therefore, the proton flux may be obtained by solving numerically the transport Equation (7.4) with the boundary condition equations.

With the condition that the oxygen ion conductivity is negligible, the transport equation may be reduced to:

$$J_p = -\frac{D_p D_h (C_p + C_h)}{D_p C_p + D_h C_h} \cdot \frac{dC_p}{dx} \tag{7.9}$$

Integration of the above equation gives the proton permeation flux as:

$$J_p = -\frac{D_p D_h [C_p(II) + C_h(II)]}{L(D_p - D_h)} \cdot \ln\left[\frac{D_p C_p(II) + D_h C_h(II)}{D_p C_p(I) + D_h C_h(I)]}\right] \tag{7.9a}$$

For tubular or hollow fibre membranes, the membrane thickness is replaced by the tube wall thickness, i.e., $(R_o - R_{in})$.

7.4.3 Transport Analysis

Theoretical studies on hydrogen permeation (or proton conduction) properties in a mixed conducting membrane were carried out by Tan *et al.* [75]. The proton conduction properties of SCYb hollow fibre membranes under different temperatures and feed concentrations were analyzed in terms of hydrogen permeation flux. The diffusional and equilibrium parameters used for calculation of the theoretical hydrogen permeation in the membrane module were obtained from literature [52, 75].

7.4.3.1 Effect of Membrane Thickness

A theoretical analysis of this effect is given in Figure 7.11, where the permeation fluxes of hydrogen and oxygen are plotted against the membrane thickness. It can be seen that both the fluxes decrease as the membrane thickness is increased. At a membrane thickness of 0.06 cm or greater, the permeation flux of both the hydrogen and oxygen is approximately linearly related to the membrane thickness indicating that the effect of concentration on conductivities and transference numbers of all the charged species is negligible. When the membrane thickness is less than 0.02 cm, the hydrogen (proton) flux is inversely proportional to the membrane thickness. However, increase in oxygen flux is still approximately linear in this region (at a membrane thickness of 0.02 cm or less) resulting in a considerable difference between the fluxes of hydrogen and oxygen. It thus follows that the SCYb oxide membrane should be prepared with a thickness of less than 0.02 cm, as in this region, a slight reduction of the membrane thickness would dramatically increase the hydrogen permeation flux.

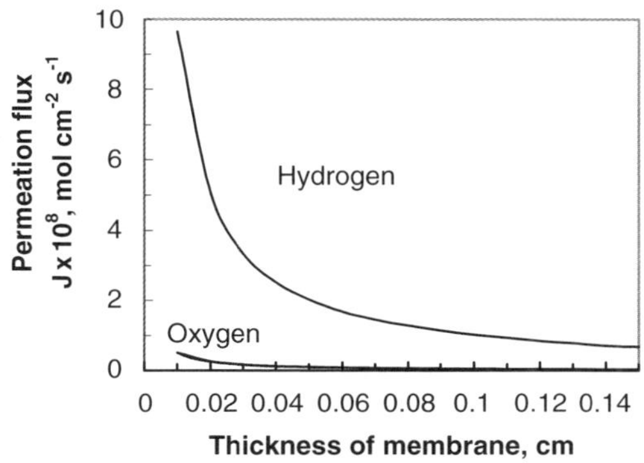

Figure 7.11 Effect of membrane thickness on the permeation flux. ($T = 1173\,K$, $P = 1$ atm, $y_{H_2}(I) = 0.01$, $y_{H_2O}(I) = 0.001$, $y_{O_2}(II) = 0.05$, $y_{H_2O}(II) = 0.01$) [75]. Copyright (2000) Elsevier, Solid State Ionics, **138** (1–2), Tan, X., Liu, S., Li, K. and Hughes, R., *Theoretical analysis of ion permeation through mixed conducting membranes and its application to dehydrogenation reactions*, p. 149–159

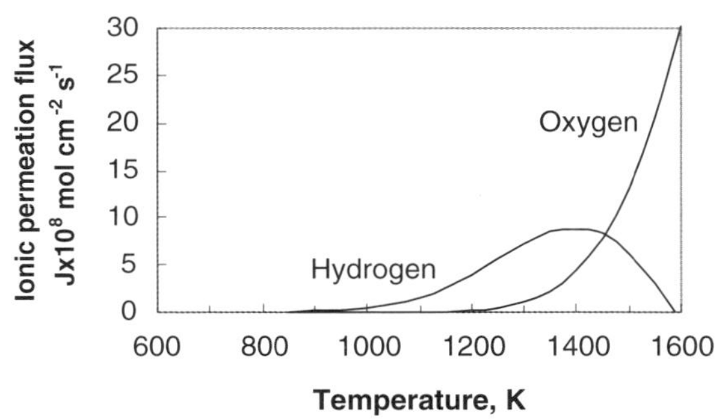

Figure 7.12 Effect of temperature on the ionic permeation flux. ($L = 0.04$ cm, $P = 1$ atm, $y_{H_2}(I) = 0.01$, $y_{H_2O}(I) = 0.001$, $y_{O_2}(II) = 0.05$, $y_{H_2O}(II) = 0.01$) [75]. Copyright (2000) Elsevier, Solid State Ionics, **138** (1–2), Tan, X., Liu, S., Li, K. and Hughes, R., *Theoretical analysis of ion permeation through mixed conducting* membranes and its application to dehydrogenation reactions, p. 149–160

7.4.3.2 Effect of Temperature

Predicted values for this effect are shown in Figure 7.12, where both the permeation fluxes of hydrogen and oxygen are plotted against the operating temperature. It can be seen that both the fluxes are very small at low temperature; for example, at a temperature of 600 K, the order of magnitude for hydrogen and oxygen ion fluxes is 10^{-12} and 10^{-19} (mole cm^{-2} s^{-1}), respectively. As the operating temperature increases to 1400 K, the fluxes for both the hydrogen and oxygen are increased to 9×10^{-8}(mole cm^{-2} s^{-1}) and 2×10^{-8}(mole cm^{-2} s^{-1}), respectively. It is interesting to note that at the operating temperature of 1400 K, which may be termed as the optimum temperature for hydrogen, a maximum value of the hydrogen flux is achievable. Any further increase in temperature after this optimum point results in a sharp decrease in hydrogen flux. On the other hand, the oxygen flux increases considerably and the membrane is eventually turned into a pure oxygen ion conductor at an operating temperature of 1600 K where the proton conducting behaviour is completely diminished.

It is generally believed that oxygen vacancies play an important role in oxygen ion conducting or proton conducting properties of perovskite materials, particularly in a higher temperature environment. The most obvious examples are $BaCeO_3$ based ceramics, such as $BaCe_{0.9}Nd_{0.1}O_{3-\alpha}$ and $BaCe_{0.9}Gd_{0.1}O_{3-\alpha}$. For $BaCe_{0.9}Nd_{0.1}O_{3-\alpha}$, the proton transport number decreases from 0.8 at 973 K to 0.2 at 1273 K, while the oxygen ion transport number grows markedly [41]. The $BaCe_{0.9}Gd_{0.1}O_{3-\alpha}$ material shows a transition from proton to oxygen ion conduction over the temperature range 873–1273 K [78], suggesting two competing processes of proton and oxygen ion conduction. Therefore, it is conceivable that $SrCe_{0.95}Yb_{0.05}O_{23-\alpha}$ may also show a mixed proton–hole or oxygen vacancy–hole conduction; which type of defect dominates the conduction would depend on the operating temperature.

7.4.3.3 Effect of Partial Pressure of Oxygen

The effect of oxygen partial pressure in compartment II at different temperatures on the fluxes of hydrogen and oxygen is illustrated in Figure 7.13. It can be seen that an increase in p_{O_2}(II) results in an increase in the hydrogen flux but the oxygen flux remains relatively unchanged at a temperature of 1200 K (as shown in Figure 7.13A). This phenomenon can be explained by the effect of oxygen pressure on the defect concentration at equilibrium. From the interface II defect concentration Equations (7.7a), (7.7b) or (7.7c), it can be inferred that a change in the oxygen partial pressure provides a negligible effect on the defect concentration at lower

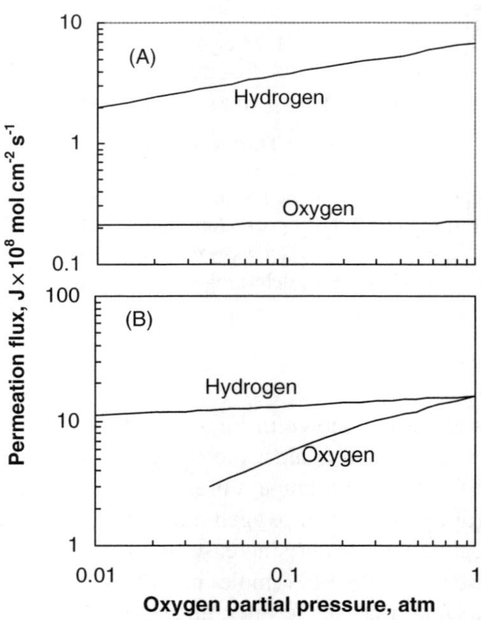

Figure 7.13 Effect of oxygen partial pressure on the ionic permeation flux. ($L = 0.04$ cm, $P = 1$ atm, $y_{H_2}(I) = 0.01$, $y_{H_2O}(I) = 0.001$, $y_{H_2O}(II) = 0.01$): A, $T = 1200$ K; B, $T = 1600$ K [75]. Copyright (2000) Elsevier, Solid State Ionics, **138** (1–2), Tan, X., Liu, S., Li, K. and Hughes, R., *Theoretical analysis of permeation through mixed conducting membranes and its application to dehydrogenation reactions,* p. 149–161

temperatures (<1300 K). However, it would result in a great increase in the proton concentration or conductivity and thus the hydrogen flux at high temperature. Figure 7.13B illustrates the effect of oxygen partial pressure on the fluxes of hydrogen and oxygen at a temperature of 1600 K. It can be seen that the effect of the oxygen partial pressure on the permeation flux of the oxygen in a higher temperature environment is more dramatic than that of hydrogen. Compared to the theoretical data shown in Figure 7.13A, the flux of the oxygen ion increases considerably as the operating temperature is increased from 1200 K to 1600 K. This suggests that the partial pressure of oxygen plays a more important role in the switch of the conducting behaviour from mixed proton–hole conduction to mixed oxygen ion–electronic conduction at a higher temperature.

Similarly, based on analysis of the interface I defect concentration equations, the impact of gas partial pressure (water vapour and hydrogen) of compartment I on the permeation can be inferred. At lower temperatures (<1300 K), the increase of water vapour or hydrogen partial pressure would also result in a significant increase in the hydrogen flux and an almost negligible oxygen flux. At higher temperatures (>1600 K), due to the domination of mixed oxygen electronic conducting behaviour, any change of gas partial pressure in both compartments can have a significant impact on the oxygen flux.

7.4.3.4 Comparison with Experimental Data

$SrCe_{095}Yb_{0.05}O_{3-\alpha}$ was previously known as a high temperature proton conducting material which has great application potential in the fields of many important electrochemical processes [8]. Its mixed proton–hole conduction was first observed and successfully applied for hydrogen permeation by Hamakawa *et al.* in 1993 [4]. One year later, the same research group studied the mixed conducting behaviour of the same material and its application to the membrane reactor in a more systematic and quantitative way [5]. Therefore, the theoretical analysis conducted by Tan *et al.* [75] was compared with the experimental work accomplished by Hamakawa *et al.* in 1994 [5]. In their experiments, no water vapour was detected in compartment I. This is obvious because the oxygen vacancy or oxygen ionic conduction is very low compared to the proton conduction as the operating temperature was less than 1200 K (Figure 7.12). The formation of water vapour due to the transfer of a vacancy is, therefore, negligible. A comparison between the theoretical data calculated by Tan *et al.* [75] and the experimental data obtained by Hamakawa *et al.* [79] is illustrated in Figure 7.14 where the

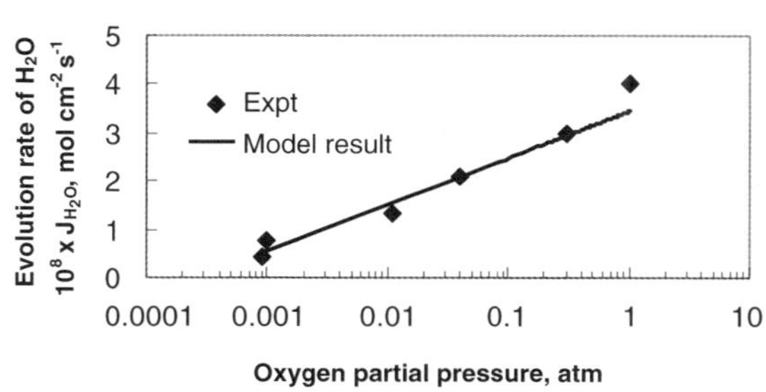

Figure 7.14 Comparison of model with experimental data on hydrogen permeation. ($L = 0.1$ cm, $T = 1173$ K, $p_{H_2} = 0.01$ atm) [75]. Copyright (2000) Elsevier, Solid State Ionics, **138** (1–2), Tan, X., Liu, S., Li, K. and Hughes, R., *Theoretical analysis of ion permeation through mixed conducting membranes and its application to dehydrogenation reactions*, p. 149–162

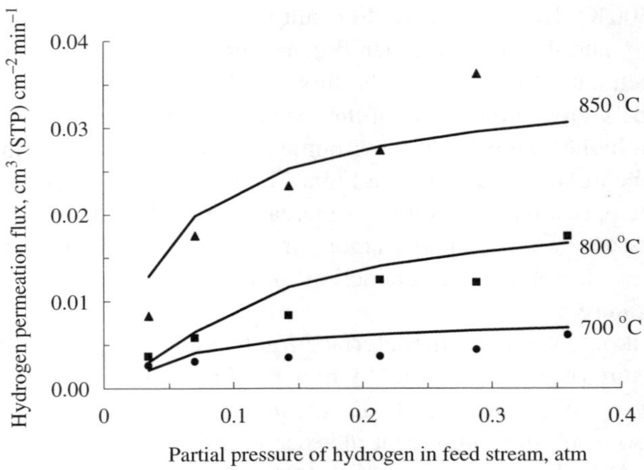

Figure 7.15 Effect of hydrogen partial pressure in the shell side on the hydrogen (proton) permeation flux at various temperatures ($p_{O_2} = 0.21$ atm) [52]. Copyright (2006) John Wiley & Sons, Ltd, American Institute of Chemical Engineers Journal, **52** (4), Liu, Y., Tan, X. and Li, K., *$SrCe_{0.95}Yb_{0.05}O_{3-\alpha}$ hollow-fiber membrane and its property in proton conduction*, p. 1577–1585

water formation rate due to the proton permeation is plotted against the oxygen partial pressure in compartment II. As can be seen from the figure the agreement between the experimental data and predicted values is exceptionally good justifying that at an operating temperature of 900 K, proton conduction is the dominating factor in the SCYb conducting membrane.

In addition to the theoretical analysis, Liu *et al.*, also carried out experimental studies on hydrogen permeation (or proton conduction) properties in a mixed conducting membrane [52]. They developed a SCYb hollow fibre membrane module containing one hollow fibre of 30 cm in length. In their measurements, a mixture of hydrogen and argon was fed into the shell side of the module while air was introduced into the lumen side of the module so that an asymmetric environment was maintained. The proton conduction properties of the SCYb hollow fibre membranes at different temperatures and feed concentrations were measured and presented in terms of hydrogen permeation flux. Again, the diffusional and equilibrium parameters used for calculation of theoretical hydrogen permeation in the membrane module were obtained from the literature [52, 75].

Figure 7.15 shows hydrogen permeation rates for the SCYb hollow fibre membrane as a function of hydrogen partial pressure under different temperatures. The oxygen partial pressure in the opposite compartment was kept 0.21 atm, i.e. air was used as the oxygen source. It can be seen that the hydrogen permeation rate increases with increasing hydrogen partial pressure and also increases with increasing temperature. The experimental data, in general, are lower than the theoretical results presented by solid lines. The reason for this might be that the resistances of surface exchange reactions, which were ignored in the mathematical model, may have an effect on hydrogen permeation. Because the hollow fibre prepared by Liu *et al.* [52] shows an asymmetric structure, i.e. dense outer surface of the SCYb hollow fibre, there may be insufficient surface area provided on the outer surface resulting in a much reduced surface exchange reaction rate. In addition, both the theoretical results and the experimental data indicate that the hydrogen permeation rate may be improved by increasing temperature rather than by increasing the hydrogen partial pressure in the feed.

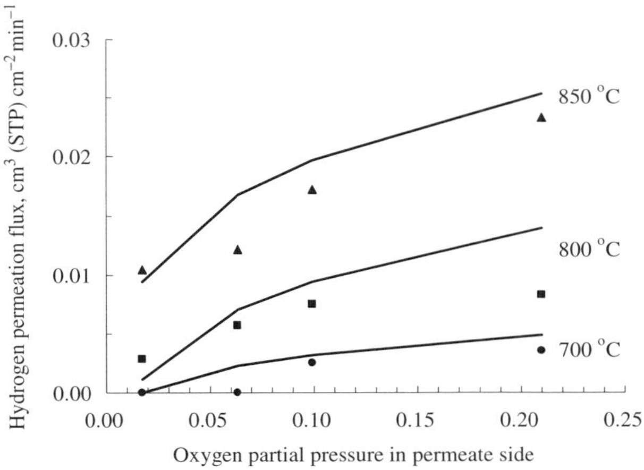

Figure 7.16 Effect of oxygen partial pressure in lumen side on the hydrogen (proton) permeation flux at various temperatures ($p_{H_2} = 0.142$ atm) [52]. Copyright (2006) John Wiley & Sons, Ltd, American Institute of Chemical Engineers Journal, **52** (4), Liu, Y., Tan, X. and Li, K., *$SrCe_{0.95}Yb_{0.05}O_{3-\alpha}$ hollow-fiber membrane and its property in proton conduction,* p. 1577–1586

Effects of oxygen partial pressure in the air stream side on the hydrogen permeation rate through the SCYb hollow fibre membrane under different temperatures are shown in Figure 7.16 where the hydrogen partial pressure was kept constant at 0.142 atm. It can be seen that as the oxygen partial pressure increases, more electron holes can be produced, resulting in an increase in the hole concentration on the membrane surface exposed to the air stream (Figure 7.10). Therefore, the permeation flux of charged defects or the hydrogen permeation rate is improved due to the increasing driving force, as reflected in Figure 7.16. However, at higher oxygen partial pressures, the formation rate of the holes shows less effect on the defect permeation process, which suggests that the permeation process is controlled by other steps, such as the formation rate of protons. As a result, the beneficial effect of further increasing oxygen concentration gradually diminishes. Again, in general, the experimental data are lower than the theoretical results as shown by solid lines. However, the general trends are clear to merit a meaningful comparison between the experimental and theoretical results.

It should be mentioned that the membrane thickness used for the calculations is the difference between the outer and inner radius of the hollow fibres, that is, 400 μm. However, if the outer dense layer is taken as the actual membrane thickness for hydrogen permeation, that is, <50 μm, the theoretical results would be much higher than that of the experimental data, as shown in Figure 7.17. For example, the hydrogen permeation flux for a membrane with a thickness of 50 μm at 850 °C may approach 0.08 cm^3(STP) cm^{-2} min^{-1}, which is similar to the experimental results obtained by Hamakawa *et al.* [49], whereas the actual permeation flux in the hollow fibre membrane is only 0.02 cm^3(STP) cm^{-2} min^{-1}. This suggests both the inner and outer surfaces of the membrane are dense. Also, densification of the middle section of the fibre further leads the membrane to have a thickness equivalent to the fibre wall thickness. It is envisaged that if both dense skin layers in the inner and outer membrane surfaces are eliminated, leaving the finger like pore structures on both the membrane surfaces for enhanced surface reactions and the central dense layer for the selective permeation, the hydrogen permeation flux would be expected to be considerably increased.

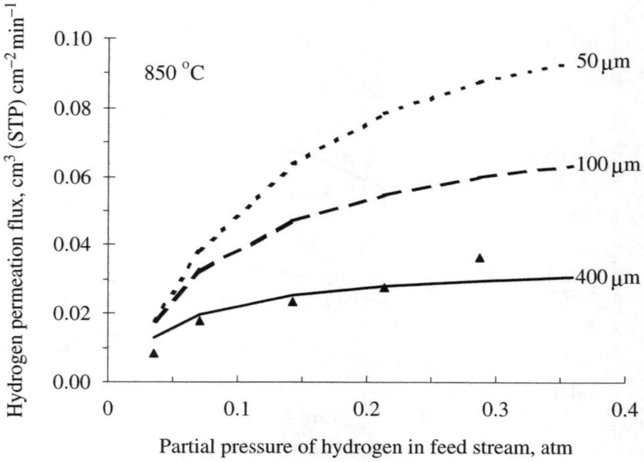

Figure 7.17 Hydrogen permeation flux under various membrane thickness ($T = 850\,°C, p_{O_2} = 0.21$ atm) [52]. Copyright (2006) John Wiley & Sons, Ltd, American Institute of Chemical Engineers Journal, **52** (4), Liu, Y., Tan, X. and Li, K., *$SrCe_{0.95}Yb_{0.05}O_{3-\alpha}$ hollow-fiber membrane and its property in proton condition,* p. 1577–1586

7.5 APPLICATIONS OF PROTON CONDUCTING CERAMIC MEMBRANES

Although mixed proton conductors have already found applications in gas sensors, some other important applications in separation and chemical reactions are being investigated [27]. Several promising applications which have attracted more attention are discussed below.

7.5.1 Hydrogen Production

The demand for hydrogen is expected to rise in the coming years, especially as hydrogen is considered to be a future clean fuel (or fuel carrier) directly for both electric power and vehicles, because of the concern over global warming. The technology for separation or enrichment of hydrogen from various hydrogen production processes will be required in the near future. Currently a few methods exist to produce hydrogen, such as methane partial oxidation ($CH_4 + 1/2O_2 \rightarrow CO + 2H_2$), steam reforming of natural gas ($CH_4 + H_2O \rightarrow CO + 3H_2$), the water gas shift reaction ($CO + H_2O \rightarrow CO_2 + H_2$) or the gasification of heavy carbonaceous materials. These generally proceed at high temperatures and/or pressures and produce hydrogen mixed with other gases. The recovery of hydrogen from high temperature, high pressure streams without the need for cooling or depressurization would enhance the efficiency and performance of systems. Membrane technologies for the removal of hydrogen from the reaction mixtures will provide the possibility to improve these methods by shifting the equilibrium towards the production of hydrogen. The purity of recovered hydrogen will also be very high, thus reducing the cost of hydrogen production. There are two main processes, water–gas shift and water dissociation, that have been investigated and thought of as having potential for hydrogen production using high temperature mixed proton conducting membranes.

The water–gas shift reaction has been studied extensively as a basis for improving the yield of hydrogen production by Argonne National Laboratory (ANL) and National Energy Technology Laboratory (NETL) in America [17, 80–82]. A membrane made from $BaCe_{0.8}Y_{0.2}O_{3-\delta}$ (BCY) was studied during the early stages, where they found the electronic conductivity of BCY is still not sufficient to support hydrogen flux [17, 80]. In order to increase the electronic conductivity, various dual phase (i.e. 40–50 % of a metal is dispersed in a ceramic matrix) membranes were developed [81]. The hydrogen permeability was increased by the increase in electronic conductivity, but the stability of the membranes still requires further investigation.

It has also been considered to produce hydrogen from water splitting into oxygen and hydrogen at high temperatures ($H_2O(g) \rightarrow H_2 + 1/2O_2$), because water resources is almost unlimited. But very low concentrations of hydrogen (0.1 %) and oxygen (0.042 %) are generated at 1600 °C, because of the small equilibrium constant of the water dissociation reaction [83]. The equilibrium can be disturbed and shifted towards the product side using either an oxygen permeable or hydrogen permeable mixed conducting membrane with the advantage of no electric power or electrical circuitry being required, compared with the use of steam electrolysis [82, 84, 85]. A ZrO_2-TiO_2-Y_2O_3 membrane was used to separate hydrogen directly from water splitting by Naito and Arashi [84]. A production rate of $0.6\,cm^3\,cm^{-2}\,min^{-1}$ of hydrogen was reported at 1683 °C. Recently, a production rate of $6\,cm^3\,cm^{-2}\,min^{-1}$ of hydrogen has been achieved at a much more moderate temperature (<900 °C) by Balachandran *et al.* [85] using a 0.13 mm thick dual phase membrane made from 60 vol% gadolinium doped CeO_2 mixed with a 40 vol% metal.

7.5.2 Dehydrogenation Reactions

A number of industrially important dehydrogenation reactions, for example, methane coupling to C_2 hydrocarbon, display equilibrium limited conversion and hence cannot completely convert all the reactants to products. As a result, commercialization of these reactions becomes unlikely and there are significant economic incentives to devise strategies to overcome this difficulty. In principle, complete conversion of a reactant can be achieved by continuously removing products (in some of them) from the reacting zone. For dehydrogenation reactions, one may simultaneously carry out direct oxidation of the hydrogen produced, with injected oxygen. Such an arrangement, however, inevitably results in many problems, such as deep oxidation of the product and difficulty in controlling the reaction temperature, all of which will decrease the product selectivity [86].

An alternative to the direct oxidation scheme mentioned above is to conduct the reaction in a membrane reactor such that the hydrogen formed may be withdrawn continuously from the reaction zone. The withdrawal of hydrogen is tantamount to a departure from equilibrium conditions and the reaction may proceed until a complete conversion is obtained. In contrast to direct oxidation, the membrane reactor method also enables the reaction to be carried out at an easily controlled temperature, minimizing the presence of side reactions and resulting in a more favorable product distribution. Due to many inherent defects, conventional hydrogen permeable palladium based membranes cannot be used in alkane dehydrogenation reactions [87]. One of the more promising materials for use in such a membrane reactor is proton or mixed proton–hole conducting ceramics since they have very good high temperature stability and can withstand the pressure differences across the membrane wall at high temperatures [88].

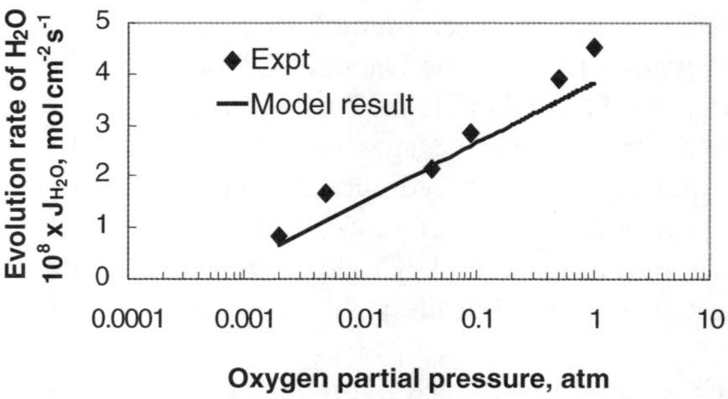

Figure 7.18 Comparison of model with experimental data on methane coupling reaction. ($L = 0.1$ cm, $T = 1173$ K, flow rate of $CH_4 = 20\,ml\,min^{-1}$, membrane area = $1.87\,cm^2$) [75]. Copyright (2000) Elsevier, Solid State Ionics, **138** (1–2), Tan, X., Liu, S., Li, K. and Hughes, R., *Theoretical analysis of ion permeation through mixed conducting membranes and its application to dehydrogenation reactions*, p. 149–162

Hamakawa *et al.* [79] have performed some experiments on the methane coupling reaction using SCYb membranes. C_2 hydrocarbon generation from methane is the result of the formation of the $CH_3^{\bullet}$ radical and a proton:

$$CH_4 \rightarrow CH_3^{\bullet} + H^+ \tag{7.10}$$

$$CH_3^{\bullet} \rightarrow C_2H_4,\ C_2H_6,\ e \tag{7.11}$$

Figure 7.18 shows a comparison between the predicted values and the experimental data for the methane coupling reaction using the SCYb membrane [79]. The water formation rate due to proton permeation through the membrane is plotted against the partial pressure of oxygen in compartment II. It can be seen that the predicted values are in good agreement with the experimental data for low ranges of oxygen partial pressure. As the oxygen pressure is increased to 1 atm, the theoretical data starts to deviate from the experimental results. However, the general trend shows fairly good agreement, especially at lower partial pressures of oxygen.

Nonoxidative coupling of methane to produce C_2 products has recently attracted attention [25, 28]. This reaction is equilibrium limited and may be improved by using membranes [4–6, 88–90]. The principle idea is that hydrogen in the form of protons produced during the dehydrogenation reaction of methane is removed continuously using a mixed proton–hole conducting membrane; the equilibrium of the reaction is thus disturbed so that a potential and significant improvement in C_2 yield, by shifting the conversion towards the product side, can be achieved. On the other hand, CO_x by-products can also be restrained since no oxygen is introduced into the methane compartment and thus, C_2 selectivity is possible to achieve up to 100 % [89].

Further research interests in applying mixed proton–hole conducting ceramic membranes in direct coupling of methane into C_2 products has been focused on preparation of membranes with ultrathin separation layers and large ratios of membrane area to volume of the membrane module, as well as the improvement of the catalytic properties of the membranes through loading an appropriate catalyst onto the membranes by various methods [25]. The preparation of ultrathin ceramic membranes is, so far, a big challenge in membrane fabrication technology. Further discussions on proton–hole ceramic membranes in dehydrogenation reactions will be covered in Chapter 8.

NOTATION

C	Defect concentration, $mol\,m^{-3}$
D	Effective diffusivity of defect, $m^2 s^{-1}$
$h^{\bullet}$	Proton hole
$H_i^{\bullet}$	Interstitial proton
J	Permeation flux of defect, $mol\,m^{-2} s^{-1}$
K_1	Equilibrium constant of Equation (7.8), $Pa^{-1/2} mol\,m^{-3}$
K_2	Equilibrium constant of Equation (7.8), $Pa^{-1} mol\,m^{-3}$
K_3	Equilibrium constant of Equation (7.8), Pa^{-1}
L	Thickness of membrane, m
$O_o^{\times}$	Lattice oxygen
$OH_O^{\bullet}$	Substitutional hydroxyl defect
p	Gas partial pressure, Pa
R_{in}	Inner radius of tubular membrane, m
R_o	Outer radius of tubular membrane, m
t	Transference number of defect
T	Temperature, K
$V_O^{\bullet\bullet}$	Oxygen vacancy
x	Location coordinate, m
Yb'_{Ce}	Substitution of Yb for Ce

Subscripts

h	Electron hole
H_2	Hydrogen
H_2O	Water
i, j	Hole, proton or vacancy
p	Proton
V	Vacancy
O_2	Oxygen

REFERENCES

1. Yuan, S. and Pal, U., *Analytic solution for charge transport and chemical-potential variation in single-layer and multilayer devices of different mixed-conducting oxides.* Journal of the Electrochemical Society, **143** (10): 3214–3222 (1996).
2. Nafe, H., *Current-voltage relation and charge distribution in mixed ionic-electronic solid conductors.* Journal of the Electrochemical Society, **144** (11): 3922–3929 (1997).
3. Heyne, L., *Electrochemstry of mixed ionic electronic conductors*, in *Solid Electrolytes, Topics in Applied Physics*, Geller, S., Editor. (1977) Berlin: Springer, p. 166–221.
4. Hamakawa, S., Hibino, T. and Iwahara, H., *Electrochemical methane coupling using protonic conductors.* Journal of the Electrochemical Society, **140** (2): 459–462 (1993).

5. Hamakawa, S., Hibino, T. and Iwahara, H., *Electrochemical methane coupling in a proton-hole mixed conductor and its application to a membrane reactor.* Journal of the Electrochemical Society, **141** (7): 1720–1726 (1994).
6. White, J. H., Schwartz, M., and Sammells, A. F., *Solid state proton and electron mediating membrane and use in catalytic membrane reactors, US5821185.* (1998).
7. Bonanos, N., Knight, K. S. and Ellis, B., *Perovskite solid electrolytes – structure, transport-properties and fuel-cell applications.* Solid State Ionics, **79**: 161–170 (1995).
8. Iwahara, H., *Proton conducting ceramics and their applications.* Solid State Ionics, **86–88**: 9–15 (1996).
9. Kreuer, K. D., *Proton conductivity: materials and applications.* Chemistry of Materials, **8**: 610–641 (1996).
10. Kreuer, K. D., *On the development of proton conducting materials for technological applications.* Solid State Ionics, **97**: 1–15 (1997).
11. Liang, K. C. and Nowick, A. S., *High-temperature protonic conduction in mixed perovskite ceramics.* Solid State Ionics, **61**: 77–81 (1993).
12. Iwahara, H., Esaka, T., Uchida, H. and Yamauch, T., *High temperature type protonic conductor based on* $SrCeO_3$ *and its application to the extraction of hydrogen gas.* Solid State Ionics, **18–19**: 1003–1007 (1986).
13. Iwahara, H., Esaka, T., Uchida, H. and Maeda, N., *Proton conduction in sintered oxides and its applications to steam electrolysis for hydrogen production.* Solid State Ionics, **3–4**: 359–363 (1981).
14. Wagner, C., *Equations for transport in solid oxides and sulfides of transition metals.* Progress in Solid State Chemistry, **10** (1): 3–16 (1975).
15. Kofstad, P., *Nonstoichiometry: Difusion and Electrical Conductivity in Binary Metal Oxides.* (1972) New York: Wiley, 1–20.
16. Lin, Y. S., Wang, W. J. and Han, J. H., *Oxygen permeation through thin mixed-conducting solid state mmembranes.* American Institute of Chemical Engineers Journal, **40**: 786–798 (1994).
17. Guan, J., Dorris, S. E., Balachandran, U. and Liu, M., *Transport properties of* $BaCe_{0.95}Y_{0.05}O_{3-\delta}$ *mixed conductors for hydrogen separation.* Solid State Ionics, **100**: 45–52 (1997).
18. Kreuer, K. D., Schonherr, E. and Maier, J., *Proton and oxygen diffusion in* $BaCeO_3$ *based compounds – a combined thermal gravimetric analysis and conductivity study.* Solid State Ionics, **70**: 278–284 (1994).
19. Condon, J. B. and Schober, T., *Modeling of point defect migration in proton conducting ceramics.* Solid State Ionics, **97**: 51–58 (1997).
20. Iwahara, H., *Technological challenges in the application of proton conducting ceramics.* Solid State Ionics, **77**: 289–298 (1995).
21. Yajima, T., Koide, K., Fukatsu, N., Osashi, T. and Iwahara, H., *A new hydrogen sensor for molten aluminum.* Sensors and Actuators B, **13–14**: 697–699 (1993).
22. Iwahara, H., Uchida, H. and Tanaka, S., *High temperature type proton conductor based on* $SrCeO_3$ *and its application to solid electrolyte fuel cells.* Solid State Ionics, **9–10**: 1021–1025 (1983).
23. Chiang, P. H., Eng, D. and Stoukides, M., *Electrocatalytic nonoxidative dimerization of methane over Ag electrodes.* Solid State Ionics, **61**: 99–103 (1993).
24. Terai, T., Li, X. H. and Fujimoto, K., *Development of catalyst-membrane system for oxidative coupling of methane by water.* Chemistry Letters, **28** (4): 323–324 (1999).
25. Liu, S., Tan, X. , Li, K. and Hughes, R., *Methane coupling using catalytic membrane reactors.* Catalysis Reviews in Science and Engineering, **43**: 147–198 (2001).
26. Lin, Y. S., *Microporous and dense inorganic membranes: current status and prospective.* Separation and Purification Technology, **25**: 39–55 (2001).
27. Iwahara, H., Asakura, Y., Katahira, K. and Tanaka, M., *Prospect of hydrogen technology using proton-conducting ceramics.* Solid State Ionics, **168**: 299–310 (2004).
28. Gellings, P. J. and Bouwmeester, H. J. M., *Solid state aspects of oxidation catalysis.* Catalysis Today, **58**: 1–53 (2000).

29. Schober, T., *Applications of oxidic high-temperature proton conductors.* Solid State Ionics, **162–163**: 277–281 (2003).
30. Norby, T., *Solid-state protonic conductors: principles, properties, progress and prospects.* Solid State Ionics, **125**: 1–11 (1999).
31. Slade, R. C. T., *The perovskite-type proton-conducting solid electrolyte $BaCe_{0.90}Y_{0.10}O_{3-\delta}$ in high temperature electrochemical cells.* Solid State Ionics, **61**: 111–114 (1993).
32. Kreuer, K. D., *Aspects of the formation and mobility of protonic charge carriers and the stability of perovskite-type oxides.* Solid State Ionics, **125**: 285–302 (1999).
33. Schober, T. and Bohn, H. G., *Water vapor solubility and electrochemical characterization of the high temperature proton conductor $BaZr_{0.9}Y_{0.1}O_{2.95}$.* Solid State Ionics, **127**: 351–360 (2000).
34. Fukui, T., Ohara, S. and Kawatsu, *Ionic conductivity of gadolinium-doped barium praseodymium oxide.* Solid State Ionics, **116**: 331–337 (1999).
35. Shin, S., Huang, H. H., Ishigame, M. and Iwahara, H., *Protonic conduction in the single crystals of $SrZrO_3$ and $SrCeO_3$ doped with Y_2O_3.* Solid State Ionics, **40**: 910–913 (1991).
36. Iwahara, H., Yajima, T., Hibino, T., Ozaki, K. and Suzuki, H., *Protonic conduction in calcium, strontium and barium zirconates.* Solid State Ionics, **61**: 65–69 (1993).
37. Yajima, T., Kazeoka, H., Yogo, T. and Iwahara, H., *Proton conduction in sintered oxides based on $CaZrO_3$.* Solid State Ionics, **47**: 271–275 (1991).
38. Sata, N., Hiramoto, K., Ishigamw, M., Hosoya, S., Nijmuma, N. and Shin, S., *Site identification of protons in $SrTiO_3$: mechanism for large protonic conduction.* Physics Reviews, **B 54** (22): 15795–15799 (1996).
39. Colomban, P., *Proton Conductors: Solids, Membranes, and Gel-Materials and Devices.* (1992) Cambridge, UK: Cambridge University Press.
40. Norwick, A. S. and Du, Y., *High-temperature protonic conductors with perovskite-related structures.* Solid State Ionics, **77**: 137–146 (1995).
41. Yajima, T., Iwahara, H. and Uchida, H., *Protonic and oxide ionic conduction in $BaCeO_3$-based ceramics – effect of partial substitution for Ba in $BaCe_{0.9}O_{3-\delta}$ with Ca.* Solid State Ionics, **47**: 117–124 (1991).
42. Yajima, T., Suzuki, H., Yogo., T. and Iwahara, H., *Protonic conduction in $SrZrO_3$-based oxides.* Solid State ionics, **51**: 101–107 (1992).
43. Matzke, T. and Cappadonia, M., *Proton conductive perovskite solid solutions with enhanced mechanical stability.* Solid State Ionics, **86–88**: 659–663 (1996).
44. Matsumoto, H., Shimura, T., Higuchi, O., Tanaka, H., Katahira, K., Otake, T., Kudo, T., Yashiro, K., Kaimai, A., Kawada, T. and Mizusaki, J., *Protonic-electronic mixed conduction and hydrogen permeation in $BaCe_{0.9-x}Y_{0.1}Ru_xO_{3-alpha}$.* Journal of the Electrochemical Society, **152** (3): A488–A492 (2005).
45. Matsumoto, H., Shimura, T., Higuchi, T., Otake, T., Sasaki, Y., Yashiro, K., Kaimai, A., Kawada, T. and Mizusaki, J., *Mixed protonic-electronic conduction properties of $SrZr_{0.9-x}Y_{0.1}Ru_xO_{3-alpha}$.* Electrochemistry, **72** (12): 861–864 (2004).
46. Song, S. J., Wachsman, E. D., Rhodes, J., Dorris, S. E. and Balachandran, U., *Numerical modeling of hydrogen permeation in chemical potential gradients.* Solid State Ionics, **164** (1–2): 107–116 (2003).
47. Qi, Xiwang and Lin, Y. S., *Electrical conduction and hydrogen permeation through mixed proton-electron conducting strontium cerate membranes.* Solid State Ionics, **130** (1–2): 149–156 (2000).
48. Li, L. and Iglesia, E., *Modeling and analysis of hydrogen permeation in mixed proton-electronic conductors.* Chemical Engineering Science, **58** (10): 1977–1988 (2003).
49. Hamakawa, S., Li, L., Li, A. and Iglesia, E., *Synthesis and hydrogen permeation properties of membranes based on dense $SrCe_{0.95}Yb_{0.05}O_{3-alpha}$ thin films.* Solid State Ionics, **148** (1–2): 71–81 (2002).
50. Mather, G. C. and Islam, M. S., *Defect and dopant properties of the $SrCeO_3$-based proton conductor.* Chemistry of Materials, **17** (7): 1736–1744 (2005).

51. Cheng, S. G., Gupta, V. K. and Lin, J. Y. S., *Synthesis and hydrogen permeation properties of asymmetric proton-conducting ceramic membranes.* Solid State Ionics, **176** (35–36): 2653–2662 (2005).
52. Liu, Y., Tan, X. and Li, K., *$SrCe_{0.95}Yb_{0.05}O_{3-alpha}$ hollow-fiber membrane and its property in proton conduction.* American Institute of Chemical Engineers Journal, **52** (4): 1577–1585 (2006).
53. Shirsat, A. N., Kaimal, K. N. G., Bharadwaj, S. R. and Das, D., *Thermodynamic stability of $SrCeO_3$.* Journal of Solid State Chemistry, **177** (6): 2007–2013 (2004).
54. Liang, K. C., Du, Y. and Norwick, A. S., *Fast high-temperature proton transport in nonstoichiometric mixed perovskites.* Solid State Ionics, **69**: 117–120 (1994).
55. Norwick, A. S. and Vaysleyb, A. V., *Isotope effect and proton hopping in high-temperature protonic conductors.* Solid State Ionics, **97**: 17–26 (1997).
56. Norwick, A. S., Du, Y. and Liang, K. C., *Some factors that determine proton conductivity in nonstoichiometric complex perovskites.* Solid State Ionics, **125**: 303–311 (1999).
57. Haugsrud, R., Larring, Y. and Norby, T., *Proton conductivity of Ca-doped Tb_2O_3.* Solid State Ionics, **176** (39–40): 2957–2961 (2005).
58. Nigara, Y., Mizusaki, J., Kawamura, K., Kawada, T. and Ishigame, M., *Hydrogen permeability in $(CeO_2)_{0.9}(CaO)_{0.1}$ at high temperatures.* Solid State Ionics, **113–115**: 347–354 (1998).
59. Nigara, Y., Kawamura, K., Kawada, T., Mizusaki, J. and Ishigame, M., *Hydrogen permeability in sintered CeO_2 at high temperatures.* Journal of the Electrochemical Society, **146** (8): 2948–2953 (1999).
60. Nigara, Y., Kawamura, K., Kawada, T. and Mizusaki, J., *Hydrogen permeability in $Ce_{0.8}Yb_{0.2}O_{1.9}$ at high temperatures.* Solid State Ionics, **136–137**: 215–221 (2000).
61. Sakai, N., Yamaji, K., Horita, H., Yokokawa, H., Hirata, Y., Sameshima, S., Nigara, Y. and Mizusaki, J., *Determination of hydrogen solubility in oxide ceramics by using SIMS analyses.* Solid State Ionics, **125**: 325–331 (1999).
62. Shimura, T., Komori, M. and Iwahara, H., *Ionic conduction in pyrochlore-type oxides containing rare earth elements at high temperature.* Solid State Ionics, **86**: 685–689 (1996).
63. Norby, T. and Christiansen, N., *Proton conduction in Ca- and Sr-substituted $LaPO_4$.* Solid State Ionics, **77**: 240–243 (1995).
64. Serra, E., Bini, A. C., Cosoli, G. and Pilloni, L., *Hydrogen permeation measurements on alumina.* Journal of the American Ceramic Society, **88** (1): 15–18 (2005).
65. Norby, T. and Larring, Y., *Mixed hydrogen ion-electronic conductors for hydrogen permeable membranes.* Solid State Ionics, **136**: 139–148 (2000).
66. Zhang, G., Dorris, S. E., Balachandran, U. and Liu, M., *Interfacial resistances of Ni-BCY mixed-conducting membranes for hydrogen separation.* Solid State Ionics, **159** (1–2): 121–134 (2003).
67. Siriwardane, R. V., Poston Jr., J. A., Fisher, E. P., Lee, T. H., Dorris, S. E. and Balachandran, U., *Characterization of ceramic hydrogen separation membranes with varying nickel concentrations.* Applied Surface Science, **167** (1–2): 34–50 (2000).
68. Balachandran, U. B., Lee, T. H., Wang, S., Picciolo, J. J., Dusek, J. T. and Dorris, S. E., *Dense cermet membranes for hydrogen separation.* Abstracts of Papers of the American Chemical Society, **225**: U864–U864 (2003).
69. Balachandran, U., Lee, T. H., Chen, L., Song, S. J., Picciolo, J. J. and Dorris, S. E., *Hydrogen separation by dense cermet membranes.* Fuel, **85** (2): 150–155 (2006).
70. Huang, H. H., Ishigame, M. and Shin, S., *Protonic conduction in the single crystals of Y-doped $SrZrO_3$.* Solid State Ionics, **47**: 251–255 (1991).
71. Hibino, T., Mizutani, K., Yajima, T. and Iwahara, H., *Characterization of proton in Y-doped $SrZrO_3$ polycrystal by IR spectroscopy.* Solid State Ionics, **58** (1–2): 85–88 (1992).
72. Norby, T. and Larring, Y., *Concentration and transport of protons and oxygen defects in oxides*, in *Ceramic Oxygen Ion Conductors and Their Technological Application*, Steele, B. C., H., Editor. (1996) The Institute of Materials.
73. Yugami, H., Matsuo, S. and Ishigame, M., *Presistent spectral hole burning of En^{3+} and Pr^{3+} in $SrZr_{1-x}Y_xO_3$ crystals.* Solid State Ionics, **77**: 195–200 (1995).

74. Kreuer, K. D., Fuchs, A., and Maier, J., *H/D isotope effect of proton conductivity and proton conduction mechanism in oxides.* Solid State Ionics, **77**: 157–162 (1995).
75. Tan, X., Liu, S., Li, K. and Hughes, R., *Theoretical analysis of ion permeation through mixed conducting membranes and its application to dehydrogenation reactions.* Solid State Ionics, **138** (1–2): 149–159 (2000).
76. Uchida, H., Yoshikawa, H. and Iwahara, H., *Dissolution of water vapour (or hydrogen) and proton conduction in $SrCeO_3$ based oxides at high temperatures.* Solid State Ionics, **35**: 229–237 (1989).
77. Schober, T., Schilling, W. and Wenzl, H., *Defect model of proton insertion into oxides.* Solid State Ionics, **86–88**: 653–658 (1996).
78. Bonanos, N., *Transport properties and conduction mechanism in high-temperature protonic conductors.* Solid State Ionics, **53–56** (Part 2): 967–974 (1992).
79. Hamakawa, S., Hibino, T. and Iwahara, H., *Electrochemical hydrogen permeation in a proton-hole mixed conductor and its application to a membrane reactor.* Journal of the Electrochemical Society, **141** (7): 1720–1725 (1994).
80. Guan, J., Dorris, S. E., Balachandran, U. and Liu, M., *The effects of dopants and A:B site nonstoichiometry on properties of perovskite-type proton conductors.* Journal of the Electrochemical Society, **145** (5): 1780–1786 (1998).
81. Guan, J., Dorris, S. E., Balachandran, U. and Liu, M., *Transport properties of $BaCe_{0.95}Y_{0.05}O_{3-\delta}$ and its application for hydrogen separation.* Solid State Ionics, **110**: 303–310 (1998).
82. Maiya, P. S., Anderson, T. J., Mieville, R. L., Dusek, J. T., Picciolo, J. J. and Balanchandran, U., *Maximizing H_2 production by combined partial oxidation of CH_4 and water gas shift reaction.* Applied Catalysis A: Genernal, **196**: 65–72 (2000).
83. Ihara, S., *Approximations for the thermodynamic prperties of high-temperature dissociated water vapour.* Bulletin of the Electrotechnical Laboratory, **41**: 259–280 (1977).
84. Naito, H. and Arashi, H., *Hydrogen production from direct water splitting at high temperature using a ZrO_2-TiO_2-Y_2O_3 membrane.* Solid State Ionics, **79**: 366–370 (1995).
85. Balachandran, U., Lee, T. H., Wang, S. and Dorris, S. E., *Use of mixed conducting membranes to produce hydrogen by water dissociation.* International Journal of Hydrogen Energy, **29** (3): 291–296 (2004).
86. Eng, D. and Stoukides, M., *Catalytic and electrocatalytic methane oxidation with solid oxide membranres* Catalysis Reviews Science and Engineering, **33** (3–4): 375–412 (1991).
87. Armor, J. N., *Applications of catalytic inorganic membrane reactors to refinery products.* Journal of the Membrane Science, **147**: 217–233 (1998).
88. Marnellos, G., Sanopoulou, O., Rizou, A. and Stoukides, M., *The use of proton conducting solid electrolyte for improved performance of hydro- and dehydrogenation reactors.* Solid State Ionics, **97**: 375–383 (1997).
89. Chiang, P. H., Eng, D. and Stoukides, M., *Electrocatalytic methane dimerization with a Yb-doped $SrCeO_3$ electrolytes.* Journal of the Electrochemical Society, **138**: L11 (1991).
90. Li, L., Borry, R. W. and Iglesia, E., *Design and optimization of catalysts and membrane reactors for the non-oxidative conversion of methane.* Chemical Engineering Science, **57**: 4594–4604 (2002).
91. Liu, J. F. and Norwick, A. S., *The incorporation and migration of protons in Nd-doped $BaCeO_3$.* Solid State Ionics, **50**: 131–138 (1992).
92. Liu, Y., Tan, X. and Li, K., *Mixed conducting ceramics for catalytic membrane processing.* Catalysis Reviews – Science and Engineering, **48** (2): 145–198 (2006).

8

Ceramic Membrane Reactors

8.1 INTRODUCTION

Selective permeation through a membrane has become a recognized chemical engineering process for the separation of fluid mixtures. During the past two decades, considerable attention has been placed on the integration of the permeation process into chemical reactors. Membrane reactors, which combine reaction and separation or combine mixing/distribution and reaction in one unit operation, are the results of this integration. These reactors have been used to enhance the reaction yield and conversion of thermodynamically limited reactions [1] or to control the reaction pathway by introducing a reactant into the reaction zone in a controlled manner [2].

Use of membrane reactors to shift the equilibrium in a reversible reaction has been recognized since the 1960s [3–5]. The combined reaction and separation process first found its application in biological systems at low temperatures [6]. In recent years, however, availability of ceramic or metallic membranes has unfolded its application to high temperature catalytic reactions [7, 8]. A comprehensive analysis of equilibrium shift isothermal reactors with a permselective wall was undertaken by Mohan and Govid [1].

Use of membrane reactors to control the addition and distribution of a reactant has been mainly studied for partial oxidative reactions of hydrocarbons [9]. The membrane, often a dense type, controls the oxygen reactant supply to the hydrocarbon compartment and avoids the direct presence of gas phase oxygen, which is often deleterious to hydrocarbon selectivity, thus to some degree, suppressing deep oxidation of the hydrocarbons. Theoretical studies of a methane coupling reaction [10–12] predicted that much better C_2 yields could be achieved in membrane reactors compared to conventional fixed bed reactors. The use of membrane reactors also has advantages in controlling hot spots in exothermic reactions [13, 14].

Objectives of this chapter are to provide readers with an extensive account of ceramic membrane reactors currently studied for various conversion limited reactions. The principles, advantages or disadvantages and crucial problems of most ceramic membrane reactors used in separation/reaction will be discussed. Studies on ceramic membranes as separators will be shown with the example of nonoxidative methane coupling reactions, while membranes used as a distributor will be treated with examples such as elimination of dissolved oxygen from water and oxidative methane coupling reactions. More emphasis will be placed on catalytic mixed conducting (both oxygen and hydrogen) membrane reactors as they have greater potential and have received less attention in the literature.

Ceramic Membranes for Separation and Reaction K. Li

8.2 MEMBRANES AS PRODUCT SEPARATORS

Membrane reactors where the membrane acts as a product separator is a process which combines membrane separation with a chemical reaction step in one unit. Due to the integration of reaction and separation, the process becomes simpler, leading to a much lower fabrication cost. In addition, a membrane reactor is capable of promoting a reaction process by selectively removing at least one of the products through the membrane from the reaction zone, making the equilibrium reaction shift to the product side. As a result, the reaction conversion can be increased beyond the thermodynamic limits (equilibrium of reactions) as illustrated in Figure 8.1 [15]. It can be seen that hydrogen produced in an equilibrium limited reaction is removed continuously through a permselective membrane, the reaction equilibrium shifts to the product side, resulting in a high conversion even at lower temperatures and thus the deactivation of catalyst and undesirable side reactions can be conceivably avoided.

As equilibrium limited reactions, usually dehydrogenation or hydrogenation, take place at high temperatures, various ceramic or metal membranes are generally the primary choice unless reaction catalysts show desirable activity and selectivity at lower temperatures (<300 °C) where polymeric membranes may be applied [16–18]. In addition, the available membranes must be able to provide sufficient permeability and/or permselectivity and always have long term stability in both mechanical and chemical properties under technical conditions [19].

8.2.1 Microporous Membrane Reactors

Because of the capablity of selective permeation of hydrogen, microporous membranes have been applied in a variety of dehydrogenation reactions. Some studies, since 1990, of dehydrogenation reactions using microporous membrane reactors are summarized in Table 8.1. As can be seen, the microporous membranes are usually applied in a packed bed membrane reactor (PBMR) configuration. In most cases, catalysts are packed in the tube side, while sweep gas such as helium or argon flowing over the membrane surface is introduced on the shell side of the reactors. All the studies reported that better yields and greater conversions can be achieved in PBMRs compared to conventional packed bed reactors under the same operating conditions. In addition, continuous removal of hydrogen by the membrane shows additional benefits in slowing down undesirable side reactions, resulting in an increase in the selectivity.

Performance of catalytic membrane reactors in dehydrogenation reactions can be controlled either by the membrane or by the catalyst. For example, in a silica membrane reactor for C_3H_8 dehydrogenation, the increase in propene yield is only significant for a relatively

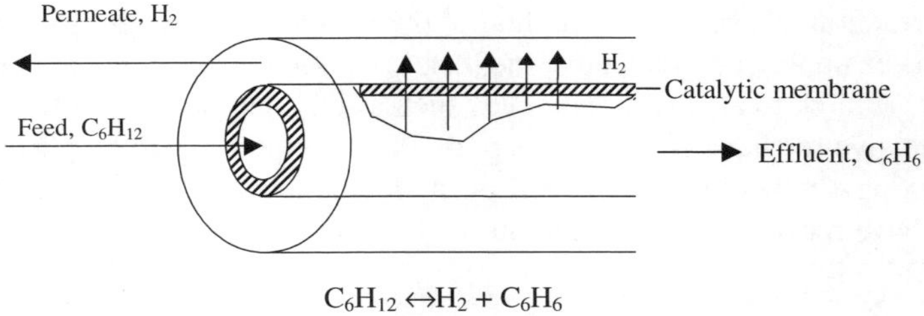

Figure 8.1 Selective permeation of H_2 in an equilibrium limited reaction

Table 8.1 Dehydrogenation reactions in microporous membrane reactors

Reaction	Membranes	Reactor	Results	Ref.
Isobutane to isobutene	MFI on α-Al_2O_3 tube	PBMR, cat. in tube	Four times of Y above FBR	[21]
	MFI on α-Al_2O_3 tube	PBMR, cat. in tube	Y enhanced remarkably	[117]
	MFI on α-Al_2O_3 tube	PBMR, cat. in tube	Two times X above FBR	[27]
	MFI on α-Al_2O_3 tube	PBMR, cat. in tube	70 % Y above FBR	[118]
	MFI on α-Al_2O_3 tube	PBMR, cat. in tube	14 % X above equilibrium	[23]
	Carbon membrane in HF	PBMR, cat. in shell	30 % X above FBR	[26]
	Silica on Vycor tube	PBMR, cat. in tube	Y and S above FBR	[118]
Cyclohexane to benzene	FAU on α-Al_2O_3 tube	PBMR, cat. in shell	123 % X above equilibrium	[32]
	Carbon membrane in HF	PBMR, cat. in shell	X above equilibrium	[119]
	Silica on alumina tube	PBMR, cat. in tube	300 % Y above FBR	[120]
Propane to propene	Silica on γ-Al_2O_3 tube	PBMR, cat. in tube	5 % Y above FBR	[30]
	Silica on γ-Al_2O_3 tube	PBMR, cat. in tube	Y higher than equilibrium	[20]
Methanol to methyl formate	Cu/SiO_2/ceramic composite membrane	CMR	33 % X above FBR; 35.5 % Y above FBR	[33]

Note: MFI = ZSM-5 or silicalite zeolite membrane; FAU = NaX or NaY zeolite membrane; cat. = catalyst Vycor = Vycor porous glass; X = conversion; Y = product yield; HF = hollow fibre; S = Selectivity; PBMR = packed bed membrane reactor; CMR = catalytic membrane reactor; FBR = fixed bed reactor

small value of propane feed loading, because at high loading of propane feed, the hydrogen formed cannot be removed fast enough through the membrane and conversion is again limited by the thermodynamic equilibrium [20]. The Pd/Ag membrane reactor demonstrated a far superior performance of the silica membrane due to its high H_2 flux as well as much higher permselectivity. Obviously, increasing the sweep flow, which may result in a higher H_2 permeation rate, will also enhance dehydrogenation reactions [21]. On the other hand, if the dehydrogenation process is kinetically controlled, change of the membrane would not give higher product yields [22, 23]. Therefore, in order to achieve an enhancement of net performance in membrane reactors, the H_2 production and the permeation (or extraction) rate must be equivalent, which can be realized by adjusting conditions such as the reactant feed flow and the membrane area [24].

The effect of the sweeping mode on the dehydrogenation reaction was studied theoretically by Casanave *et al.* [25] where dehydrogenation of isobutane in a packed bed zeolite membrane reactor with a Pt-In catalyst was used as the model reaction. It was found that the dehydrogenation reaction was limited by transport properties of the membrane in cocurrent mode while in countercurrent mode limitation by kinetics was predominant. As a result, the yield of reaction in these two sweeping modes was almost the same although the separation factor was higher in countercurrent than in cocurrent. Naturally, countercurrent flow operation mode is preferred for membrane permeation controlling reactions [26].

In microporous membrane reactors, if the membrane exhibits a low hydrogen permselectivity, the sweep gas may be introduced into the reaction side through the membrane with the result of dilution of the feed. This can also lead to an increase in the conversion [26, 27]. Such a conversion increase is not desired since separation of the sweep gas in subsequent steps is not easy and expensive. Using vacuum as a driving force for membrane transport can avoid dilution of the feed, but the transport selectivity is poorer compared to sweep gas operation with N_2 where mutual blocking of counter diffusion by nitrogen and hydrocarbons takes place [28].

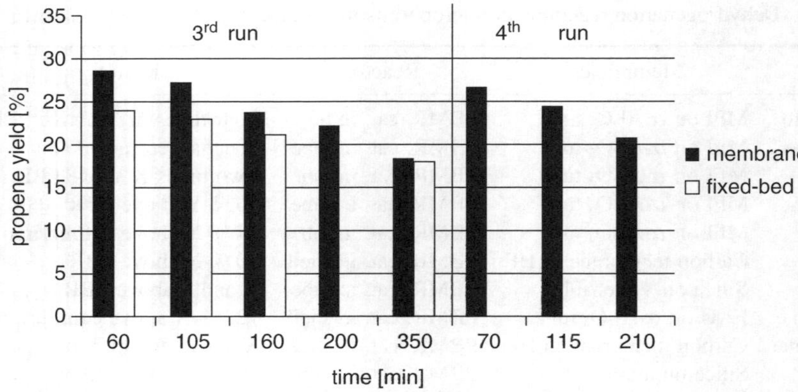

Figure 8.2 Yields of catalytic propane dehydrogenation in a silica membrane reactor and a fixed bed reactor as a function of time ($T = 535\,°C$; WHSV $= 1.1\,h^{-1}$; $V_{propane} = 30\,ml\,min^{-1}$) [30]. Reprinted from Separation and Purification Technology, **25** (13), Schafer *et al.*, *Development of . . .*, p. 3–9. Copyright (2001) with permission from Elsevier

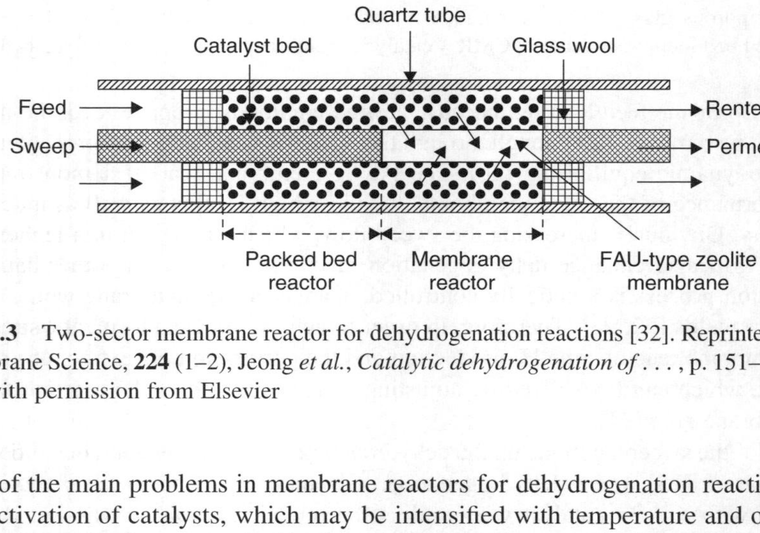

Figure 8.3 Two-sector membrane reactor for dehydrogenation reactions [32]. Reprinted from Journal of Membrane Science, **224** (1–2), Jeong *et al.*, *Catalytic dehydrogenation of . . .*, p. 151–158. Copyright (2003) with permission from Elsevier

One of the main problems in membrane reactors for dehydrogenation reactions is coking and deactivation of catalysts, which may be intensified with temperature and olefin content, and accelerated by the removal of hydrogen [29, 30]. As a result, the performance of membrane reactors often deteriorates gradually as a function of time, as shown in Figure 8.2, where the yield of the catalytic propane dehydrogenation is plotted against operating time. Although the deactivated catalyst can be regenerated by burning off the coke in an O_2 containing atmosphere, followed by reactivation in pure H_2 gas [20], a suitable catalyst with resistance to coking has to be designed owing to the specific reactive environment in membrane reactors.

It is generally accepted that deactivation of catalyst by coking in dehydrogenation reactions can be inhibited by hydrogen. In order to restrain coke formation on the catalyst, hydrogen concentration in the feed side should be kept at a certain value [31]. This leads to a design of membrane reactor, as shown in Figure 8.3, where the permselective membrane is only located at the rear section of the reactor [32].

Another main problem with microporous membranes is their long term stability. Amorphous silica microporous membranes have been shown to be too sensitive to steam and coke in the presence of light alkanes and olefins. For zeolite membranes, it is difficult to control the membrane thickness, hence the permeance, since for optimal PBMR design, one must strive to balance membrane flux with reaction rate.

One of the advantages of microporous membranes is the incorporation of catalyst, making them a true catalytic membrane. A Cu/SiO_2/ceramic composite membrane using an ion exchange method was prepared by Guo *et al.* [33], who used the membrane for dehydrogenation of methanol to methyl formate. A 57.3 % conversion of methanol and 50.0 % yield of methyl formate were achieved in the catalytic membrane reactor at 240 °C, while only 43.1 % conversion of methanol and 36.9 % yield of methyl formate could be obtained in a fixed bed reactor. Another example is the incorporation of metal ions such as iron into MFI zeolite membranes [34, 35]. Such incorporation significantly influences the membranes' physico-chemical characteristics as well as the adsorption properties, which can prevent carbon deposition on the membranes. This is beneficial to reactions such as dehydrogenation of ethylbenzene to styrene.

8.2.2 Dense Ceramic Membrane Reactors

So far, only H_2 permeable ceramic membranes have been applied as product separators in the coupling of methane to produce valuable C_2 products (ethane or ethylene):

$$2CH_4 \leftrightarrow C_2H_6 + H_2 \tag{8.1}$$

The above reaction is a difficult one, only taking place at high temperatures (>650 °C). Compared to palladium or palladium alloy membranes, H_2 permeable ceramic membranes have good thermochemical stability at high temperatures and also show some catalytic activities in methane coupling, and thus are more suitable for this reaction, as dense metal membranes are only suitable for use a at temperature range of 300–600 °C. Table 8.2 summarizes the applications of the H_2 membrane reactors in methane coupling reactions.

In actual experiments, methane (usually mixed with an inert gas) and oxygen (usually air) are respectively introduced into two compartments separated by the membrane at high temperatures as shown in Figure 8.4a. As methane flows over a proton–hole conducting membrane surface at high temperature, it is catalytically dissociated into methane radicals by combining with a proton hole transported from the oxygen side:

$$CH_4 + h^{\bullet} \rightarrow \cdot CH_3 + H^{\bullet} \tag{8.2}$$

Table 8.2 Proton conducting membrane reactors for methane coupling reactions

Membrane	Membrane reactor	T (°C)	Results	Ref.
$SrCe_{0.95}Yb_{0.05}O_{3-\alpha}$	Disk, 0.75 mm	650–750	Y = 0.17%	[121, 122]
$SrCe_{0.95}Yb_{0.05}O_{3-\alpha}$	Tube, 1.0 mm	900		[38, 123]
$SrCe_{0.95}Yb_{0.05}O_{3-\alpha}$	Disk, 1.0 mm	815–820		[124]
$BaCe_{0.95}Mn_{0.05}O_{3-\alpha}$	Tube, 1 mm, no catalyst	950		[125]

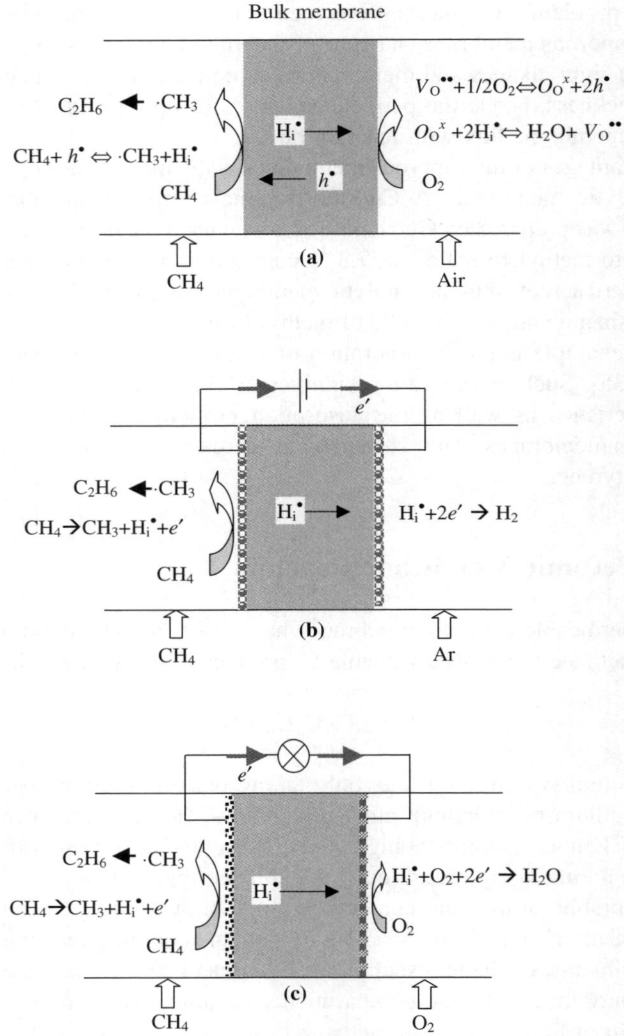

Figure 8.4 Different operating modes for proton conducting membrane reactors in methane coupling reactions

Note that Reaction (8.2) is actually the combination of the following two steps:

$$CH_4 \rightarrow \cdot CH_3 + H^{\bullet} + e' \tag{8.3}$$

$$h^{\bullet} + e' \leftrightarrow \text{nil} \tag{8.4}$$

The proton is transported to the oxygen side under the electrochemical potential gradient and is oxidized by oxygen to form water. Meanwhile, the methane radicals are released into the gas phase, where the coupling reaction takes place:

$$2\cdot CH_3 \leftrightarrow C_2H_6 \tag{8.5}$$

Following the same mechanism, ethylene may be produced from ethane by the formation of ethane radicals:

$$C_2H_6 + h^\bullet \rightarrow \cdot C_2H_5 + H^\bullet \tag{8.6}$$

$$\cdot C_2H_5 \Leftrightarrow C_2H_4 + H^\bullet \tag{8.7}$$

It should be mentioned that the higher carbon products ($>C_2$) can be further produced following a similar mechanism [36]. Furthermore, if some particular catalyst is applied to the membrane surface, even benzene and other valuable aromatic hydrocarbons may be formed [37].

As described above, the use of proton conducting membranes appears advantageous since no oxygen is brought into direct contact with hydrocarbons to form CO_x and the C_2 products and water are formed in separate compartments.

The primary shortcoming of the reported H_2 permeable (mainly proton conducting) membranes is low hydrogen permeability partly due to the low electronic conductivity. In this case, an external voltage may be imposed on both sides of the membrane so that the electrons released from the hydrocarbons can be pumped out leading to hydrogen permeability enhancement as shown in Figure 8.4b. Therefore, argon may be used as the sweep gas instead of oxygen. Hamakawa *et al.* [38] tested a $SrCe_{0.95}Yb_{0.05}O_{3-\alpha}$ (SCYb) membrane reactor with external electrical supply by two porous Ag electrodes attached to the membrane. It was found that the C_2 formation rate in the electrochemical membrane reactor was seven times higher than that in the absence of an external electrical supply. In addition, due to the electrochemical pumping, the catalytic activity of the membrane can be dramatically changed and the reaction rate enhancement can be orders of magnitude higher than the rate of proton transport through the electrolyte. This phenomenon, known as the non Faradaic electrochemical modification of catalytic activity (NEMCA) effect, has been reported in the literature [39–42].

Another operating mode of a membrane reactor is the 'fuel cell' configuration as shown in Figure 8.4c. In this case, different catalysts are applied to both sides of the membrane as anode and cathode of the cell, respectively. The electrons formed on the anode pass through the external circuit to the cathode. As a result, C_2 products, water and electricity can be simultaneously produced in the membrane reactor. This operating mode was experimentally tested by Asano *et al.* and Hibino *et al.* [43, 44]. The main problems with this operating mode are the high ohmic resistance of the electrolyte membrane and the deposition of carbon on the membrane as well.

Considerable research interest in applying proton conducting ceramic membranes in dehydrogenation reactions has been focused on preparation of membranes with an ultrathin separation layer and a large ratio of membrane area to volume of membrane elements, as well as the improvement of catalytic properties of the membranes through loading an appropriate catalyst on the membranes using various methods [45]. Liu *et al.* and Li *et al.* [37, 46] studied two asymmetric proton conducting ceramic membranes of $SrZr_{0.95}Y_{0.05}O_{3-\alpha}$ and $SrCe_{0.95}Y_{0.05}O_{3-\alpha}$, coated with an active catalyst of Mo/H-ZSM5 for the methane coupling reaction, as shown Figure 8.5. The membrane is a thin dense film (10–30 μm) coated on the top of a porous support made of the same material so as to match the thermal expansion coefficient of the materials. In their experiments, benzene and other valuable aromatic hydrocarbons were formed in addition to ethylene.

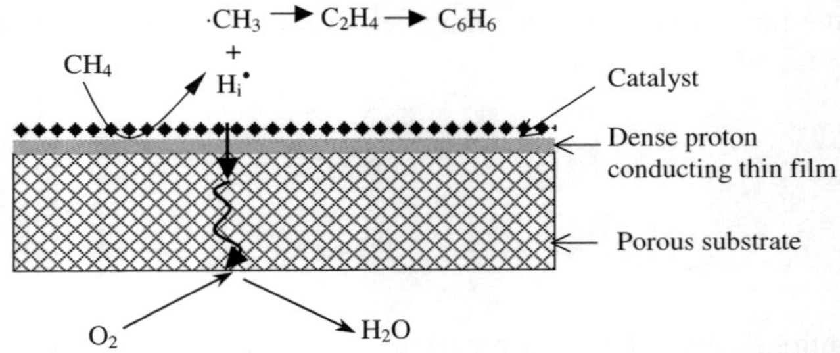

Figure 8.5 Catalyst attached proton conducting asymmetric membrane reactor for methane coupling reaction

A $SrTi_{0.4}Mg_{0.6}O_{3-\alpha}$ catalyst was used by Terai *et al.* [47] in a $SrCe_{0.95}Yb_{0.05}O_{3-\alpha}$ (SCYb) membrane reactor for the oxidative coupling of methane using water as the oxidant since it was previously shown to be an effective catalyst for this reaction:

$$2CH_4 + H_2O \leftrightarrow C_2H_6 + H_2 + H_2O \tag{8.8}$$

It was found that the yield of C_2 hydrocarbons was significantly increased due to the use of the membrane.

The dehydrogenation of C_3H_8 in a $SrCe_{0.95}Yb_{0.05}O_{3-\alpha}$ proton conducting membrane reactor operated in the hydrogen pumping mode was studied by Karagiannakis *et al.* [48], in which a polycrystalline palladium film deposited on the membrane surface was used as a working electrode catalyst. It was found that the presence of water vapour in the feed gas strongly influences the reaction rate. For example, upon increasing the H_2O partial pressure from 0.3 to 2.8 kPa, a four fold increase on the reaction rate was observed. This is mainly due to the presence of water vapour, which can restrict the formation and deposition of carbon on the electrodes so that the catalytic activity is sustained [48]. In other work, Karagiannakis *et al.* [49] compared the catalytic and electrocatalytic activities of platinum and palladium and found that platinum exhibited higher overall reaction rates, while the product selectivities were similar to those on palladium. On the other hand, the electrocatalytic effect on palladium electrodes was much more pronounced with substantial carbon deposition.

Recently, a $SrCe_{0.95}Yb_{0.05}O_{3-\alpha}$ hollow fibre membrane reactor was developed by Liu *et al.* [50] who conducted the methane coupling reaction in the reactor with no catalysts. At a temperature of 950°C, the C_2 yield was up to 13.4 %, which is much higher than the previously reported values. It was explained that the asymmetric structure of the membrane was responsible for C_2 yield improvement. In addition, higher hydrocarbons (C_3–C_5) were also produced, although their formation rates were much lower than those of the C_2 products. Detailed analysis of this novel reactor for the methane coupling reaction is given below.

8.2.2.1 Experimental Investigation of a Dense Ceramic Membrane Reactor for Methane Coupling Reactions

A ceramic hollow fibre membrane reactor fabricated from $SrCe_{0.95}Yb_{0.05}O_{3-\alpha}$ (SCYb) has been developed by Liu *et al.* [50]. The reactor, consisting of one hollow fibre housed in a ceramic tube, was used for methane coupling experiments. A soft silicone rubber tube was used on

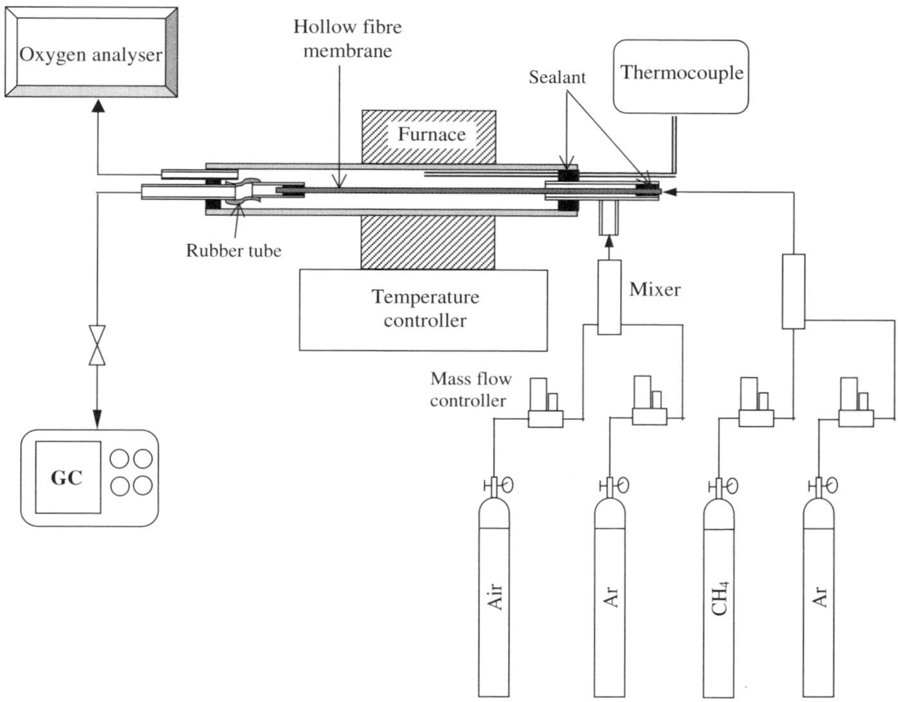

Figure 8.6 Experimental setup for methane coupling in the SCYb hollow fibre membrane reactor [50]. Copyright (2006) American Chemical Society, Industrial and Engineering Chemistry Research, **45** (11), Liu, Y., Tan, X. and Li, K., *Nonoxidative methane coupling in a* $SrCe_{0.95}Yb_{0.05}O_{3-\alpha}$ *(SCYb) hollow fiber membrane reactor*, p. 3782–3790

one end of the fibre to offset the thermal expansion mismatch between the hollow fibre membrane and the module shell at high temperatures, as shown in Figure 8.6, where it can also be be seen that the membrane reactor was located in a tubular furnace with a much shorter heating zone. Therefore, the sealing joints and the silicone rubber tube could be kept away from the furnace heating zone, preventing the silicone rubber tube and seals from being burned away. The feed, methane mixed with argon, was introduced into the lumen side of the reactor while the mixture of air and argon was fed cocurrently into the shell side. An oxygen meter is used to monitor the change of oxygen concentration in the lumen gas stream, while the hydrocarbon concentrations in the product stream were analyzed using a gas chromatography. Detailed information for the experiments can be found elsewhere [50].

Some experimental results for the performance of the reactor for methane coupling reactions are given in Figures 8.7–8.9. Figure 8.7 shows the experimental results where the C_2 formation rates at various temperatures are plotted against feed concentrations of methane ((A)–(D)) and oxygen ((a)–(d)), respectively. As is expected, both ethane and ethylene formation rates increase with increasing methane feed concentration. However, the formation rate of ethane remains relatively unchanged as oxygen feed concentration increases. This suggests that increasing the permeation of hydrogen by means of increasing the oxygen concentration promotes the formation of ethylene more than that of ethane, especially at higher temperatures. On the other hand, it can also be seen that as the temperature increases, the increase

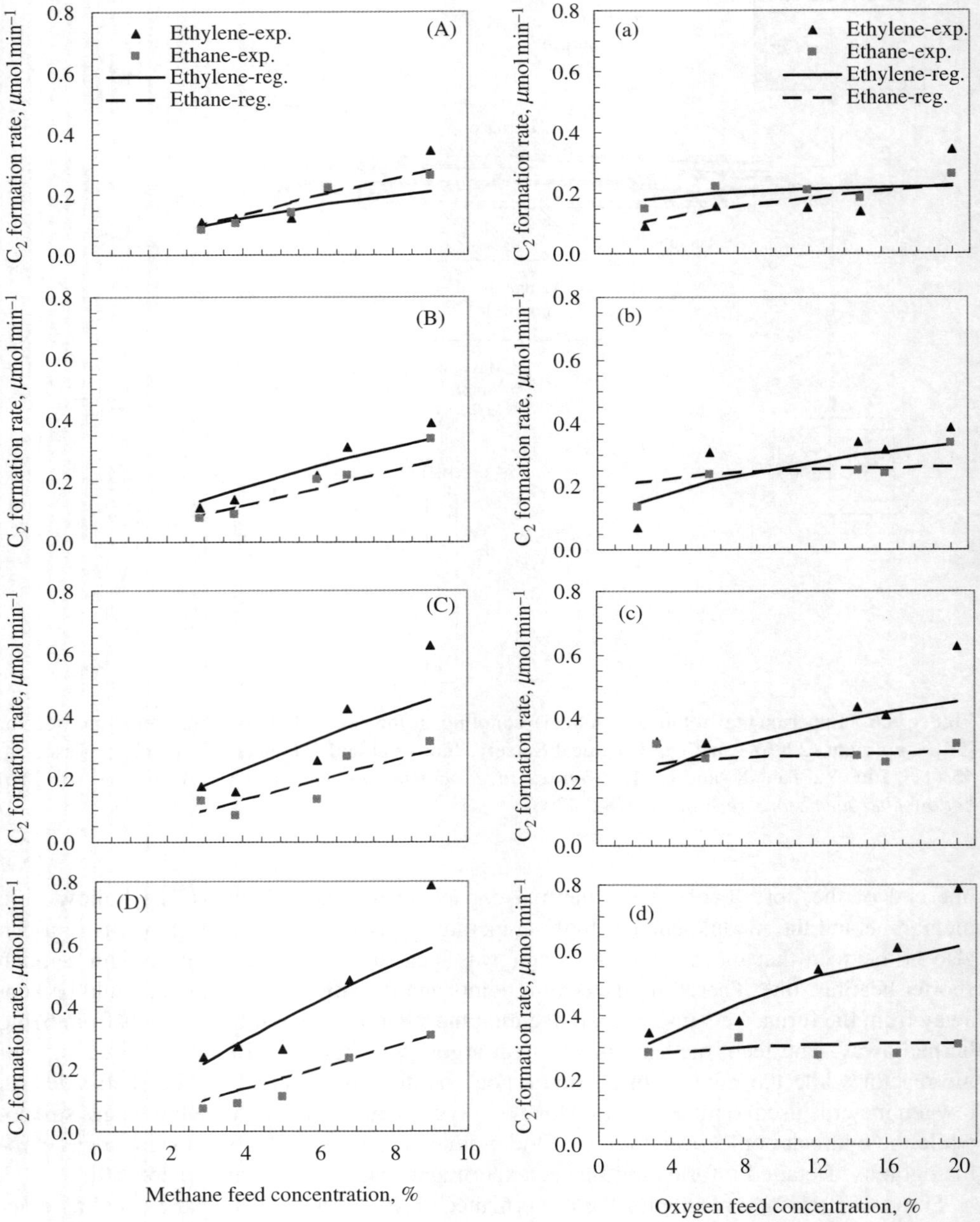

Figure 8.7 C_2 formation rates against feed concentrations at different temperatures (feed flow rate of methane stream = 4.3 $cm^3\,min^{-1}$, feed flow rate of air stream = 40 $cm^3\,min^{-1}$): (A), (a) 800 °C; (B), (b) 850 °C; (C), (c) 900 °C; (D), (d) 950 °C [50]. Copyright (2006) American Chemical Society, Industrial and Engineering Chemistry Research, **45** (11), Liu, Y., Tan, X. and Li, K., *Nonoxidative methane coupling in a $SrCe_{0.95}Yb_{0.05}O_{3-\alpha}$ (SCYb) hollow fiber membrane reactor*, p. 3782–3791

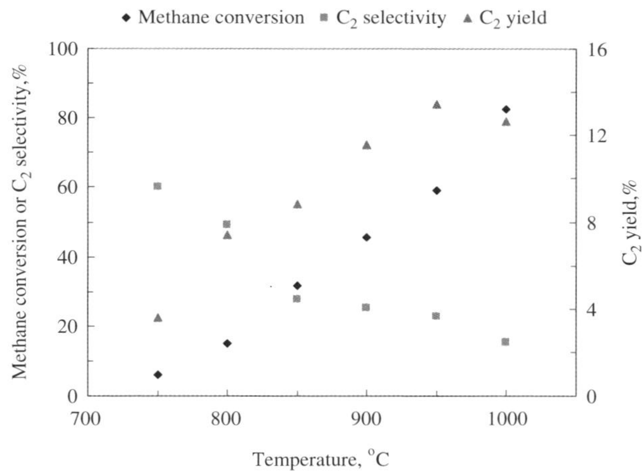

Figure 8.8 Effect of temperature on the OCM conversion, C_2 selectivity and C_2 yield (feed flow rate of methane stream = 3.96 $cm^3 min^{-1}$, methane feed concentration = 9.22 %, air feed flow rate = 40 $cm^3 min^{-1}$) [50]. Copyright (2006) American Chemical Society, Industrial and Engineering Chemistry Research, **45** (11), Liu, Y., Tan, X. and Li, K., *Nonoxidative methane coupling in a $SrCe_{0.95}Yb_{0.05}O_{3-\alpha}$ (SCYb) hollow fiber membrane reactor*, p. 3782–3792

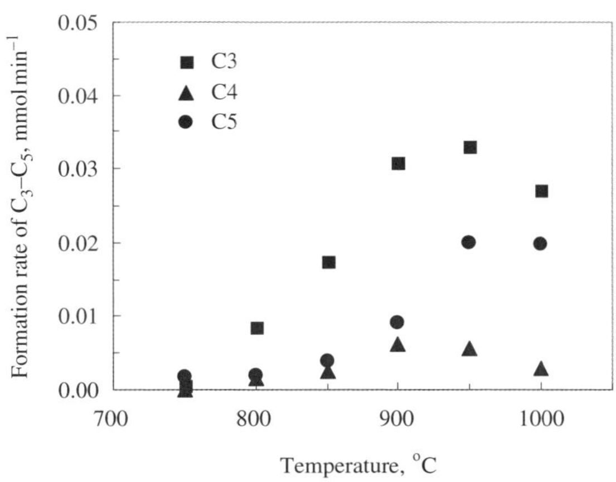

Figure 8.9 C_3–C_5 formation rates as a function of temperature (feed flow rate of methane stream = 4.3 $cm^3 min^{-1}$, methane feed concentration = 9.22 %, air feed flow rate = 40 $cm^3 min^{-1}$) [50]. Copyright (2006) American Chemical Society, Industrial and Engineering Chemistry Research, **45** (11), Liu, Y., Tan, X. and Li, K., *Nonoxidative methane coupling in a $SrCe_{0.95}Yb_{0.05}O_{3-\alpha}$ (SCYb) hollow fiber membrane reactor*, p. 3782–3793

in ethylene formation rate is more pronounced than that of ethane. Therefore, higher temperature favors the formation of ethylene rather than ethane, which suggests that the activation energy for ethylene formation is higher than that for ethane formation.

Figure 8.8 shows the effect of operating temperature on methane conversion, C_2 selectivity and yield at given feed flow rates. As expected, the methane conversion increases with increasing temperature until it reaches 82.7 % at 1000°C. However, as the temperature increases, the C_2 selectivity is gradually decreased. The maximum C_2 selectivity at 750°C is only about 60 %. This value is much less than expected as no oxygen takes part in the methane coupling reaction. The reasons for this may be due to the carbonization of methane in the absence of oxygen on the membrane surface at high temperatures. Furthermore, since SCYb also exhibits some oxygen ion conduction at very high temperatures [51], oxygen permeation into the methane side may further decrease the C_2 selectivity. As a result, only 15.3 % C_2 selectivity is achieved at a temperature of 1000°C. Nevertheless, the C_2 yield still increases from 3.6 % to 13.4 % as the temperature increases from 750°C to 950°C, but then is decreased with further increasing temperature. Although the obtained C_2 yield is still much lower than the desired industrial values (usually considered to be above 30 %), it is better than most literature values reported so far [52, 53].

On the other hand, C_2 products can be further converted into higher hydrocarbons such as C_3, C_4 and C_5, following a similar mechanism to the formation of C_2 compounds. Obviously, the formation of these higher hydrocarbons would contribute to the low C_2 selectivity although the formation rates are one to two orders of magnitude lower than the C_2 formation rate, as shown in Figure 8.9, which plots the formation rates of C_3, C_4 and C_5 hydrocarbons as a function of temperature. It can be seen that the C_3 formation rate is the highest, but the C_5 formation rate is higher than C_4. Furthermore, as temperature increases, the formation rates of these higher hydrocarbons increase up to a maximum value at around 950°C and then reduce with further increase in temperature. Obviously, a temperature higher than 950°C promotes the carbonization of methane with the result of a decrease in hydrocarbon production.

8.3 MEMBRANES AS REACTANT DISTRIBUTORS

Dehydrogenation of organics can also be performed by means of partial oxidation reactions, i.e. catalytic oxidative dehydrogenation. However, the catalytic oxidative dehydrogenation of organics is difficult to implement because, in general, the intermediate products are more reactive towards oxygen than the original hydrocarbons, resulting in a very low selectivity. As a result, total oxidation of the original substance is often reached instead of partial oxidation. One way to increase the selectivity towards the intermediate products is to control the oxygen concentration along the reactor length. This can be conveniently implemented by means of the membrane reactor concept. The use of a membrane allows for the oxygen and the hydrocarbon reactants to be fed into different compartments. The most preferable configuration is the catalytic membrane reactor (CMR) where the membrane also acts as a catalyst providing a reactive interface for the reaction to take place, while avoiding a long contact time between the desired products and oxygen [54]. However, packed bed membrane reactors (PBMR) are, so far, still the most preferred design because of the lower catalytic activity of the membranes. With the use of PBMRs, the oxygen concentration along the reactor length can be carefully controlled to favorably influence the reaction selectivity. By applying membrane reactors, it is possible to optimize the selectivity for different products in a given reaction network [55]. This is, of course, not possible in conventional fixed bed reactors, where the oxygen concentration is at a maximum at the reactor entrance, but decreases monotonically along the length of the reactor. As a result, the selectivity is typically low at the reactor inlet although the reaction rate is the highest, while the selectivity is increased gradually, but the conversion decreased, along the reactor length. Obviously, this negatively impacts on the overall yield. In membrane reactors, since the oxidant and the organics are fed into the reactors separately, it creates reactor conditions less prone to explosion and other undesirable safety effects, which are typically associated with the oxidation of gaseous hydrocarbons, and thus potentially broadens the range of feasible operations. In addition, the use of membranes to control the distribution of oxygen feed to the reaction side could significantly reduce hot spots as encountered in co-feed reactors [56].

In recent years, a number of studies have been carried out on membrane reactors for a variety of catalytic oxidative dehydrogenation reactions. These studies also have made use of either porous or dense membranes to control the addition of oxygen. The porous membranes include zeolite and alumina either intact or impregnated by a variety of catalysts. Dense membranes are typically made of metallic silver or its alloys, and various oxide ceramics that are capable of transporting oxygen selectively. They have good permselectivity towards oxygen, which allows for the use of air during the partial oxidation reaction. The permselectivity of porous membranes is not as high as dense membranes, but its permeability is much higher.

8.3.1 Porous Membrane Reactors

Porous membranes have several advantages over dense membranes, which motivate their use as oxygen distributors in catalytic dehydrogenation processes. Apart from the high gas permeability, the membrane cost of porous membranes either in membrane materials or in membrane production is more attractive than that of the dense membranes. It is known that large scale production of porous membranes with consistent quality is relatively mature. This makes the membranes affordable in industrial applications. Furthermore, most reactions take place in the temperature range 400–700 °C, where the dense membranes cannot provide enough oxygen permeation flux [57]. In addition, porous membranes generally have more thermochemical resistance to the harsh operating environment based on the nature of the membrane materials. The main problem related to porous membranes is the low permselectivity towards oxygen. Therefore, many efforts have been focused on the improvement of membrane permselectivity as well as retaining high permeability, which is generally achieved by modification of the membrane pores.

8.3.1.1 Techniques in Modification of Membrane Pores

The microporous membranes for oxidative dehydrogenation reactions are mainly of alumina, which are usually obtained by reducing the pore size of mesoporous membranes. As illustrated in Figure 8.10, several chemical vapour and liquid deposition methods have been used in depositing a solid inside the pores of mesoporous membranes [58]. If the deposition follows a heterogeneous mechanism, the modification will result in narrowing the pore size of the mesoporous membranes as shown in Figure 8.10(a). This is undesirable because in this case the pore size is reduced at the expense of a substantial loss of the porosity (and the permeability) of the membrane. Furthermore, the pore size distribution and noncircular pore cross section of most mesoporous membranes make it even less effective to reduce the pore size of these membranes by a heterogeneous mechanism in order to obtain microporous membranes. The ideal approach is to deposit solid of a fractal structure, as shown in Figure 8.10b, in the pores of mesoporous ceramic membranes. In general, along permeation flow direction, the deposit should be as narrow as possible (Figure 8.7c) in order to minimize the mass transfer resistance.

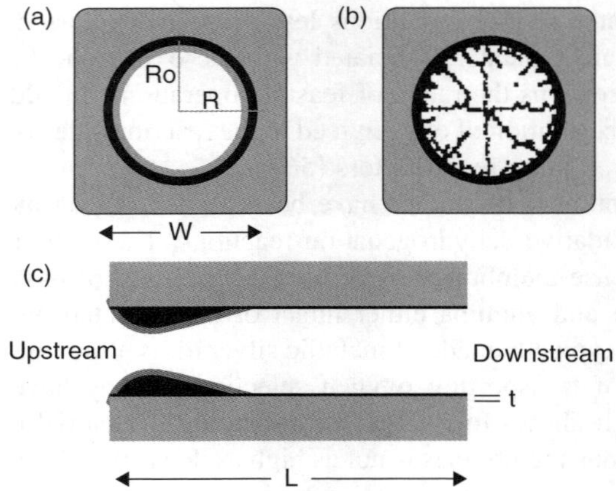

Figure 8.10 Schematic illustration of modifying the pore size of mesoporous membranes to obtain microporous inorganic membranes [58]. Reprinted from Separation and Purification Technology, **25** (1–3), Lin, *Microporous and . . .*, p. 39–55. Copyright (2001) with permission from Elsevier

The atomic layer chemical vapour deposition (ALCVD) method described by Klaus *et al.* [59] provides a flexibility in depositing a solid in mesoporous ceramic membranes in a controlled manner. To apply ALCVD for membrane modification, the pore surface of the mesoporous ceramic membranes containing –OH groups is exposed to an aluminum or silicon vapor precursor (e.g. $AlCl_3$) which will be chemsorbed to form the intermediate species on the pore surface (e.g. $–O–AlCl_2$). The pores are then evacuated to remove the entire aluminum or silicon precursor in the gas phase in the pores, and subsequently exposed to water vapour which reacts with the surface intermediate species to form one atomic layer of alumina or silica (e.g. –O–Al–OH).

Pan *et al.* [60] used a modified ALCVD method to decrease the pore size of mesoporous γ-alumina membranes in a more effective manner. In a modification of ALCVD on sol-gel derived γ-alumina by two half reactions between H_2O and $Al(CH_3)_3$ (at 180°C), the membrane pores with surface –OH groups were exposed to $Al(CH_3)_3$ vapour at 26 mbar for 5 min (to form the surface intermediate species: $–O–Al(CH_3)_2$ and were evacuated to leave some residual $Al(CH_3)_3$ molecules (at 1 mbar) in the membrane pores. Water vapour was introduced into the membranes to react with the surface intermediates to form one atomic layer of alumina (heterogeneous deposition). It also reacted with the residual $Al(CH_3)_3$ molecules in the pore space to form solid alumina particles in the gas phase (homogeneous deposition) that were then deposited on the membrane wall. This operation was repeated to grow additional layers of alumina.

As described before, microporous membranes may be easily catalytically modified by incorporating catalysts in the membrane pores. This makes the membrane itself possess catalytic activity. One of the advantages of catalytic membrane reactors (CMRs) is that the permeation flux in a CMR contributes to the rapid removal of products from the vicinity of the catalyst, resulting in an increase in selectivity in partial oxidation reactions. There are two approaches to perform the modification. One is to deposit catalyst layers by various techniques, such as the sol-gel method, and the other is to load catalysts inside the membrane pores by methods such as ion exchange. As an example, an oxidative dehydrogenation of propane catalyst (V–Mg–O or Ni based) was supported on commercial porous alumina membranes to form a catalytic membrane for the oxidative dehydrogenation of propane [61].

The sol-gel process is a widely used method for the synthesis of porous inert or catalytic membranes [62]. It is also attractive for multilayer depositions which can lead to controlled structure, composition and activity for the membrane. The main problem with sol-gel derived membranes is the thermal and hydrothermal instability (pore growth, grain coarsening, phase transition, defect development) above 400–500 °C. Mixed oxide systems or stabilizing agents can improve the membrane stability in some cases [63], where a precise control of the process parameters (synthesis of designed precursors, control of the hydrolysis–condensation reactions etc.) is required.

8.3.1.2 Applications of Porous Ceramic Membrane Reactors

Table 8.3 summarizes the studies of porous membranes as oxygen distributors either in packed bed membrane reactors (PBMRs) or catalytic membrane reactors (CMRs). It can be seen that all the studies reported that the use of membrane reactors in selective oxidation reactions can result in an increase in olefin selectivity and a suppression of deep oxidation. In most of the studies, the hydrocarbon was introduced into one side of the membrane where catalysts are packed, while the oxidant was fed on the other side and then permeated through the membrane into the reaction zone. Thus, the oxidant was added continuously along the catalyst zone.

In membrane reactors, not only the concentration profiles of the reactants differ from the profiles in classical fixed bed reactors, but there exists additional differences in residence time behaviour as well. For example, moderate nonuniform membranes outperform uniform membranes because they creates a pattern of moderately decreasing permeability in the direction of flow along the membrane length, which lowers the oxygen partial pressure [45]. As a result, the selectivity of the desired product is increased. For the selective oxidation mode, a combination of catalytically active membranes and catalyst particles is recommended [64, 65]. A key factor for the improvement of selectivity is the membrane permeability, which must be tailored to provide a homogeneous distribution of oxygen along the reactor and a suitable value for the permeation flux [66]. In addition to a higher selectivity, membrane reactors give more homogeneous temperature profiles, and easier control of operation.

A better operating mode can be found in the fluidized bed membrane reactor (FLBMR) [67], because in addition to high productivity, the FLBMR can provide a broader favorable operating range with respect to the oxygen–hydrocarbon ratio, which indicates a lower sensitivity against oscillations and disturbances in the reactant feed, corresponding to a higher safety of operation. Furthermore, because of excellent heat transfer characteristics, FLBMRs may be easily scaled up, especially for strongly exothermic reactions. Ostrowski *et al.* [68] compared the results of catalytic partial oxidation of methane (CPOM) to syngas on an Ni/Al_2O_3 catalyst in fixed bed and fluidized bed reactors with and without membranes by means of mathematical modeling and simulations. It was concluded that in both reactor types, the conversion of methane and yield of syngas could be significantly improved by means of integrated product separation on applying porous membranes. Due to the higher selectivity of separation the membrane fluidized bed was superior compared to the fixed bed.

For CMRs, the structural characteristics of the membrane, and in particular, the relative laminar and Knudsen contributions to permeation, have a strong influence on the reactor

Table 8.3 Porous membrane reactors for oxidative dehydrogenation reactions

Reactions	Membrane, membrane reactor and catalyst	Comments	Ref.
Ethane → ethylene	Inert ceramic alumina composite membrane tube; PBMR; $VO_x/\gamma\text{-}Al_2O_3$	33 % yield compared to 22 % in the FBR	[64, 65]
	Two layer porous stainless steel tube; FLBMR; $VO_x/\gamma\text{-}Al_2O_3$	37 % C_2 yield; better than PBMR; selectivity significantly improved	[67]
n-butane → butadiene and butene	Porous $\alpha\text{-}Al_2O_3$ tube; PBMR; V/MgO	Up to 61.1 % selectivity and 42.2 % yield of butadiene and butene	[45]
	Porous $\alpha\text{-}Al_2O_3$ tube; CMR; V/MgO	21 % yield obtained, higher than PBR	[22]
Methanol → formaldehyde	Porous 316L SS tube; PBMR-outside; Fe-Mo oxide catalyst packed	Formaldehyde selectivity up to 100 %	[126, 127]
Methanol → H_2	Composite multilayered $\alpha\text{-}Al_2O_3$ tube; CMR; Pt	Conversion up to 84 %, hydrogen selectivity between 50 and 55 %	[128]
Propane to propylene	Macroporous $\alpha\text{-}Al_2O_3$ coated by $\gamma\text{-}Al_2O_3$ or ZSM-5; CMR; $V/\gamma\text{-}Al_2O_3$ or V/ZSM-5	$V/\gamma\text{-}Al_2O_3$ CMR gives a high propylene selectivity	[69]
	Porous Al_2O_3 tube; PBMR; V/MgO	Selectivity remarkably increased over FBR	[66]
	Porous $\alpha\text{-}Al_2O_3$ tube; CMR; V/MgO or V/Al_2O_3	Propene yield up to 15 %; V/MgO is more selective but less active	[129]
	Porous $\alpha\text{-}Al_2O_3$ tube; CMR; MFI and V-MFI	Propene selectivity up to 15 %	[130]
Propane to Acrolein	Porous $\alpha\text{-}Al_2O_3$ tube; PBMR; $Ag_{0.01}Bi_{0.85}V_{0.54}Mo_{0.45}O_4$	Acrolein selectivity and yield 2–4 times higher than conventional reactors	[131]
Methane → C_2, (OCM)	Porous $\gamma\text{-}Al_2O_3$ tube; PBMR; $Mn\text{-}W\text{-}Na/SiO_2$	C_2 yields up to 27.5 %, 10 % higher than the co-feed reactor (30 % higher C_2 selectivity)	[132]
	Porous Vycor membrane tube; PBMR; Sm_2O_3	C_2 selectivity enhanced	[133]
Ethylene epoxidation	Porous 316L SS tube; PBMR; Cs-doped $Ag/\alpha\text{-}Al_2O_3$	PBMR-E > FBR > PBMR-O in performance	[134]

performance [22]. An appropriate thickness of the catalytic layer is necessary to enhance the selectivity [69].

Apart from porous membranes used as an oxygen distributors, they have also been used as distributors for other reactants, such as hydrogen. Li and coworkers used porous membranes to distribute hydrogen in water for elimination of dissolved oxygen (DO) from water [70, 71] and their performance was compared to conventional ion exchange columns. The membrane reactors have a shell and tube type construction. A catalyst (palladium doped anion exchange resin based on styrene and divinyl benzene) was packed in either the shell side or the tube side of the reactors. Hydrogen as a reactant was distributed through the porous membrane into water, where the reaction between the saturated DO and hydrogen takes place at the catalyst surface. A detailed analysis of these reactors for DO elimination is given below. Obviously, the models presented below can be easily applied to other reactants so long as the kinetic information of the reactions are known and assumptions made are identical.

8.3.1.3 Analysis of Membrane Reactors for Elimination of DO from Water

Analysis of porous membrane reactors, as shown in Figure 8.11 for dissolved oxygen removal, were carried out by Tan and Li [70]. Water saturated with dissolved oxygen was

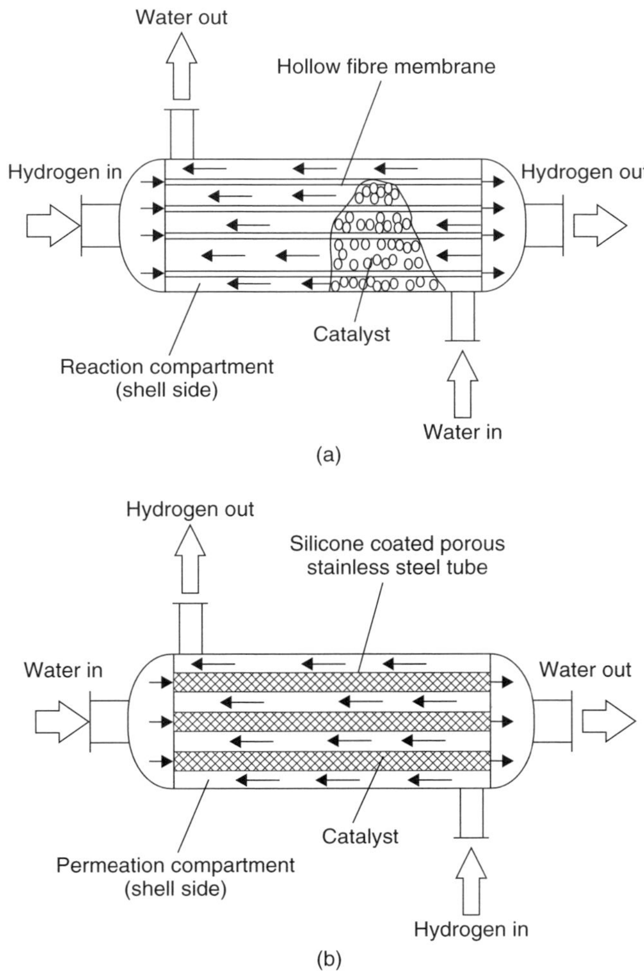

Figure 8.11 Diagrams of the membrane reactors: (a) shell packed reactor; (b) tube packed reactor [70]. Copyright (2000) Elsevier, Chemical Engineering Science, **55** (7), Tan, X. and Li, K., *Investigation of novel membrane reactors for removal of dissolved oxygen from water*, p. 1213–1224

used as the feed while purified hydrogen was used as both a sweep gas and a reducing agent. The following assumptions were adopted during the formulation of the model by Tan and Li [70]:

(1) Oxygen and hydrogen concentrations in the liquid phase are small so that the density and the flow velocity in the liquid stream are constant along the tube length.

(2) Dispersed plug flow conditions are expected to prevail as the reactors are packed with catalyst particles.

(3) Equilibrium is attained instantaneously between the bulk gas flow and the liquid interface.

(4) Gas phase pressure is constant.

(5) The concentration of hydrogen in bulk liquid is the same as that on the catalyst surface.

(6) The chemical reaction between the dissolved oxygen and hydrogen on the catalyst surface is considered to be elementary.

(7) The system is operated isothermally and in steady state.

Assumption (1) approximates the system realistically and simplifies the mass balance equations considerably. Assumption (5) is admissible since the catalyst has a strong affinity to the dissolved hydrogen [72] and the resistance of the hydrogen transfer between the bulk liquid and the active sites of the catalyst is thus negligible. Reaction between the dissolved oxygen and hydrogen is exothermic. However, in this case, both the concentrations of the dissolved oxygen and hydrogen in water are low and the assumption (7) can be justified as no noticeable temperature changes occurred during the entire experimental runs. Based on the above assumptions, the general conservation equations for both the dissolved hydrogen and oxygen in the liquid phase can be written as:

Tube packed membrane reactor:

$$-u\frac{\mathrm{d}C_i}{\mathrm{d}x}+D_{Li}\frac{\mathrm{d}^2C_i}{\mathrm{d}x^2}+k_{\mathrm{G}i}\cdot\frac{2}{R_t}\cdot(p_i-K_{\mathrm{H}i}C_i)-\upsilon_i(1-\varepsilon_\mathrm{b})k_\mathrm{R}C_A^2C_B=0 \qquad i=A,B \qquad (8.9a)$$

Shell packed membrane reactor:

$$-u\frac{\mathrm{d}C_i}{\mathrm{d}x}+D_{Li}\frac{\mathrm{d}^2C_i}{\mathrm{d}x^2}+k_{\mathrm{G}i}\cdot\frac{2nR_\mathrm{o}}{R_\mathrm{s}^2-nR_\mathrm{o}^2}\cdot(p_i-K_{\mathrm{H}i}C_i)-\upsilon_i(1-\varepsilon_\mathrm{b})k_\mathrm{R}C_A^2C_B=0 \qquad i=A,B \qquad (8.9b)$$

with the boundary conditions:

$$x=0 \qquad C_i=C_{i0}+\frac{D_{Li}}{u}\cdot\frac{\mathrm{d}C_i}{\mathrm{d}x} \qquad (8.9c)$$

$$x=Z \qquad \frac{\mathrm{d}C_i}{\mathrm{d}x}=0 \qquad (8.9d)$$

where the various symbols are defined in the notation section at the end of this chapter.

The conservation equation at the gas phase side can also be formulated and for the countercurrent flow arrangement, the equation can be written as:

Tube packed membrane reactor:

$$\frac{G\times10^{-3}}{RT}\cdot\frac{\mathrm{d}p_i}{\mathrm{d}x}=k_{Gi}\cdot2n\pi R_\mathrm{t}\cdot(p_i-K_{\mathrm{H}i}C_i) \tag{8.10a}$$

Shell packed membrane reactor:

$$\frac{G\times10^{-3}}{RT}\cdot\frac{\mathrm{d}p_i}{\mathrm{d}x}=k_{Gi}\cdot2n\pi R_\mathrm{o}\cdot(p_i-K_{\mathrm{H}i}C_i) \tag{8.10b}$$

with the boundary condition:

$$x=Z \qquad p_A=p_{A0} \qquad p_B=0 \tag{8.10c}$$

The extent of the DO removal is generally defined as:

$$\sigma_\mathrm{t}=1-C_B(x=Z)/C_{B0} \tag{8.11}$$

Obviously, Equation (8.11) is the overall DO removal by both the chemical reaction and membrane permeation, the latter of which is given by:

$$\sigma_\mathrm{p}=\frac{G\times10^{-3}\cdot p_B(\mathrm{out})}{RT\cdot LC_{B0}} \tag{8.12}$$

By performing the following dimensionless transform:

$$y_A=p_A/p_{A0} \qquad y_B=p_B/K_{\mathrm{H}B}C_{B0} \tag{8.13a}$$

$$x_A=K_{\mathrm{H}A}C_A/p_{A0} \qquad x_B=C_B/C_{B0} \tag{8.13b}$$

$$z=x/Z \tag{8.13c}$$

Equations (8.9a–d) and (8.10a–c) above then become:

$$\frac{\mathrm{d}x_A}{\mathrm{d}z}=\frac{1}{\gamma P_e}\cdot\frac{\mathrm{d}^2x_A}{\mathrm{d}z^2}+\beta N_\mathrm{m}(y_A-x_A)-N_\mathrm{R}x_A^2x_B \tag{8.14a}$$

$$\frac{\mathrm{d}x_B}{\mathrm{d}z}=\frac{1}{P_e}\cdot\frac{\mathrm{d}^2x_B}{\mathrm{d}z^2}+N_\mathrm{m}(y_B-x_B)-N_\mathrm{c}N_\mathrm{R}x_A^2x_B \tag{8.14b}$$

$$\frac{\mathrm{d}y_A}{\mathrm{d}z}=\frac{\beta N_\mathrm{m}}{\alpha N_\mathrm{F}}(y_A-x_A) \tag{8.15a}$$

$$\frac{\mathrm{d}y_B}{\mathrm{d}z}=\frac{N_\mathrm{m}}{N_\mathrm{F}}(y_B-x_B) \tag{8.15b}$$

and the boundary conditions of Equations (8.14a–b) and (8.15a–b) are rewritten as:

$$z=0 \qquad x_A=\frac{1}{\gamma P_e}\cdot\frac{dx_A}{dz} \qquad x_B=1+\frac{1}{P_e}\cdot\frac{dx_B}{dz} \tag{8.16a}$$

$$z=1 \qquad \frac{dx_A}{dz}=0 \qquad \frac{dx_B}{dz}=0 \tag{8.16b}$$

$$z=1 \qquad y_A=1 \qquad y_B=0 \tag{8.16c}$$

The extents of dissolved oxygen removal in dimensionless form are:

$$\sigma_t=1-x_B\,|_{z=1} \tag{8.17}$$

$$\sigma_p=N_F y_B\,|_{out} \tag{8.18}$$

The dimensionless parameters in Equations (8.14a), (8.14b), (8.15a) and (8.15b) are defined in Table 8.4, and are classified into two categories: operating condition modulus and system characteristic modulus.

It should be noted that the dimensionless parameters defined in Table 8.4 have physical significance. The mass transfer modulus, N_m, is a ratio of the maximum permeation rate for oxygen across the membrane to the liquid space velocity. It is proportional to the mass transfer area of the membrane per unit of the reactor volume. The reaction modulus, N_R, is the multiplication of reaction rate with the residence time of liquid water and is proportional to

Table 8.4 Definitions of dimensionless parameters

Dimensionless parameter	Expression
Mass transfer modulus	$N_m=\frac{2k_{GB}/R_t}{u/Z}\cdot K_{HB}$ for tube-packed reactor
	$N_m=\frac{k_{GB}K_{HB}}{u/Z}\cdot\frac{2R_t}{R_s^2-nR_t^2}$ for shell-packed reactor
Reaction modulus	$N_R=\frac{\upsilon_B(1-\varepsilon_b)k_R}{u/Z}\left(\frac{p_{A0}}{K_{HA}}\right)\cdot C_{B0}$
Flow modulus	$N_F=\frac{G\times10^{-3}\times K_{HB}}{LRT}$
Concentration modulus	$N_C=\frac{\upsilon_B p_{A0}/K_{HA}}{\upsilon_A C_{B0}}$
Peclet number	$P_e=\frac{uZ}{D_{LB}}$
Ratio of the solubilities	$\alpha=K_{HA}/K_{HB}$
Ratio of the mass transfer rate	$\beta=\frac{k_{GA}K_{HA}}{k_{GB}K_{HB}}$
Ratio of the axial dispersion coefficients	$\gamma=D_{LB}/D_{LA}$

the amount of catalyst packed in an unit volume of the reactor. It indicates the importance of the chemical reaction on the dissolved oxygen removal. The flow modulus, N_F, may be rewritten as:

$$N_F = \frac{\frac{G \times 10^{-3}}{RT} \cdot K_{HB} C_{B0}}{L C_{B0}} \tag{8.19}$$

Therefore it is the ratio of the equilibrium oxygen flow in the gas phase and the initial dissolved oxygen flow in liquid phase. Obviously, it reflects the effect of relative flows of hydrogen gas and water on dissolved oxygen removal and thus may be employed to assess the minimum amount of reducing agent needed for the process. The concentration modulus, N_c, is the concentration ratio between the dissolved hydrogen and oxygen at the inlet conditions. It shows the effect of hydrogen pressure on the dissolved oxygen removal. Peclet number, P_e, indicates the degree of the axial dispersion. The bigger the Peclet number, the less the axial dispersion. With the P_e number increasing from 0 to ∞, the flow pattern transforms from perfect mixing to plug flow.

Equations (8.14) and (8.15) are a group of highly nonlinear differential equations and can only be solved numerically except for some special cases. By transforming these equations into a group of first order ordinary differential equations, they can be easily integrated by conventional methods. Some initial values are not known at either end to start the integration. Therefore, the prediction–calibration technique is used to overcome this difficulty. The tolerance function is:

$$Er = \sqrt{\left[\left(\frac{dx_A|_{z=1}}{dz}\right)^2 + \left(\frac{dx_B|_{z=1}}{dz}\right)^2 + (y_A|_{z=1} - 1)^2 + (y_B|_{z=1})^2\right]} \tag{8.20}$$

Detailed procedures for solving the above equations and the solution algorithm can be found elsewhere [70].

When the amount of hydrogen introduced into the reactor is so large that its concentration in water is approximately constant everywhere within the reactor, the chemical reaction may be considered as pseudo first order with respect to the dissolved oxygen concentration. Accordingly, only the mass balance for oxygen needs to be formulated, that is:

$$\frac{dx_B}{dz} = \frac{1}{P_e} \cdot \frac{d^2 x_B}{dz^2} + N_m (y_B - x_B) - N_c N_R x_B \tag{8.21}$$

Together with Equation (8.15b) and the boundary conditions, an analytical solution using the Laplace transform technique may be obtained as:

$$x_B = \sum_{k=1}^{3} \frac{P_1(\mu_k)}{Q'(\mu_k)} \cdot e^{\mu_k z} \tag{8.22a}$$

$$y_B = \sum_{k=1}^{3} \frac{P_2(\mu_k)}{Q'(\mu_k)} \cdot e^{\mu_k z} \tag{8.22b}$$

where:

$$Q(s) = s^3 - (P_e + N_{\rm m}/N_F)\cdot s^2 + P_e(N_{\rm m}/N_{\rm F} - N_{\rm m} - N_{\rm c}N_{\rm R})\cdot s + P_eN_{\rm c}N_{\rm R}N_{\rm m}/N_{\rm F} \tag{8.23a}$$

$$Q'(s) = 3s^2 - 2(P_e + N_{\rm m}/N_{\rm F})s + P_e(N_{\rm m}/N_{\rm F} - N_{\rm m} - N_{\rm c}N_{\rm R}) \tag{8.23b}$$

$$P_1(s) = x_0 s^2 - (N_{\rm m}/N_{\rm F}\cdot x_0 + P_e x_0 - x_0')s - N_{\rm m}/N_{\rm F}\cdot(x_0' - P_e x_0) - P_e N_{\rm m} y_0 \tag{8.23c}$$

$$P_2(s) = y_0 s^2 - (P_e y_0 - N_{\rm m}/N_{\rm F}\cdot x_0)s - P_e(N_{\rm m} + N_{\rm c}N_{\rm R})y_0 - (x_0' - P_e x_0)N_{\rm m}/N_{\rm F} \tag{8.23d}$$

and μ_k is the root of the following equation:

$$s^3 - (P_e + N_{\rm m}/N_{\rm F})\cdot s^2 + P_e(N_{\rm m}/N_{\rm F} - N_{\rm m} - N_{\rm c}N_{\rm R})\cdot s + P_eN_{\rm c}N_{\rm R}N_{\rm m}/N_{\rm F} = 0 \tag{8.24}$$

The boundary values of x_0 y_0 and x_0' may be obtained by substituting Equations (8.22a) and (8.22b) into the boundary conditions:

$$\sum_{k=1}^{3}\left(1 - \frac{\mu_k}{P_e}\right)\cdot\frac{P_1(\mu_k)}{Q'(\mu_k)} = 1 \tag{8.25a}$$

$$\sum_{k=1}^{3}\frac{\mu_k P_1(\mu_k)\exp(\mu_k)}{Q'(\mu_k)} = 0 \tag{8.25b}$$

$$\sum_{k=1}^{3}\frac{P_2(\mu_k)\exp(\mu_k)}{Q'(\mu_k)} = 0 \tag{8.25c}$$

The DO removal can be expressed as:

$$\sigma_{\rm t} = 1 - \sum_{k=1}^{3}\frac{P_1(\mu_k)\cdot e^{\mu_k}}{Q'(\mu_k)} \tag{8.26}$$

$$\sigma_{\rm p} = N_{\rm F}\sum_{k=1}^{3}\frac{P_2(\mu_k)}{Q'(\mu_k)} \tag{8.27}$$

In the absence of axial dispersion, $P_e \to \infty$, then Equations (8.26) and (8.27) reduce to:

$$\sigma_{\rm t} = 1 - \frac{1}{\lambda_1 - \lambda_2}[\lambda_1 e^{\lambda_1} - \lambda_2 e^{\lambda_2} + (N_{\rm m} y_0 - N_{\rm m}/N_{\rm F})(e^{\lambda_1} - e^{\lambda_2})] \tag{8.28}$$

$$\sigma_{\rm p} = N_{\rm m}\cdot\frac{e^{\lambda_1} - e^{\lambda_2}}{\lambda_1 e^{\lambda_1} - \lambda_2 e^{\lambda_2} + (N_{\rm m} + N_{\rm c}N_{\rm R})(e^{\lambda_1} - e^{\lambda_2})} \tag{8.29}$$

where :

$$\lambda_{1,2} = \frac{-(N_m + N_c N_R - N_m/N_F) \pm \sqrt{(N_m + N_c N_R - N_m/N_F)^2 + 4N_c N_R N_m/N_F}}{2} \quad (8.31)$$

$$y_0 = \frac{N_m}{N_F} \cdot \frac{e^{\lambda_1} - e^{\lambda_2}}{\lambda_1 e^{\lambda_1} - \lambda_2 e^{\lambda_2} + (N_m + N_c N_R)(e^{\lambda_1} - e^{\lambda_2})} \quad (8.32)$$

Of course, when the gas flow is so high that the oxygen concentration in gas phase approximately equals to 0, Equation (8.28) becomes:

$$\sigma_t = 1 - \exp[-(N_m + N_c N_R)] \quad (8.33)$$

Equations (8.28) and (8.33) are useful for optimization of the operating conditions of the membrane reactors studied.

Some simulation and experimental data are presented in the following sections. In the first, the effect of hydrogen flow rate and pressure on DO removal was studied theoretically using the basic parameters listed in Table 8.5. All the parameters employed are in the range of the experimental conditions. In the second, the simulation study was extended to the axial dispersion effect and the magnitude of physical stripping and chemical reaction in the reactor. Finally, the experimental results were compared with the modelling values to confirm the validity of the model developed.

Effects of Operating Pressure and Flow Rate of Hydrogen

For the membrane reactors designed and the reaction system employed, the overall reaction rate constant, k_R can be written as:

$$\frac{1}{k_R} = \frac{R_p C_A^2}{3k_S} + \frac{1}{k_r} \quad (8.34)$$

where k_S is the mass transfer coefficient between the bulk liquid and the surface of the catalyst and k_r is the reaction rate constant. For reaction kinetic controlled processes, i.e., $1/k_r >> R_p C_A^2/3k_S$, an increase in hydrogen pressure will favor dissolved oxygen removal as the high

Table 8.5 Basic parameters used for the simulation

Mass-transfer modulus, N_m	0.5
Reaction modulus, N_R	1
Feed flow modulus, N_F	10
Feed concentration modulus, N_c	1.5
Peclet number, P_e	(∞ no axial dispersion)
Ratio of the solubility	1.645
Ratio of the mass transfer rate	2.343
Ratio of the axial dispersion coefficients	1

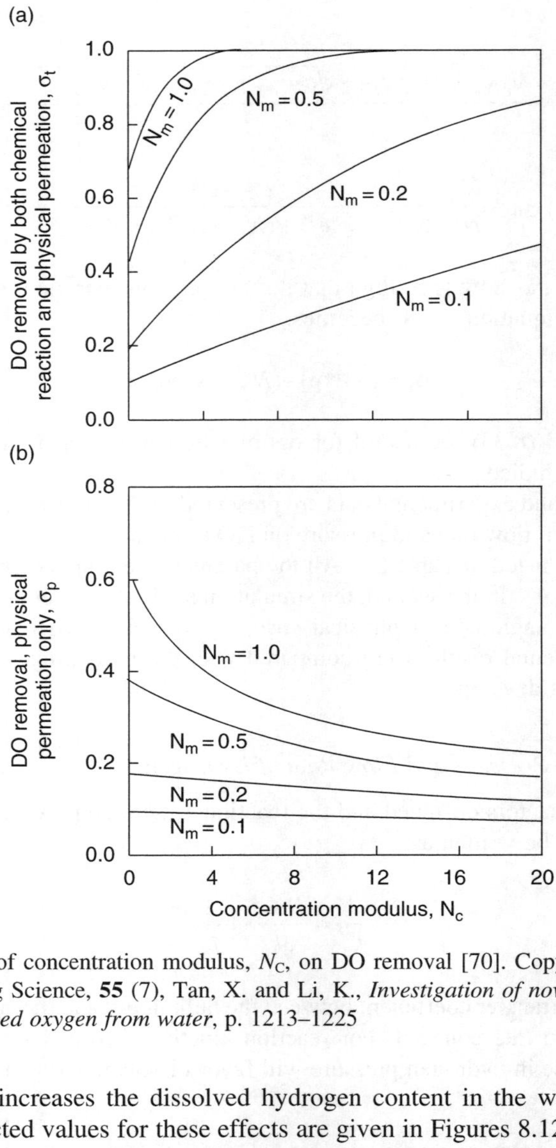

Figure 8.12 Effect of concentration modulus, N_C, on DO removal [70]. Copyright (2000) Elsevier, Chemical Engineering Science, **55** (7), Tan, X. and Li, K., *Investigation of novel membrane reactors for removal of dissolved oxygen from water*, p. 1213–1225

operating pressure increases the dissolved hydrogen content in the water giving a higher reaction rate. Predicted values for these effects are given in Figures 8.12 and 8.13. In Figure 8.12a, the hydrogen pressure, illustrated as the concentration modulus, is plotted against the dissolved oxygen removal for different values of N_m, which is related to the membrane area employed per unit volume of the reactor. As can be seen, the dissolved oxygen can be removed completely from water by means of increasing the system pressure or partial pressure of hydrogen at a N_m value greater than 0.5. Smaller values of the mass transfer modulus N_m due to either insufficient membrane area or high membrane resistance result in a poor performance of the reactor in terms of the dissolved oxygen removal. Figure 8.12b indicates that at higher N_m values ($N_m = 1$), the dissolved oxygen removal due to the physical permeation decreases greatly when the hydrogen partial pressure (i.e. the concentration modulus, N_C) is increased. However, the overall dissolved oxygen removal still increases (Figure 8.12a). This is because more hydrogen is transferred into the water due to the high hydrogen pressure and it reacts

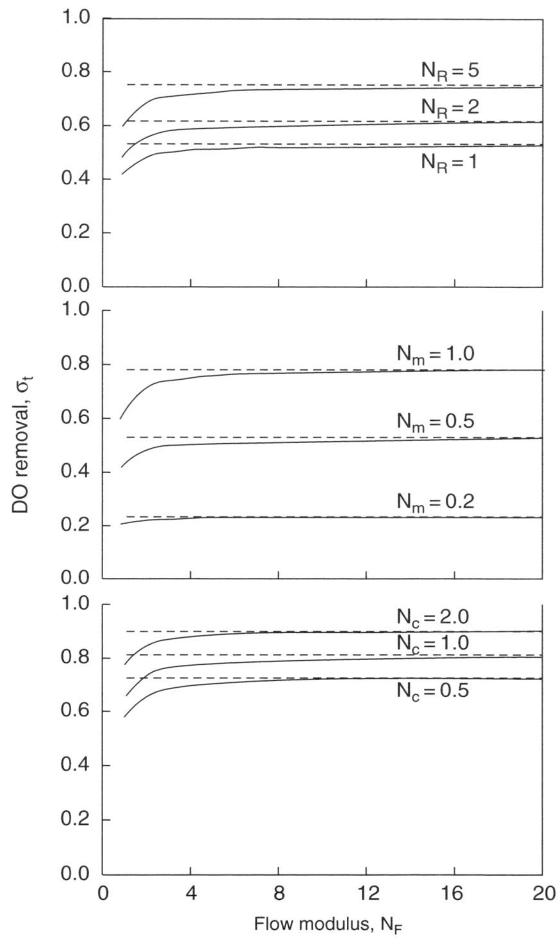

Figure 8.13 Effect of flow modulus, N_F, on DO removal [70]. Copyright (2000) Elsevier, Chemical Engineering Science, **55** (7), Tan, X. and Li, K., *Investigation of novel membrane reactors for removal of dissolved oxygen from water*, p. 1213–1226

with the dissolved oxygen resulting in considerable depletion of DO in water, hence reducing the driving force for the oxygen permeating through the membrane.

When the overall rate of the reaction is controlled by the liquid film resistance adjacent to the catalyst surface, i.e., $R_pC_A^2/3k_S >> 1/k_r$, the dissolved oxygen removal becomes independent of the hydrogen pressure. As can be seen from Equations (8.34) and (8.9), the reaction kinetic becomes first order with respect to the dissolved oxygen. Such kinetic behaviour has been observed experimentally by Suppiah *et al.* [73] who carried out the experiments under conditions where the hydrogen is present in excess of the stoichiometric requirement.

The effect of the hydrogen flow rate, illustrated as flow modulus, N_F, on dissolved oxygen removal is shown in Figure 8.13 for different operating and system conditions. It can be seen that as the N_F value increases, the dissolved oxygen removal is improved but only up to a certain value of the flow modulus, N_F, which may be termed as the effective flow modulus. Any further increase in the flow modulus after this effective flow modulus does not result in better DO removal as shown in the figure. This behaviour is true for different operating and system conditions and it is interesting to note that the effective flow modulus, in this case about $N_F = 10$, remains almost identical and does not seem to be dependent on the various operating and system conditions such as N_R, N_m and N_c. In Figure 8.13, the theoretical data obtained at the condition $N_F \to \infty$ is also plotted (dashed lines). As can be seen, at $N_F = 10$, the DO removal approaches the dashed line indicating that the maximum driving force of the oxygen is obtained and further increase in hydrogen flow rate results in no effect on the DO removal.

Effect of Axial Dispersion

In order to study the effect of axial dispersion on DO removal, the hydrogen is assumed to be in excess in the reactor and thus Equation (8.21) can be employed for the simulation. The result is shown in Figure 8.14 for different operating and system conditions. The dashed lines are the theoretical data obtained in the absence of axial dispersion. It can be seen that axial dispersion has a negligible effect on the DO removal when the Peclet number, P_e, is greater than 35. At $P_e << 35$, the effect of axial dispersion becomes considerable, for example, at $P_e = 5$, the differences in DO removal calculated from models with and without consideration

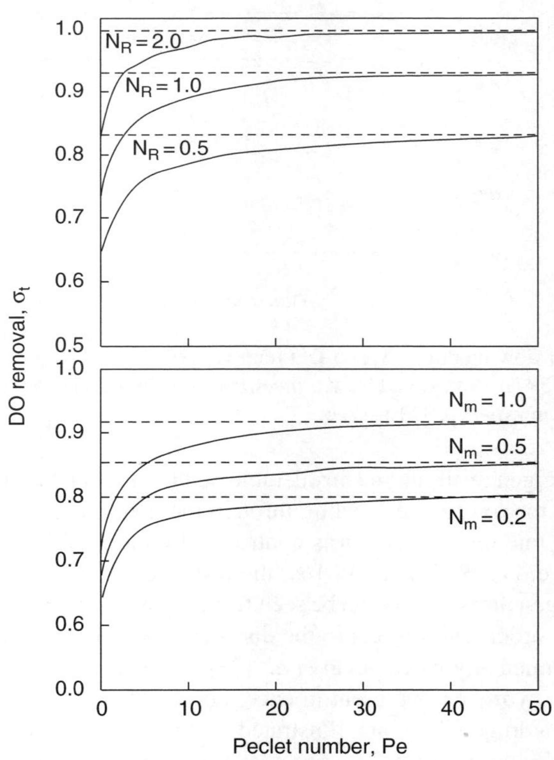

Figure 8.14 Effect of the Peclet number on DO removal [70]. Copyright (2000) Elsevier, Chemical Engineering Science, **55** (7), Tan, X. and Li, K., *Investigation of novel membrane reactors for removal of dissolved oxygen from water*, p. 1213–1227

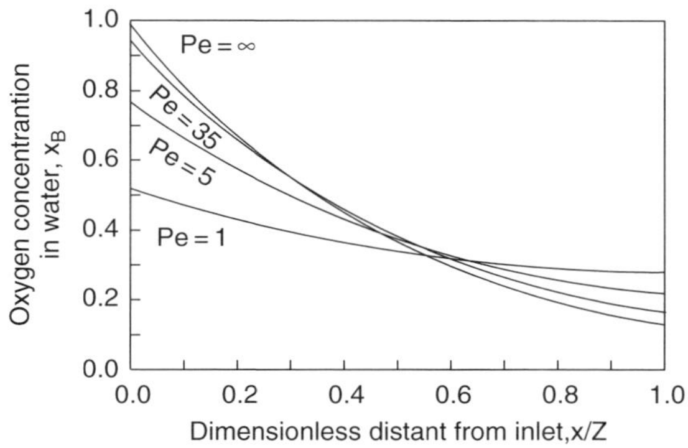

Figure 8.15 Concentration distributions at different Peclet numbers [70]. Copyright (2000) Elsevier, Chemical Engineering Science, **55** (7), Tan, X. and Li, K., *Investigation of novel membrane reactors for removal of dissolved oxygen from water*, p. 1213–1228

of the axial dispersion is more than 15 %. Figure 8.15 shows the effect of Peclet number on the DO concentration profiles in water. It can be seen, again, that a Peclet number greater than 35 is a reliable value to indicate a negligible effect of the axial dispersion. It should be noted that $P_e > 35$ obtained from the membrane reactors studied is slightly higher than that obtained from the normal fixed bed reactors [74].

Magnitude of Physical Stripping and Chemical Reaction

As described above, dissolved oxygen removal is a result of both physical stripping and chemical reaction. Their contributions to DO removal are related to the dimensionless moduli, N_m and N_R, respectively. Obviously, they are dependent on the amount of catalyst and the membrane area used. In case of excess hydrogen operation, the DO removal can be estimated by Equations (8.28) to (8.33). In general cases where the membrane resistance to hydrogen transfer is considerable, i.e. a hydrogen concentration gradient in the water is present, the relationships between the DO removal and mass transfer/reaction moduli must be calculated numerically. In Figure 8.16, the reaction modulus, N_R is plotted against the transfer modulus, N_m for different DO removal. It can be seen from the figure that there must be enough membrane area in order to remove the dissolved oxygen to a certain extent, which also depends on values of the concentration modulus, N_c or the hydrogen partial pressure. If the transfer modulus is too small, i.e. $N_m < 0.6$, DO removal over 99.9 % is not attainable even if the reaction modulus, N_R approaches infinity. On the other hand, it is also important to have a higher reaction modulus, N_R for the complete removal of DO once N_m reaches a certain extent, otherwise, a higher hydrogen partial pressure, i.e. higher N_c values, must be provided. Accordingly, Figure 8.16 may also be used as a diagram for the preliminary membrane reactor design.

The contributions of physical and chemical DO removal in the membrane reactors are illustrated in Figure 8.17, where the percentage of physical removal is plotted against the surface area ratio between the catalyst and the membrane, A_s/A_t. As can been seen, for a given membrane area, the physical contribution decreases as the catalyst surface is increased. When A_s/A_t is greater than 1.0, more than 95 % of the DO removal is achieved by the chemical method. It thus indicates that, at a high area ratio, A_s/A_t the removal of DO by physical

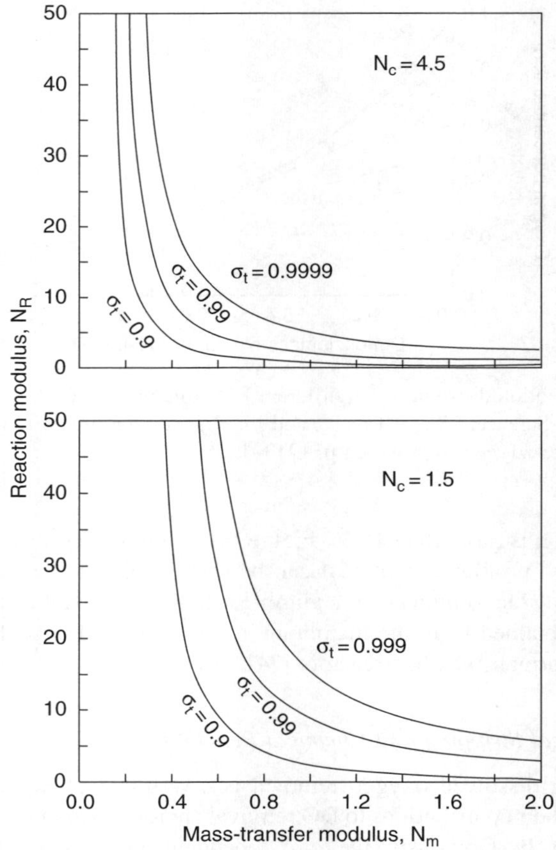

Figure 8.16 Mass transfer modulus N_m and reaction modulus N_R required for different DO removal at two different values of N_c [70]. Copyright (2000) Elsevier, Chemical Engineering Science, **55** (7), Tan, X. and Li, K., *Investigation of novel membrane reactors for removal of dissolved oxygen from water*, p. 1213–1229

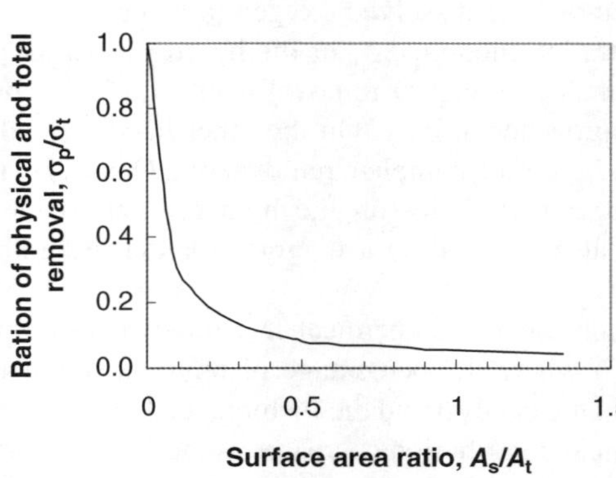

Figure 8.17 Contribution to the DO removal

stripping becomes small and negligible and the membranes used in these reactors become, somewhat, gas distributors, to ensure the continuous supply of the reducing agent, i.e. H_2. It has been shown by Imaoka *et al.* [75] and Sato *et al.* [76] that chemical reduction of the dissolved oxygen is usually more effective than the physical method, therefore, a high area ratio, A_s/A_t may be selected for design purposes. However, the membrane area may also be an important factor as it directly relates to the hydrogen supply and its distribution within the reactor. Furthermore, use of membranes in the reactors provides an additional advantage, as the gas and liquid flows can be operated independently and operating constraints such as flooding and loading, usually found in packed columns, do not exist in the membrane reactors investigated.

8.3.2 Dense Ceramic Membranes

Dense membranes have good permselectivity towards oxygen, which allows for the use of air during partial oxidation reactions. Although silver and its alloys with other metals like vanadium are also of unique oxygen permselectivity, most recent studies have been primarily focused on dense ceramic membranes whose oxygen permselectivities are based on the oxygen ionic conduction performance of solid electrolytes. In such a membrane reactor, oxygen is added discretely along the length of the reactor into the catalyst bed so that the local hydrocarbon to oxygen ratio in the reaction zone can be kept high, leading to much high selectivity. As a result, the product yield can be greatly increased, since the total amount of oxygen participating in the reaction is not reduced. Furthermore, the use of air instead of pure oxygen as the oxidant in practical industrial processes would decrease the cost remarkably, since the separation of nitrogen from subsequent product streams becomes unnecessary.

As mentioned in Chapter 6, a large number of oxygen permeable membrane materials or oxide conducting electrolyte ceramics have been developed and studied for their transport properties, chemical and thermal stabilities and potential applications. Increasing both the ion and electron conductivity and improving the stability of the materials are the two major challenges [77–81]. For example, dual phase membranes that are usually composed of a continuous phase for ion conduction and a dispersed one for improving electron conduction, have been developed so that enhanced oxygen permeation flux can be achieved [82–84]. In these cases, the two phase materials can be either ceramic–metal [84] or ceramic–ceramic combinations [85].

8.3.2.1 Configurations of the Dense Ceramic Membrane Reactors

In most studies, the membrane reactor is built in the configurations shown in Figure 8.18 where a disc shaped membrane made from mixed conducting ceramics is attached to a ceramic or quartz tube with a glass or gold ring as the sealant. The hydrocarbon stream, usually diluted by an inert gas, and the oxygen containing stream are fed into the opposite sides of the membrane, respectively. At high temperature, the hydrocarbon reacts with the oxygen permeated from the air side to form products. Sometimes, a catalyst specifically for a reaction is used, as shown in Figure 8.18b, which is directly packed or coated as a porous layer on the membrane surface. Another configuration is shown in Figure 8.18c where suitable electrodes attached to the membrane surfaces are connected to an external circuit for the

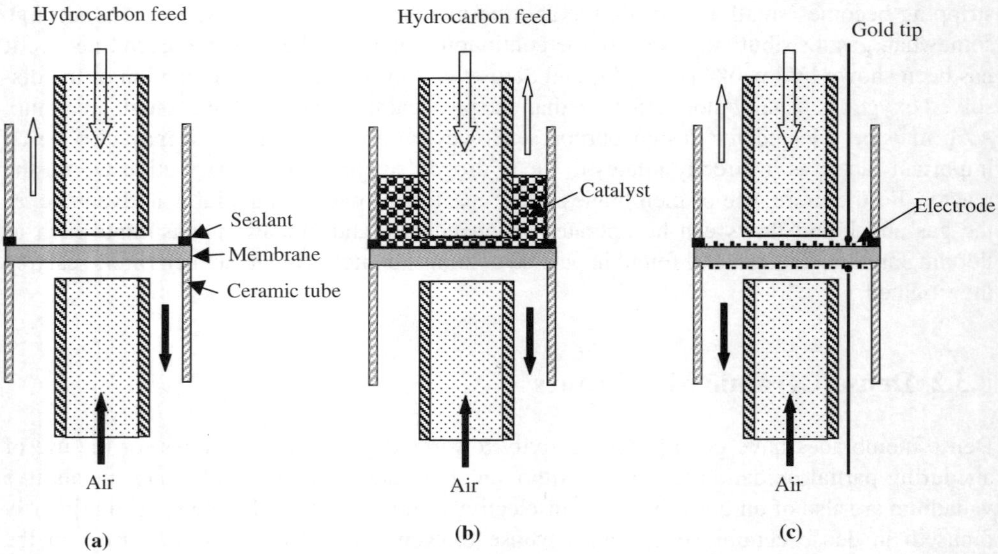

Figure 8.18 Configuration of dense membrane reactors with disc shaped membranes

electric current with gold wires. This case is suitable for the oxide electrolytes with less electron conductivity [86].

The disc shaped membrane reactors are easily fabricated, and thus are especially suitable for demonstration of various reactions. However, the disc membranes have limited membrane areas (around 5 cm^2), leading to low conversions and yields. In order to increase the membrane area for oxygen permeation, tubular reactors can be applied. As shown in Figure 8.19a, the membrane reactor is composed of a dense oxide membrane tube enclosed in a ceramic tube acting as the reactor shell. The hydrocarbon and air streams are separately introduced into the tube side and the shell side, respectively. Similar to the disc membrane reactors, the catalyst may be packed in the membrane tube (hydrocarbon side) to promote a desired reaction, as shown in Figure 8.19b. It should be mentioned that the catalyst can also be packed in the shell side. The disadvantages of these two structures include the difficulties of connecting and high temperature sealing. Since the two ends of the membrane tube are fixed in the reactor, the thermal expansion of the tube should also be taken into consideration. However, use of a dead end tubular membrane makes the thermal expansion of the membrane tube at high operating temperatures no longer a problem as shown in Figure 8.19c. As a result, the construction of the membrane reactor can be simplified.

Another reactor configuration is hollow fibre membrane reactors, which possess the largest membrane area for oxygen permeation and surface reactions compared with the aforesaid configurations. Figure 8.20 illustrates schematically a hollow fibre membrane reactor where a bundle of hollow fibre membranes is glued to two short ceramic tubes with a sealant and located inside a ceramic tube [87]. Two short silicone rubber tubes are used on both sides of the fibre bundle to offset the thermal expansion of the hollow fibres at high temperatures. Since both the sealing and connecting joints are located outside of the heating zone, normal sealant can be used, and therefore, the problem of high-temperature sealing may be easily avoided.

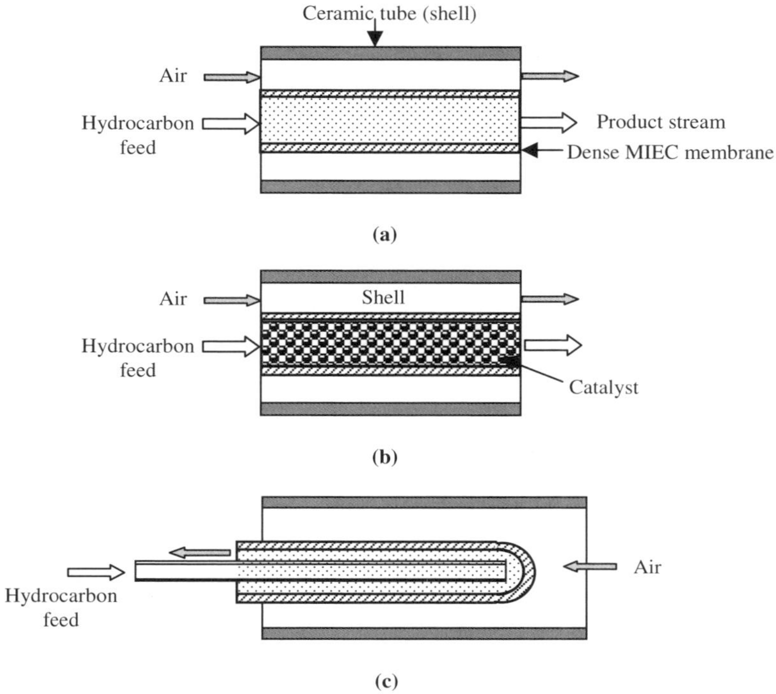

Figure 8.19 Configuration of the tubular membrane reactor

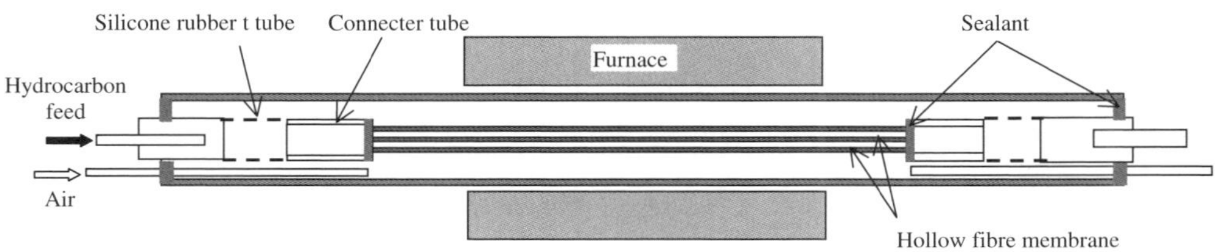

Figure 8.20 Hollow fibre membrane reactor. Copyright (2005) John Wiley & Sons, Ltd, AIChE J., **51** (7), Tan, X., Liu, Y. and Li, K., *Mixed conducting ceramic . . .* , p. 1991–2000

8.3.2.2 Applications of Dense Ceramic Membrane Reactors

As described above, dense ceramic membranes made from mixed oxygen ionic and electronic conducting oxides allow separation of oxygen from air at elevated temperatures (usually > 700 °C). Therefore, by using dense ceramic membrane reactors, expensive oxygen separation from air by cryogenic separation is no longer required, which could significantly reduce the costs for the dehydrogenation reactions [88]. Considering that oxygen can only be transported at high temperatures, these dense ceramic membranes have so far been limited to a few reactions such as methane conversions.

The conversion of methane into other valuable hydrocarbons has significant industrial importance. So far there are two different main methods that have been studied in converting methane to chemicals and liquid fuels, namely direct or indirect routes. The direct route is the oxidative coupling of methane (OCM) into C_2 products (ethane and ethylene) and the indirect route refers to the production of syngas (H_2 and CO mixture) by steam reforming or partial oxidative reaction of methane followed by a further conversion into higher hydrocarbons by the Fischer–Tropsch process.

Partial Oxidation of Methane (POM) to Syngas

Of all the potential applications for mixed conducting ceramic membranes, the partial oxidation of methane to syngas ($CO + H_2$) is thought to be one of the most commercially important applications [89, 90]. Recent progress in applying mixed conducting membranes to partial oxidation of methane to syngas has been reported in the literature [88, 91].

In general, there are two types of mechanism that have been proposed for the partial oxidation of methane to syngas on catalysts [92]. In the direct partial oxidation (DPO) mechanism, CH_4 and O_2 react to form CO and H_2 as primary products:

$$CH_4 + \frac{1}{2}O_2 \rightarrow CO + 2H_2 \tag{8.35}$$

On the other hand, in the indirect mechanism, one part of CH_4 is oxidized to form CO_2 and H_2O:

$$CH_4 + 2O_2 \rightarrow CO_2 + 2H_2O \tag{8.36}$$

with the kinetic equation:

$$R_1 = A_1 p_{CH_4} p_{O_2} \exp(-E_1/RT) \tag{8.36a}$$

Then the CO_2 and H_2O formed are converted with CH_4 by dry reforming and steam reforming, respectively.

$$CH_4 + CO_2 \leftrightarrow 2CO + 2H_2 \tag{8.37}$$

with the kinetic equation:

$$R_2 = A_2 p_{CH_4} p_{CO_2} \exp(-E_2/RT)\left(1 - \frac{p_{CO}^2 p_{H_2}^2}{K_3 p_{CH_4} p_{CO_2}}\right) \tag{8.37a}$$

$$CH_4 + H_2O \rightarrow CO + 3H_2 \tag{8.38}$$

with the kinetic equation:

$$R_3 = A_3 p_{CH_4} p_{H_2O} \exp(-E_3/RT)\left(1 - \frac{p_{CO} p_{H_2}^3}{K_3 p_{CH_4} p_{H_2O}}\right) \tag{8.38a}$$

Therefore, the indirect pathway is also called the combustion and reforming reactions (CRR) mechanism.

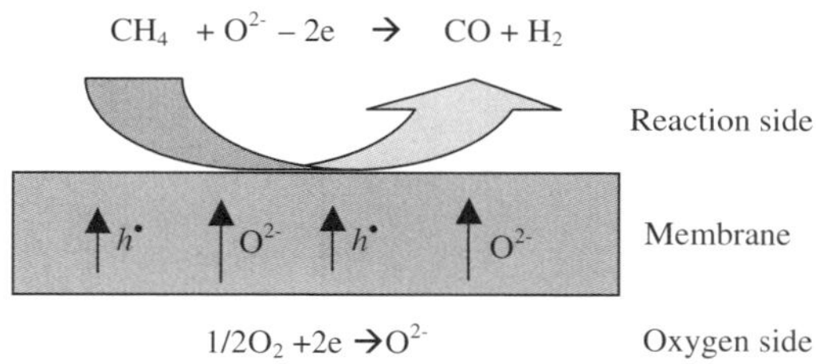

Figure 8.21 Schematic diagram of the membrane reactor for POM reaction

In solid oxide membrane reactors, the mechanism for the POM reaction is schematically illustrated in Figure 8.21. The steps may be described as follows: (1) oxygen is first adsorbed on the membrane surface of the high oxygen pressure side, then the adsorbed oxygen was dissociated into oxygen ions and electron holes; (2) the O^{2-} and electron holes are diffused to the other side of the membrane surface via bulk diffusion under the oxygen partial pressure gradient; (3) the oxygen ions and electron holes recombined to form molecular oxygen via the reverse process of (1); (4) molecular oxygen is then diffused via the gas phase to the surface of the catalyst; (5) molecular oxygen is dissociated into oxygen ionic or other types of oxygen species on the catalyst, and then reacted with methane to form syngas or CO_2.

Developing a mixed conducting membrane with a high oxygen flux and long term phase stability would be the main challenge for successful application in the syngas production process. Although membranes made from the La–Sr–Co–Fe family of materials exhibit high ionic and electronic conductivity, hence the high oxygen permeation rate, considerable doubts on these cobalt containing perovskite membranes still remain due to their poor phase stability in a reducing environment. Partial decomposition of the membrane surface exposed to the reducing environment has been observed [88, 93, 94]. Recently, much effort has been placed on the improvement of the oxygen permeability and the chemical stability of mixed conducting materials [95–98], especially several new materials including Ba–Sr–Co–Fe by Shao *et al.* [99, 100] and Ba–Co–Fe–Zr by Tong *et al.* [101], which have been developed and found to be of good stability in syngas generation. Membranes made from $A_2B_2O_5$ brownmillerite structure materials have been tested and a high yield of syngas and long term operation have been reported [102]. The performance of these recently studied materials is summarized in Table 8.6.

Problems relating to limited thermodynamic stability and poor dimensional stability of candidate materials were addressed by Hendriksen *et al.* [86]. It was suggested that improvement of the electrode performance, by catalyst addition or enhancement of the surface area, has to be introduced considering its effect on the stability of the membrane. Slow kinetics at the anode may protect the material from the syngas environment, and improvement of the anode performance has to be matched by improvement of cathode performance (and possibly a reduction of membrane thickness) to assure the membrane operates within the stability limits of the material. Reduction of membrane thickness to the level where a support is needed may be difficult for the perovskite materials due to their dimensional instability in a reducing environment.

Oxidative Coupling of Methane (OCM)

Oxidative coupling of methane (OCM) to C_2 hydrocarbons (ethane and ethylene) is a promising process to upgrade natural gas. Dense mixed conducting ceramic membranes have an

Table 8.6 Mixed conducting oxide membrane reactors for partial oxidation of methane to syngas

Membrane	Temp. (°C)	Stability	Comments X(CH_4 conversion) S_{CO} (CO selectivity)	Ref.
$La_{0.6}Sr_{0.4}Co_{0.2}Fe_{0.8}O_{3-\alpha}$: Tube; Ni/Al_2O_3	825–885	Fail, 7 hrs	$X > 96$, $S_{CO} > 97$	[96, 135]
$La_{0.2}Sr_{0.8}Fe_{0.2}Co_{0.8}O_{3-\alpha}$ and $SrFeCo_{0.5}O_3$: Tube with Rh based catalyst	850	1000 h	$X > 99$ %; $S_{CO} > 98$ %; $J_{O_2} = 0.3\,ml\,cm^{-2}\,min^{-1}$;	[77, 90]
$La_{0.6}Sr_{0.4}Co_{0.2}Fe_{0.8}O_{3-\alpha}$: Disk			$X = 5–25$ %	[136]
$La_{0.8}Sr_{0.2}Ga_{0.6}Fe_{0.4}O_{3-\alpha}$: Disk		good	$X = 5–42$ %; $S_{co} = 100$ %	
$La_{0.5}Sr_{0.5}Fe_{0.8}Ga_{0.2}O_{3-\alpha}$: Tube with catalyst	850		$X = 97$ %; $S_{CO} = 100$ % $J_{O_2} = 0.35\,ml\,cm^{-2}\,min^{-1}$	[98]
$Sr_4(Fe_{1-x}Co_x)_6O_{13\pm\delta}$: Tube	850	1000 h	$X > 98$, $J_{O_2} = 10\,ml\,cm^{-2}\,min^{-1}$	[77]
$A_2B_2O_5$ Brownmillerite structure: Tube	900	1 y	$X = 90$; $S_{CO} > 96$ $J_{O_2} = 10–12\,ml\,cm^{-2}\,min^{-1}$	[102]
$BaCo_{0.4}Fe_{0.4}Zr_{0.2}O_{3-\alpha}$; Disk; $LiLaNiO/\gamma\text{-}Al_2O_3$	850	2200 h	$X = 98$; $Sco = 100$ $J_{O_2} = 5.6\,ml\,cm^{-2}\,min^{-1}$	[101]
$Ba_{0.5}Sr_{0.5}Co_{0.8}Fe_{0.2}O_3$ Disk; $LiLaNiO/\gamma\text{-}Al_2O_3$.	875	500 h	$X > 97$ %; $S_{CO} > 95$ % $J_{O_2} = 11.5\,ml\,cm^{-2}\,min^{-1}$	[99, 137–139]
$Ba_{0.5}Sr_{0.5}Co_{0.8}Fe_{0.2}O_{3-\alpha}$: Tube packed with $LiLaNiO/\gamma\text{-}Al_2O_3$.	875	500 h	$X = 94$ %; $S_{CO} > 95$ % $J_{O_2} = 8.0\,ml\,cm^{-2}\,min^{-1}$	[80]

$SrFeCo_{0.5}O_y$: tube packed with Ni/γ-Al_2O_3	900		X = 98 %, S_{CO} = 98 %, H_2/CO = 1.8, r_{CO} = 16 mL min^{-1} cm^{-2}	[140]
YSZ-$SrCo_{0.4}Fe_{0.6}O_{3-\alpha}$; Disc, Ni/$\gamma$-$Al_2O_3$		270 h	X = 64 %, S_{CO} = 100 % J_{O_2} = 4.5 ml cm^{-2} min^{-1}	[141]
$Ca_{0.8}Sr_{0.2}Ti_{1-x}Fe_xO_{3-\alpha}$: Disc, electrochemical cell; $Ni_{1.0}$/$Ca_{0.8}Sr_{0.2}Ti_{1-x}Fe_xO_{3-\alpha}$				[142]
8 % YSZ; Disc, electrochemical cell; $Ni_{1.0}$/$Ca_{0.8}Sr_{0.2}Ti_{1-x}Fe_xO_{3-\alpha}$			S_{CO} = 100 %	[143]
$YBa_2Cu_3O_{7-x}$; disc; Ni/ZrO_2	900	Not good	X = 100 %, S_{CO} = 95 %, J_{O_2} = 1.5 ml cm^{-2} min^{-1}	[144]
$Sm_{0.4}Ba_{0.6}Co_{0.2}Fe_{0.8}O_{3-\alpha}$; Disc; Rh/MgO	900		X = 90 %, S_{CO} = 98 %,	[92]
$La_{1-x}Sr_xGa_{1-y}M_yO_{3-\alpha}$ (M = Fe, Co, or Ni)	1000		H_2/CO ratio close to 2	[136]
$SrFe_{0.7}Al_{0.3}O_{3-\alpha}$ and $La_{0.3}Sr_{0.7}Co_{0.8}Ga_{0.2}O_{3-\alpha}$: Disc	950		H_2/CO ratio close to 2, CO selectivity at 1223 K achieved 65 % and 48 %,	[145]
$Ba_{0.5}Sr_{0.5}Co_{0.8}Fe_{0.2}O_{3-\alpha}$; Disk; LiLaNiOx/ γ-Al_2O_3	700–850		X = 92 %; S_{CO} = 90 %, J_{O_2} = 15.5 ml cm^{-2} min^{-1}; Syngas rate = 79 ml cm^{-2} min^{-1}	[146]
$La_{0.2}Ba_{0.8}Fe_{0.8}Co_{0.4}O_{3-\alpha}$: Disk; 5 % Ni/$Al_2O_3$	850	850 h	X = 80 %; S_{CO} = 98 %	[147]
$SrCo_{0.4}Fe_{0.5}Zr_{0.1}O_{3-\alpha}$: Disk; NiO/$Al_2O_3$	900		X = 84 %; S_{CO} = 95 %	[148]

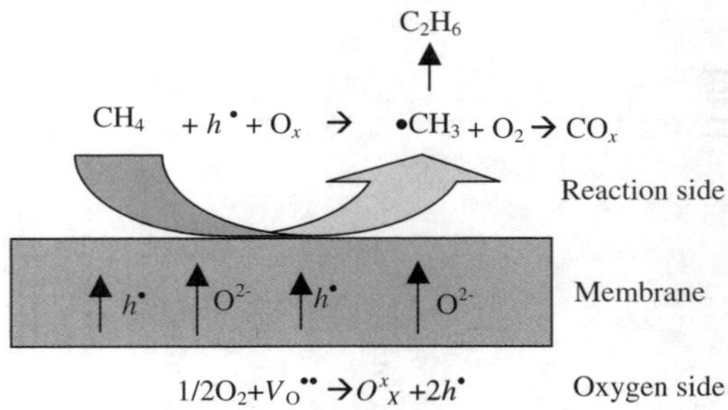

Figure 8.22 Schematic diagram of the dense membrane reactor for OCM reaction

excellent permselectivity towards oxygen. Therefore, air is allowed to be used directly as an oxidant during the reaction without any nitrogen being introduced into the reaction compartment. On the other hand, the distinct oxygen transport mechanism of mixed conducting ceramic membranes delivers the oxygen into the reaction compartment in a form of dissociated and ionized oxygen. The ionized oxygen reacting with methane on the membrane surface would follow a different reaction mechanism so that the formation of CO_x from side reactions due to the presence of gas phase oxygen could be minimized, and thus an increased selectivity for C_2 hydrocarbons could be expected.

In addition to the oxygen permeability properties, some mixed ionic and electronic conducting membranes (MIECMs) are also catalytically active and selective for OCM [103, 104]. For example, $La_{0.8}Sr_{0.2}CoO_3$ (LSC) possesses OCM catalytic properties similar to those of Li/MgO in terms of C_2 yield and selectivity at temperatures of around 800 °C. The mechanism for OCM on the dense membranes is schematically illustrated in Figure 8.22. It generally suggests that the coupling reaction takes place in the gas phase.

The kinetic equations for OCM on a catalyst operated in the conventional mode are derived based on the following mechanism written in Kroger–Vink notation [105]:

$$CH_4 + h^{\bullet} + 0.5O_O^x \rightarrow \cdot CH_3 + 0.5H_2O + 0.5V_O^{\bullet\bullet} \tag{8.39}$$

$$2\cdot CH_3 \rightarrow C_2H_6 \tag{8.40}$$

In an integral form:

$$2CH_4 + 2h^{\bullet} + O_O^x \rightarrow C_2H_6 + H_2O + V_O^{\bullet\bullet} \tag{8.41}$$

The methane radical will also be oxidized to carbon dioxide:

$$\cdot CH_3 + xO_2 \rightarrow CO_x + H_2O \tag{8.42}$$

Meanwhile, the ethane produced will be further converted into ethylene following a similar mechanism:

$$C_2H_6 + 2h^{\bullet} + O_O^x \rightarrow C_2H_4 + H_2O + V_O^{\bullet\bullet} \tag{8.43}$$

$$\cdot C_2H_5 + xO_2 \rightarrow CO_x + H_2O \tag{8.44}$$

A great deal of effort has been placed on oxidative methane coupling using dense oxygen permeable ceramic membranes to improve the methane coupling reaction. Hazbun [106, 107] conducted a methane oxidative coupling reaction for C_2 compound conversion using a two layer tubular membrane, a 10 %Y_2O_3-89 %ZrO_2-1 %TiO_2 layer for oxygen permeation and a LiO/MgO/ZrO_2 layer as the catalyst. A product yield of 20–25 % C_2 compounds achieved with 50–60 % selectivity and 35–45 % conversion at 700–750 °C showed the feasibility of using a dense membrane reactor for oxidative coupling of methane. Since then, a variety of mixed conducting perovskite type oxides have been reported as promising OCM catalysts, which are very active and selective in converting methane into C_2 (C_2H_4, C_2H_6) products [103, 108, 109]. $La_{0.2}Sr_{0.8}CoO_3$ showed good OCM catalytic properties in terms of C_2 yield (>14 %) and selectivity (>50 %), while $SrCo_{1-x}Fe_xO_3$ showed very poor OCM catalytic properties [103]. $La_{1-x}Sr_xCo_{1-y}Fe_yO_{3-\delta}$ series materials were also investigated in OCM reactions and higher yields of C_2 products were observed [110, 111]. Zeng *et al.* [111] found that the surface catalytic properties of LSCF membranes for OCM strongly depend on the oxygen activity of the membrane surface exposed to the methane stream. A more comprehensive analysis of the performance of oxygen permeable membranes made from perovskite oxides formulae $(La_{1-x}A_x)(Co_{1-y}Fe_y)O_{3-\delta}$ for methane coupling was performed by Xu and Thomson [112].

More recently, other mixed conducting oxides were studied for OCM, including $Ba_{0.5}Sr_{0.5}Co_{0.8}Fe_{0.2}O_{3-\delta}$ (BSCF) [99] and $BaCe_{0.8}Gd_{0.2}O_{3-\delta}$ [113], and showed promising performances in OCM reactions. The C_2 hydrocarbon selectivity of $Ba_{0.5}Sr_{0.5}Co_{0.8}Fe_{0.2}O_{3-\delta}$ was from 40–70 %, while the maximum yield of C_2 attained from $BaCe_{0.8}Gd_{0.2}O_{3-\delta}$ was 16 %. Some reported results are summarized in Table 8.7.

Compared with mixed conducting perovskite materials, membranes made from other materials such as $Bi_{1.5}Y_{0.3}Sm_{0.2}O_{3-\delta}$ [114] also showed good performance in methane coupling reactions. A 35 % yield of C_2 product was obtained using a $Bi_{1.5}Y_{0.3}Sm_{0.2}O_{3-\delta}$ membrane reactor. The material, $Bi_{1.5}Y_{0.3}Sm_{0.2}O_{3-\delta}$, was found to be not only highly oxygen permeable and catalytically active for OCM, but also chemically and mechanically stable under OCM conditions.

There is still a gap to be closed between the currently achieved yield results and desirable industrial values. It is possible to increase the yield by matching the oxygen permeation rate of the membrane and surface catalytic activation of the methane on the membrane surface because the overall C_2 yield is determined by the combined effects of both important factors. When the oxygen permeation rate is greater than the rate consumed in the methane coupling, deep oxidation of methyl radical and C_2 products to CO_x will occur and the C_2 yield will be reduced. The surface catalytic activation of methane must be improved while the oxygen permeation rate is increased, and this may be performed by fixing a catalyst on the membrane surface.

Other Oxidation Reactions

Apart from the POM and OCM reactions, dense ceramic membrane reactors have also been used for other dehydrogenation reactions, which are summarized in Table 8.8. Under reaction conditions, the oxygen permeation flux through the dense membrane is over an order of magnitude higher than those under oxygen permeation conditions with He as the purge. The oxygen permeation mechanism switches from p to n type electronic conduction under selective oxidation of ethane (SOE) reaction conditions [115].

Table 8.7 Mixed conducting oxide membrane reactors for oxidative coupling of methane

Membrane; membrane reactor	Temperature/°C	Catalyst	Comments	Ref.
$La_{0.8}Sr_{0.2}CoO_{3-\alpha}$: Disk	800–850	None	$Y_{C_2} = 12–14$; $S_{C_2} = 40–56$ %	[103]
$La_{0.8}Sr_{0.2}Co_{0.6}Fe_{0.4}O_{3-\alpha}$: Disk 1.85 mm,	850	None	$Y_{C_2} = 10–18$ %; $S_{C_2} = 70–90$ %	[111]
$La_{0.6}Sr_{0.4}Co_{0.8}Fe_{0.2}O_{3-\alpha}$: Disk 0.5–2 mm		None	$Y_{C_2} = 13$ %, $S_{C_2} = 70$ %	[110]
$La_{0.6}Sr_{0.4}Co_{0.8}Fe_{0.2}O_{3-\alpha}$: Disk 0.5–2 mm	800–1000		X = 1–3 %, $S_{C_2} = 70$ %	
$BaCe_{0.8}Gd_{0.2}O_{3-\alpha}$: Tube 0.7 mm		None	$Y_{C_2} = 16.5$ %	[113]
$Ba_{0.5}Sr_{0.5}Co_{0.8}Fe_{0.2}O_{3-\alpha}$: Disk		None	X = 40–70; $S_{C_2} = 0.5–0.3$	[99]
$Ba_{0.5}Sr_{0.5}Co_{0.8}Fe_{0.2}O_{3-\alpha}$: Tube	850	La-Sr/CaO	$Y_{C_2} = 15$ %	[149]
$La_{0.2}Sr_{0.8}CoO_{3-\alpha}$	800	With Cat.	$Y_{C_2} = 14$ %; $S_{C_2} = 40$ %	[103]
$SrCo_{0.8}Fe_{0.2}O_{3-\alpha}$	850	With Cat.	$Y_{C_2} = 6.3$ %; $S_{C_2} = 30.4$ %	[103]
$Bi_{1.5}Y_{0.3}Sm_{0.2}O_{3-\alpha}$; Tube	870–930		$Y_{C_2} = 35$ %; $S_{C_2} = 54$ % (at 900°C); Duration = 6 days	[150–152]
8 % YSZ; $SrFeO_{3-\alpha}$, Au/YSZ/Ag disc cell		None	$Y_{C_2} = 3.1$ %; S_{C_2} higher than FBR	[153]
$BaCe_{0.8}Gd_{0.2}O_{3-\alpha}$; Disc cell	900		C_2 formation rate enhanced; S_{C_2} two times the co-feed operation	[154]
Substituted $SrCoO_{3-\alpha}$ perovskite oxide	750		$S_{C_2} > 98$ %, Duration = 6 days	[108]
$La_{0.6}Sr_{0.4}Co_{0.8}Fe_{0.2}O_{3-\alpha}$; Hollow fibre	950	None	$Y_{C_2} = 15.3$ %; Duration = 4 weeks	[87]

Table 8.8 Mixed conducting oxide membrane reactors for oxidative dehydrogenation reactions

Reaction	Membrane; MR; catalyst	Temperature °C	Comments	Ref.
Propane	$Ba_{0.5}Sr_{0.5}Co_{0.8}Fe_{0.2}O_{3-\alpha}$; tube;	700–750	44.2 % selectivity much higher than that (15 %) in FBR	[155]
Ethane to ethylene	$Ba_{0.5}Sr_{0.5}Co_{0.8}Fe_{0.2}O_{3-\alpha}$ disc; V/MgO micron grain or Pd nanocluster modified surfaces	770	$Y_{C_2} = 75$ %	[156]
C_2–C_4 alkanes	YSZ, disc, Mo/V catalyst film			[157]
Ethane to ethylene	$Bi_{1.5}Y_{0.3}Sm_{0.2}O_{3-\alpha}$ (BYS); tube	825–875	$Y_{C_2} = 56$ %; $S_{C_2} = 80$ %; Duration = 2 days	[115]
Propane to propene	Bi_2O_3–25 % Er_2O_3 (BE25), BE25–Ag cermet, and $Bi_2V_{0.9}Co_{0.1}O_{5.35}$ (BICOVOX.10), disc			[158]
Ethane to syngas	$Ba_{0.5}Sr_{0.5}Co_{0.8}Fe_{0.2}O_{3-\alpha}$; tube; Li/$\gamma$-$Al_2O_3$	800–900	X = ~100 %; $S_{CO} > 91$ %; $J_{O_2} = 10$–$11\,cm^3\,cm^{-2}\,min^{-1}$; Stability > 100 h	[159]
Ethane to ethylene	$Ba_{0.5}Sr_{0.5}Co_{0.8}Fe_{0.2}O_{3-\alpha}$; tube; no catalyst	650	$S_{C_2} = 90$ %;	[160]
Heptane to syngas	$Ba_{0.5}Sr_{0.5}Co_{0.8}Fe_{0.2}O_{3-\alpha}$; tube; LiNaNiO/$\gamma$-$Al_2O_3$	850	X = 100 %; $S_{CO} = 91$–93 %; $S_{H_2} = 95$–97 %; $J_{O_2} = 11.5\,cm^3\,cm^{-2}\,min^{-1}$; Stability > 100 h	[161]

8.3.2.3 Experimental Investigation of a Dense Membrane Reactor for Oxidative Methane Coupling (OMC)

Tan and Li [87] developed a ceramic hollow fibre membrane reactor (HFMR) for OMC. Figure 8.23 schematically illustrates the HFMR as well as the experimental setup for OCM. As can be seen, two short soft rubber tubes were used in the HFMR to offset the thermal expansion mismatch between the reactor shell and the hollow fibre membranes at high temperature. The HFMR was placed in a tubular furnace where a high temperature zone is much shorter than that of the reactor so that preventing the soft rubber tubes being burnt out during the high temperature operation. Before the experiments, a leakage test of the HFMR (possibly at the connecting and sealing joints as well as defects at high temperature) was carried out. For this purpose, air and argon were introduced into the shell side and lumen side of the HFMR, respectively. Any leakage of the HFMR can be detected by a GC if nitrogen is presented in the argon stream. Such leakage measurements were conducted from room temperature to 900 °C.

Some experimental results for the OMC are shown in Figures 8.24–8.27. Figure 8.24 illustrates the experimental results of OCM in a HFMR at different temperatures. As can be seen, the methane conversion rate increases with increasing operating temperature (Figure 8.24(a)). Figure 8.24(b) depicts the effect of the operating temperature on the C_2 yield. It can be seen that the yield of ethylene increases slowly from 0.6 to 2.1 % as the temperature is increased from 800 to 900 °C and thereafter it shows a sharp increase (to 7.6 %) when the temperature is further increased to 975 °C. However, for the formation of ethane, its yield reaches a

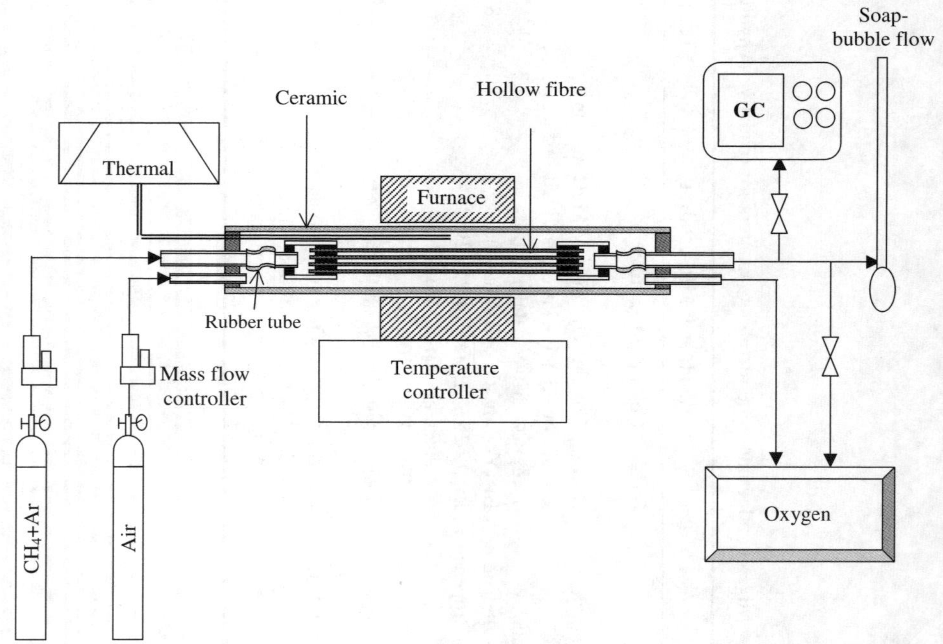

Figure 8.23 Experimental setup for OCM experiments [87]. Copyright (2006) American Chemical Society, Industrial and Engineering Chemistry Research, **45** (1), Tan, X. and Li, K., *Oxidative coupling of methane in a perovskite hollow-fiber membrane reactor*, p. 142–149

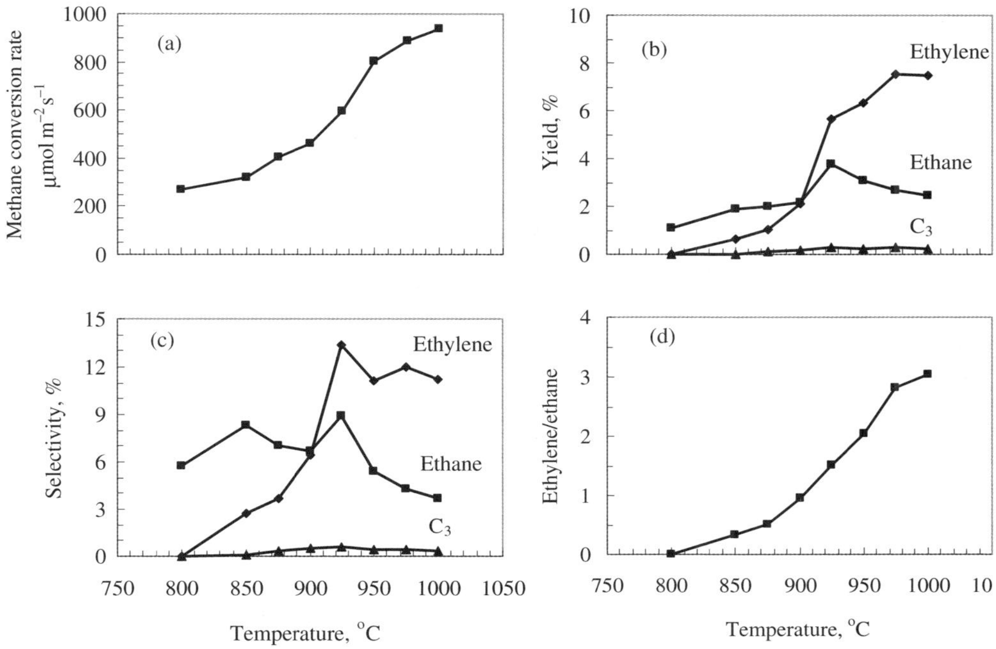

Figure 8.24 Effect of temperature on OCM in the HFMR. (Air feed flow rate = 22.96 cm^3(STP) min^{-1}, methane feed flow rate = 17.2 cm^3(STP) min^{-1}) [87]. Copyright (2006) American Chemical Society, Industrial and Engineering Chemistry Research, **45** (1), Tan, X. and Li, K., *Oxidative coupling of methane in a perovskite hollow-fiber membrane reactor*, p. 142–149

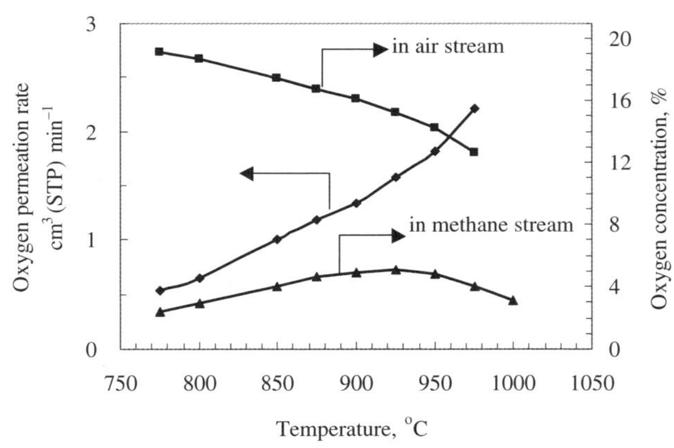

Figure 8.25 Oxygen permeation rate and oxygen concentrations in the air stream and in the methane stream at various temperatures. (Air feed flow rate = 22.96 cm^3(STP) min^{-1}, methane feed flow rate = 17.2 cm^3(STP) min^{-1}) [87]. Copyright (2006) American Chemical Society, Industrial and Engineering Chemistry Research, **45** (1), Tan, X. and Li, K., *Oxidative coupling of methane in a perovskite hollow-fiber membrane reactor*, p. 142–149

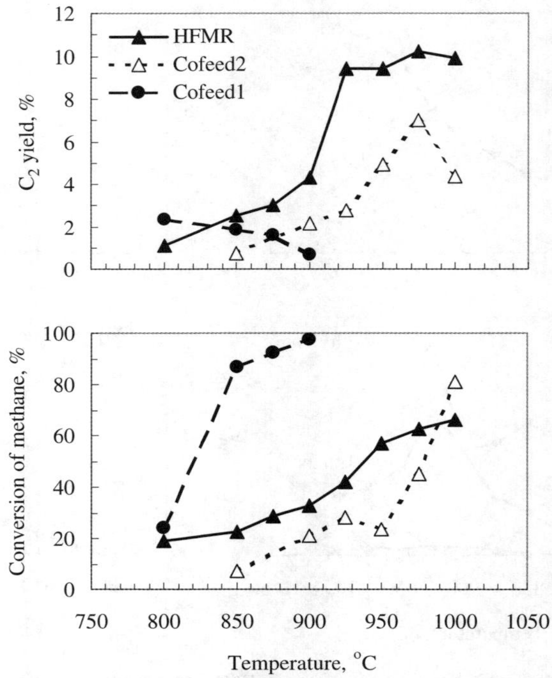

Figure 8.26 Comparison of the HFMR operation with the co-feed operations. (Air feed flow rate = 22.96 cm^3(STP) min^{-1}, methane feed flow rate = 17.2 cm^3(STP) min^{-1}) [87]. Copyright (2006) American Chemical Society, Industrial and Engineering Chemistry Research, **45** (1), Tan, X. and Li, K., *Oxidative coupling of methane in a perovskite hollow-fiber membrane reactor*, p. 142–149

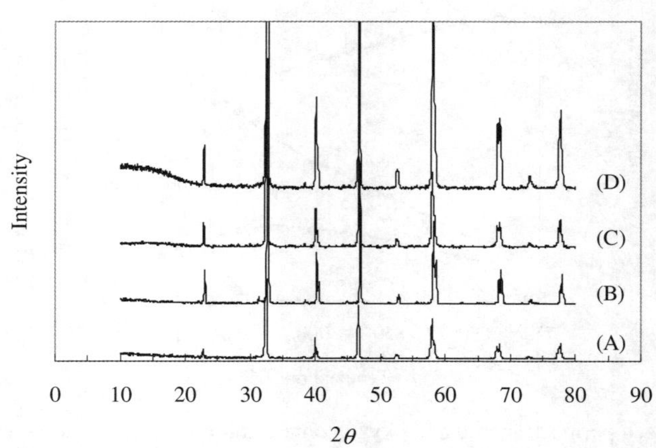

Figure 8.27 XRD patterns of the LSCF raw material and the hollow fibre membrane: (A) membrane material in powder; (B) fresh membrane in powder; (C) reacted membrane in powder; (D) reacted membrane in fibre on the outer surface [87]. Copyright (2006) American Chemical Society, Industrial and Engineering Chemistry Research, **45** (1), Tan, X. and Li, K., *Oxidative coupling of methane in a perovskite hollow-fiber membrane reactor*, p. 142–149

maximum value of 3.8 % at 925 °C. Compared to a disc shaped membrane reactor, where only 1–2 % of the C_2 yield can be realized [110], much higher C_2 yields ($C_2H_4 + C_2H_6$) are obtained in HFMR. This remarkable increase in C_2 yields results from the better structures of the hollow fibre membranes prepared by Tan *et al.* [116]. For example, due to the higher oxygen permeation in LSCF hollow fibre membranes compared to that in disc shape membranes, a methane conversion of as high as 70 % has been achieved in the HFMR while only 1–3 % was obtained in the disc shape membrane reactor [110]. However, the observed maximum selectivities for ethylene and ethane in the HFMR at 925 °C are only 13.4 % and 9 %, respectively, as shown in Figure 8.24(c), which are much lower than that of 70 % C_2 obtained in the disc shaped membrane reactor [110]. Figure 8.24(d) illustrates that higher temperature favors the formation of ethylene compared to that of ethane. As can be seen, the ethylene/ethane ratio increases from zero at 800 °C to 3.1 at 1000 °C. However, for the total production of C_2, the operating temperature should be lower than 950 °C since the C_2 selectivity is drastically decreased as the temperature becomes greater than 950 °C, as shown in Figure 8.24(c).

Figure 8.25 shows the oxygen permeation rate as well as the oxygen concentrations in the gas effluents from the HFMR at different operating temperatures. It can be seen that the oxygen concentration in the methane stream increases with increasing temperature from 775–925 °C and then decreases with temperature in the range 925–1000 °C. This suggests that the oxygen permeation rate increases faster than the oxygen reaction rate with methane in the temperature range to 775–925 °C. Thereafter, the reverse becomes true. In addition, the experimental data further indicates that the oxygen permeation rate increases from 0.26 to 1.06 cm^3(STP) min^{-1} as the temperature is increased from 775 to 975 °C, corresponding to an oxygen permeation flux of 0.065–0.265 cm^3(STP) cm^{-2} min^{-1}, if the membrane area used is calculated based only on the constant temperature zone length. In fact, the oxygen concentration in the methane stream was always above 3 % over the whole temperature range, as shown in Figure 8.25, indicating that the oxygen permeated from the air stream is more than that required for the OCM reaction. It thus follows that the OCM reaction in the HFMR is not controlled by oxygen permeation but by the reaction activity of the LSCF membranes. It is envisaged that if the right catalyst is deposited on the membrane surface of the methane compartment, a much better conversion/yield could be achieved.

Figure 8.26 compares the HFMR with co-feed operations under the same feed flow rates. As can be seen, in general, much higher C_2 yields have been achieved by the HFMR operation than by either co-feed operation 1 (where the methane–argon mixture and air are co-fed into the lumen and shell sides) or co-feed operation 2 (where the mixture of methane, argon and air is fed only into the lumen side while the shell side is blocked). Figure 8.26 further illustrates that at lower temperatures, i.e., 800 °C, the C_2 yield in co-feed operation 1 is even higher than that obtained in the HFMR, but almost no C_2 compounds are produced in co-feed operation 2. This may be explained according to the following OCM reaction mechanism on the perovskite catalyst, expressed in Kroger–Vink notation [105, 114]:

$$\frac{1}{2}O_2 + V_O^{\bullet\bullet} \Leftrightarrow O_O^x + 2h^{\bullet} \qquad \text{(on the membrane surface exposed to } O_2\text{)} \tag{8.45}$$

$$CH_4(g) \leftrightarrow CH_4(s) \tag{8.46}$$

$$CH_4(s) + \frac{1}{2}O_O^x + h^{\bullet} \rightarrow \cdot CH_3 + \frac{1}{2}H_2O + \frac{1}{2}V_O^{\bullet\bullet} \tag{8.47}$$

$$2\cdot CH_3 \rightarrow C_2H_6 \quad \text{(coupling reaction)} \tag{8.48}$$

$$\cdot CH_3 + \frac{x}{2}O_2 \rightarrow CO, CO_2 \tag{8.49}$$

In co-feed operation 1, the membrane surface area for reactions is twice that either in co-feed operation 2 or in the membrane operation. In addition, the reactant residence time in co-feed operation 1 is the longest among these operations. Therefore, the formation rate and concentration of $\cdot CH_3$ radicals in co-feed operation 1 are much higher than that in the other two operations. At lower temperatures, the formation of carbon oxides (Reaction 8.49) has not yet dominated the elimination of $\cdot CH_3$ radicals. Therefore, the C_2 yield mainly depends on the formation rate and the concentration of $\cdot CH_3$ in the gas stream. As temperature increases, the $\cdot CH_3$ radicals tend to react with oxygen to form carbon oxides other than the coupling reaction (Reaction 8.48), resulting in a decrease in C_2 yields. In co-feed operation 2, the reactions take place only on the inner surface of the membrane, but the reactant residence time is the shortest in these operations. As a result, the concentration of $\cdot CH_3$ radicals is the lowest at lower temperatures, leading to a much lower C_2 yield. As the temperature increases, the formation rate of $\cdot CH_3$ radicals is promoted with the result of higher C_2 yield. However, if the temperature is further increased to above 975 °C, more $\cdot CH_3$ radicals react with oxygen to form carbon oxides than couple to form C_2. Consequently, the C_2 yield is lower at 1000 °C than that at 975 °C. In the HFMR operation, the locally lower oxygen concentration in the methane stream restrains the oxidation of $\cdot CH_3$ radicals into carbon oxides, leading to higher C_2 yields. The methane conversions in the membrane operation are higher than that in co-feed operation 2 within the temperature range 850–975 °C because the residence time of reactants in the membrane reactor is much longer than that in co-feed operation 2. All these experimental results indicate the C_2 yield depends not only on the oxygen concentration but also on the temperature and the residence time of the reactants in the reactor.

Figure 8.27 shows the XRD patterns of the membrane material (A), the fresh membrane (B) and the reacted membranes (C and D), where C is for the reacted membrane ground into powders and D for the outer surface of the reacted membrane, respectively. It can be seen that the perovskite structure was kept unchanged both after the LSCF powders were made into hollow fibres by the spinning/sintering processes and after OCM reactions. Since XRD measurement of the reacted membranes was conducted after finishing all the OCM reactions including co-feed operations, which lasted for about four weeks, it suggests that the LSCF material is very stable in the perovskite structure. Furthermore, no obvious decrease in either the OCM performance or the mechanical strength (a three point bending test [36] performed using Intron Model 4466) of the LSCF hollow fibre membranes has been observed after the four week reaction experiments.

NOTATION

A_t	Membrane area, m^2
A_s	Surface area of catalyst, m^2
C	Oxygen concentration in water feed, $mol\,m^{-3}$
D_L	Axial dispersion coefficient, $m^2\,s^{-1}$
$D_{i\text{-}H_2O}$	Molecular diffusivity of dissolved gas in water, $m^2\,s^{-1}$

G	Rate of hydrogen gas, $m^3 s^{-1}$
k_G	Total mass transfer coefficient in gas phase, $mol\,m^{-2} s^{-1} atm^{-1}$
K_H	Henry constant, $atm\,ml\,mol^{-1}$
k_g	Mass transfer coefficient of the gas film, $mol\,m^{-2} s^{-1} atm^{-1}$
k_L	Mass transfer coefficient of the liquid film, $m\,s^{-1}$
k_m	Mass transfer coefficient of the membrane, $mol\,m^{-2} s^{-1} atm^{-1}$
k_r	Reaction rate constant, $(m^3)^2 mol^{-2} s^{-1}$
k_R	Overall reaction rate constant, $(m^3)^2 mol^{-2} s^{-1}$
k_S	Mass transfer coefficient of the liquid film (adjacent to the catalyst surface), $m\,s^{-1}$
L	Rate of water feed, $m^3 s^{-1}$
n	Number of membrane tubes
N_c	Feed concentration modulus, in Table 8.5
N_F	Feed flow modulus, in Table 8.5
N_m	Membrane transport modulus, in Table 8.5
N_R	Chemical reaction modulus, in Table 8.5
p	Partial pressure, atm
P_e	Peclet number, $P_e = uL/D_{LB}$
R	Ideal gas constant, $0.08206\,atm\,l\,mol^{-1} K^{-1}$
R_p	Average radius of the catalyst particle, m
Re	Reynolds number, $Re = 2R_p u \rho_w / \upsilon$
R_s	Inner radius of shell, m
R_t	Inner radius of membrane tube/fibre, m
R_o	Outer radius of membrane tube/fibre, m
s	Laplace transform variable
T	System temperature, K
u	Water velocity, $m\,s^{-1}$
x	Distance co-ordinate, m
x_i	Dimensionless concentration in liquid phase
x_0	Dimensionless oxygen concentration in water at $z = 0$
x_0'	Dimensionless oxygen concentration gradient at $z = 0$
y_0	Dimensionless oxygen concentration in gas phase at $z = 0$
y_i	Dimensionless concentration in gas phase
Z	Length of membrane tube, m
z	Dimensionless location co-ordinate

Greek Letters

α	Ratio of the solubility of oxygen to that of hydrogen
β	Ratio of the mass transfer rate of oxygen to hydrogen
γ	Ratio of the axial dispersion coefficients
ϕ	Packing fraction of membrane tubes
η	Catalyst efficiency factor
υ_i	Chemical reaction stoichiometric number, 1 for hydrogen and 1/2 for oxygen
μ_k	Root of Equation (8.24)
$\lambda_{1,2}$	Defined by Equation (8.31)
ε_b	Bed void fraction, dimensionless

σ_p	DO removal by permeation only
σ_t	Overall oxygen removal by both physical stripping and chemical reaction

Subscripts

A	Hydrogen
B	Oxygen
i	Shortcut for A and B
o	Inlet condition

REFERENCES

1. Mohan, K. and Govind, R., *Analysis of equilibrium shift in isothermal reactors with a permselective wall.* American Institute of Chemical Engineers Journal, **34** (9): 1493–1503 (1988).
2. Coronas, J., Menendez, M. and Santamaria, J., *Methane oxidative coupling using porous ceramic membrane reactors. 2. Reaction Studies.* Chemical Engineering Science, **49** (12): 2015–2025 (1994).
3. Pfefferle, W. C., *Process for dehydrogenation.* (1966) US Patent No. 3,290,406.
4. Wood, B. J., *Dehydrogenation of cyclohexane on a hydrogen-porous membrane.* Journal of Catalysis, **11**: 30 (1968).
5. Michaels, A. S., *New separation technique for the CPI.* Chemical Engineering Progress, **64** (6): 31 (1968).
6. Chang, H. N. and Furusaki, S., *Membrane bioreactors: present and prospects.* Advances in Biochemical Engineering/Biotechnology. **44**: 27–64 (1991).
7. Shu, J., Grandjean, B. P. A., Vanneste, A. and Kaliaguine, S., *Catalytic palladium-based membrane reactors – a review.* Canadian Journal of Chemical Engineering, **69** (5): 1036–1060 (1991).
8. Hsieh, H. P., *Inorganic membrane reactors.* Catalysis Reviews – Science and Engineering, **33** (1–2): 1–70 (1991).
9. Eng, D. and Stoukides, M., *Catalytic and electrocatalytic methane oxidation with solid oxide membranres.* Catalysis Reviews in Science and Engineering, **33** (3–4): 375–412 (1991).
10. Reyes, S. C., Iglesia, E. and Kelkar, C. P., *Kinetic-transport models of bimodal reaction sequences. 1. Homogeneous and heterogeneous pathways in oxidative coupling of methane.* Chemical Engineering Science, **48** (14): 2643–2661 (1993).
11. Reyes, S. C., Kelkar, C. P. and Iglesia, E., *Kinetic transport models and the design of catalysts and reactors for the oxidative coupling of methane.* Catalysis Letters, **19** (2–3): 167–180 (1993).
12. Cheng, S. and Shuai, X., *Simulation of a catalytic membrane reactor for oxidative coupling of methane.* American Institute of Chemical Engineers Journal, **41** (6): 1598–1601 (1995).
13. Tsai, C. Y., Ma, Y. H., Moser, W. R. and Dixon, A. G., *Modeling and simulation of a nonisothermal catalytic membrane reactor.* Chemical Engineering Communications, **134**: 107–132 (1995).
14. Coronas, J., Gonzalo, A., Lafarga, D. and Menendez, M., *Effect of the membrane activity on the performance of a catalytic membrane reactor.* American Institute of Chemical Engineers Journal, **43** (11): 3095–3104 (1997).
15. Fogler, H. S., *Elements of Chemical Reaction Engineering.* (1992) New Jersey: Prentice-Hall. p. 163.
16. Frisch, H. L., Maaref, S. and Deng-Nemer, H., *Low-temperature dehydrogenation reaction-separation membranes using zeolite 13X polyethylacrylate.* Journal of Membrane Science, **154** (1): 33–40 (1999).

17. Ziegler, S., Theis, J. and Fritsch, D., *Palladium modified porous polymeric membranes and their performance in selective hydrogenation of propyne.* Journal of Membrane Science, **187** (1–2): 71–84 (2001).
18. Liu, C., Xu, Y., Liao, S., Yu, D., Zhao, Y. and Fan, Y., *Selective hydrogenation of propadiene and propyne in propene with catalytic polymeric hollow-fibre reactor.* Journal of Membrane Science, **137** (1–2): 139–144 (1997).
19. Armor, J. N., *Applications of catalytic inorganic membrane reactors to refinery products.* Journal of Membrane Science, **147**: 217–233 (1998).
20. Weyten, H., Luyten, J., Keizer, K., Willems, L. and Leysen, R., *Membrane performance: the key issues for dehydrogenation reactions in a catalytic membrane reactor.* Catalysis Today, **56** (1–3): 3–11 (2000).
21. Ciavarella, P., Casanave, D., Moueddeb, H., Miachon, S., Fiaty, K. and Dalmon, J. A., *Isobutane dehydrogenation in a membrane reactor: Influence of the operating conditions on the performance.* Catalysis Today, **67** (1–3): 177–184 (2001).
22. Alfonso, M. J., Menendez, M. and Santamaria, J., *Oxidative dehydrogenation of butane on V/MgO catalytic membranes.* Chemical Engineering Journal, **90** (1–2): 131–138 (2002).
23. van Dyk, L., Miachon, S., Lorenzen, L., Torres, M., Fiaty, K. and Dalmon, J. A., *Comparison of microporous MFI and dense Pd membrane performances in an extractor-type CMR.* Catalysis Today, **82** (1–4): 167–177 (2003).
24. Bernstein, L. A., Reo, C. M. and Lund, C. R. F., *A batch membrane reactor for laboratory studies.* Journal of Membrane Science, **118** (1): 93–100 (1996).
25. Casanave, D., Ciavarella, P., Fiaty, K. and Dalmon, J. A., *Zeolite membrane reactor for isobutane dehydrogenation: experimental results and theoretical modelling.* Chemical Engineering Science, **54** (13–14): 2807–2815 (1999).
26. Sznejer, G. and Sheintuch, M., *Application of a carbon membrane reactor for dehydrogenation reactions.* Chemical Engineering Science, **59** (10): 2013–2021 (2004).
27. Illgen, U., Schafer, R., Noack, M., Kolsch, P., Kuhnle, A. and Caro, J., *Membrane supported catalytic dehydrogenation of iso-butane using an MFI zeolite membrane reactor.* Catalysis Communications, **2** (11–12): 339–345 (2001).
28. Sheintuch, M. and Efremenko, I., *Analysis of a carbon membrane reactor: from atomistic simulations of single-file diffusion to reactor design.* Chemical Engineering Science, **59** (22–23): 4739–4746 (2004).
29. Schafer, R., Noack, M., Kolsch, P., Stohr, M. and Caro, J., *Comparison of different catalysts in the membrane-supported dehydrogenation of propane.* Catalysis Today, **82** (1–4): 15–23 (2003).
30. Schafer, R., Noack, M., Kolsch, P., Thomas, S., Seidel-Morgenstern, A. and Caro, J., *Development of a H_2-selective SiO_2-membrane for the catalytic dehydrogenation of propane.* Separation and Purification Technology, **25** (1–3): 3–9 (2001).
31. Jeong, B.-H., Sotowa, K.-I. and Kusakabe, K., *Modeling of an FAU-type zeolite membrane reactor for the catalytic dehydrogenation of cyclohexane.* Chemical Engineering Journal, **103** (1–3): 69–75 (2004).
32. Jeong, B.-H., Sotowa, K.-I. and Kusakabe, K., *Catalytic dehydrogenation of cyclohexane in an FAU-type zeolite membrane reactor.* Journal of Membrane Science, **224** (1–2): 151–158 (2003).
33. Guo, Y. L., Lu, G. Z., Mo, X. H. and Wang, Y. S., *Vapor phase dehydrogenation of methanol to methyl formate in the catalytic membrane reactor with Cu/SiO_2/ceramic composite membrane.* Catalysis Letters, **99** (1–2): 105–108 (2005).
34. Au, L. T. Y., Chau, J. L. H., Ariso, C. T. and Yeung, K. L., *Preparation of supported Sil-1, TS-1 and VS-1 membranes: effects of Ti and V metal ions on the membrane synthesis and permeation properties.* Journal of Membrane Science, **183** (2): 269–291 (2001).
35. Zhang, X., Wang, J., Liu, H., Liu, C. and Yeung, K., *Factors affecting the synthesis of hetero-atom zeolite Fe-ZSM-5 membrane.* Separation and Purification Technology, **32** (1–3): 151–158 (2003).

36. Liu, Y., Tan, X. and Li, K., *$SrCe_{0.95}Yb_{0.05}O_{3\text{-alpha}}$ hollow-fiber membrane and its property in proton conduction.* American Institute of Chemical Engineers Journal, **52** (4): 1577–1585 (2006).
37. Liu, Z., Li, L. and Iglesia, E., *Catalytic pyrolysis of methane on Mo/H-ZSM5 with continuous hydrogen removal by permeation through dense oxide films.* Catalysis Letters, **82** (3–4): 175–180 (2002).
38. Hamakawa, S., Hibino, T. and Iwahara, H., *Electrolytic synthesis of ethane using a proton conducting ceramic based on $SrCeO_3$.* Denki Kagaku, **62** (4): 346–347 (1994).
39. Bebelis, S. and Kotsionopoulos, N., *Non-Faradaic electrochemical modification of the catalytic activity for propane combustion of Pt/YSZ and Rh/YSZ catalyst-electrodes.* Solid State Ionics, **177** (26–32): 2205–2209 (2006).
40. Kokkofitis, C., Karagiannakis, G., Zisekas, S. and Stoukides, M., *Catalytic study and electrochemical promotion of propane oxidation on Pt/YSZ.* Journal of Catalysis, **234** (2): 476–487 (2005).
41. Baranova, E. A., Foti, G. and Comninellis, C., *Current-assisted activation of Rh/TiO_2/YSZ catalyst.* Electrochemistry Communications, **6** (4): 389–394 (2004).
42. Poulidi, D., Castillo-del-Rio, M. A., Salar, R. and Metcalfe, I. S., *Electrochemical promotion of catalysis using solid-state proton-conducting membranes.* Solid State Ionics, **162–163**: 305–311 (2003).
43. Asano, K., Hibino, T. and Iwahara, H., *A novel solid oxide fuel-cell system using the partial oxidation of methane.* Journal of the Electrochemical Society, **142** (10): 3241–3245 (1995).
44. Hibino, T., Asano, K. and Iwahara, H., *Solid oxide fuel-cell which can work in uniform-gas phase using partial oxidation of methane.* Nippon Kagaku Kaishi, (7): 600–604 (1994).
45. Ge, S. H., Liu, C. H. and Wang, L. J., *Oxidative dehydrogenation of butane using inert membrane reactor with a non-uniform permeation pattern.* Chemical Engineering Journal, **84** (3): 497–502 (2001).
46. Li, L., Borry, R. W. and Iglesia, E., *Design and optimization of catalysts and membrane reactors for the non-oxidative conversion of methane.* Chemical Engineering Science, **57** (21): 4595–4604 (2002).
47. Terai, T., Li, X. H., Tomishige, K. and Fujimoto, K., *Development of catalyst-membrane system for oxidative coupling of methane by water.* Chemistry Letters, (4): 323–324 (1999).
48. Karagiannakis, G., Zisekas, S., Kokkofitis, C. and Stoukides, M., *Effect of H_2O presence on the propane decomposition reaction over Pd in a proton conducting membrane reactor.* Applied Catalysis A: General, **301** (2): 265–271 (2006).
49. Karagiannakis, G., Kokkofitis, C., Zisekas, S. and Stoukides, M., *Catalytic and electrocatalytic production of H_2 from propane decomposition over Pt and Pd in a proton-conducting membrane-reactor.* Catalysis Today, **104** (2–4): 219–224 (2005).
50. Liu, Y., Tan, X. and Li, K., *Nonoxidative methane coupling in a $SrCe_{0.95}Yb_{0.05}O_{3\text{-alpha}}$ (SCYb) hollow fiber membrane reactor.* Industrial and Engineering Chemistry Research, **45** (11): 3782–3790 (2006).
51. Tan, X., Liu, S., Li, K. and Hughes, R., *Theoretical analysis of ion permeation through mixed conducting membranes and its application to dehydrogenation reactions.* Solid State Ionics, **138** (1–2): 149–159 (2000).
52. Hamakawa, S., Hibino, T. and Iwahara, H., *Electrochemical hydrogen permeation in a proton-hole mixed conductor and its application to a membrane reactor.* Journal of the Electrochemical Society, **141** (7): 1720–1725 (1994).
53. Langguth, J., Dittmeyer, R., Hofmann, H. and Tomandl, G., *Studies on oxidative coupling of methane using high-temperature proton-conducting membranes.* Applied Catalysis A: General, **158**: 287–305 (1997).
54. Jose, G., Marcano, S. and Tsotsis, T. T., *Catalytic membranes and membrane reactors.* (2002) Weinheim: Wiley-VCH Verlag GmbH.

55. Tota, A., Hamel, C., Thomas, S., Joshi, M., Klose, F. and Seidel-Morgenstern, A., *Theoretical and experimental investigation of concentration and contact time effects in membrane reactors.* Chemical Engineering Research and Design, **82** (A2): 236–244 (2004).
56. Assabumrungrat, S., Rienchalanusarn, T., Praserthdam, P. and Goto, S., *Theoretical study of the application of porous membrane reactor to oxidative dehydrogenation of n-butane.* Chemical Engineering Journal, **85** (1): 69–79 (2002).
57. Julbe, A., Farrusseng, D. and Guizard, C., *Porous ceramic membranes for catalytic reactors – overview and new ideas.* Journal of Membrane Science, **181** (1): 3–20 (2001).
58. Lin, Y. S., *Microporous and dense inorganic membranes: current status and prospective.* Separation and Purification Technology, **25** (1–3): 39–55 (2001).
59. Klaus, J. W., Sneh, O. and George, S. M., *Growth of SiO_2 at room temperature with the use of catalyzed sequential half-reactions.* Science, **278** (5345): 1934–1936 (1997).
60. Pan, M., Cooper, C., Lin, Y. S. and Meng, G. Y., *CVD modification and vapor/gas separation properties of nanoporous alumina membranes.* Journal of Membrane Science, **158** (1–2): 235–241 (1999).
61. Pantazidis, A., Dalmon, J. A. and Mirodatos, C., *Oxidative dehydrogenation of propane on catalytic membrane reactors.* Catalysis Today, **25** (3–4): 403–408 (1995).
62. Julbe, A., Guizard, C., Larbot, A., Cot, L. and Giroir-Fendler, A., *The sol-gel approach to prepare candidate microporous inorganic membranes for membrane reactors.* Journal of Membrane Science, **77** (2–3): 137–153 (1993).
63. Yeung, K. L., Sebastian, J. M. and Varma, A., *Mesoporous alumina membranes: synthesis, characterization, thermal stability and nonuniform distribution of catalyst.* Journal of Membrane Science, **131** (1–2): 9–28 (1997).
64. Klose, F., Wolff, T., Thomas, S. and Seidel-Morgenstern, A., *Operation modes of packed-bed membrane reactors in the catalytic oxidation of hydrocarbons.* Applied Catalysis A: General, **257** (2): 193–199 (2004).
65. Klose, F., Wolff, T., Thomas, S. and Seidel-Morgenstern, A., *Concentration and residence time effects in packed bed membrane reactors.* Catalysis Today, **82** (1–4): 25–40 (2003).
66. Ramos, R., Menendez, M. and Santamaria, J., *Oxidative dehydrogenation of propane in an inert membrane reactor.* Catalysis Today, **56** (1–3): 239–245 (2000).
67. Ahchieva, D., Peglow, M., Heinrich, S., Morl, L., Wolff, T. and Klose, F., *Oxidative dehydrogenation of ethane in a fluidized bed membrane reactor.* Applied Catalysis A: General, **296** (2): 176–185 (2005).
68. Ostrowski, T., Giroir-Fendler, A., Mirodatos, C. and Mleczko, L., *Comparative study of the catalytic partial oxidation of methane to synthesis gas in fixed-bed and fluidized-bed membrane reactors: Part I: A modeling approach.* Catalysis Today, **40** (2–3): 181–190 (1998).
69. Bottino, A., Capannelli, G. and Comite, A., *Catalytic membrane reactors for the oxidehydrogenation of propane: experimental and modelling study.* Journal of Membrane Science, **197** (1–2): 75–88 (2002).
70. Tan, X. and Li, K., *Investigation of novel membrane reactors for removal of dissolved oxygen from water.* Chemical Engineering Science, **55** (7): 1213–1224 (2000).
71. Sinha, V. and Li, K., *Alternative methods for dissolved oxygen removal from water: a comparative study.* Desalination, **127** (2): 155–164 (2000).
72. Bayer, A. G., *Catalytic removal of dissolved oxygen from water.* (1990) Leverkusen: Organic Chemical Division, Forschung/Anwendungstechnik Lewatit, Ionenaustauscher/Katalysatoren, Geb. B106, pp.1–6.
73. Suppiah, S., Kutchcoskie, K. J., Balakrishnan, P. V. and Chuang, K. T., *Dissolved-oxygen removal by combination with hydrogen using wetproofed catalysts.* Canadian Journal of Chemical Engineering, **66** (5): 849–857 (1988).
74. Butts, T. J., *Reaction Kinetics and Reactor Design.* (1980) Englewood Cliffs, NJ: Prentice-Hall.

75. Imaoka, T., Yagi, Y., Kasama, Y., Sugiyama, I., Isagawa, T. and Ohmi, T. *Advanced ultrapure water systems for ULSI processing*, in *10th Annual Semiconductor Pure Water Conference*. (1991) Santa Clara, CA.
76. Sato, H., Hashimoto, N., Shinoda, T. and Takino, K. *Advanced ultrapure water systems for ULSI processing*, in *10th Annual Semiconductor Pure Water Conference*. (1991) Santa Clara, CA.
77. Tsai, C. Y., Dixon, A. G., Ma, Y. H., Moser, W. R. and Pascucci, M. R., *Dense perovskite,* $La_{(1-x)}A'_xFe_{1-y}Co_yO_{3-delta}$ *(A′ = Ba, Sr, Ca), membrane synthesis, applications, and characterization.* Journal of the American Ceramic Society, **81** (6): 1437–1444 (1998).
78. Mitchell, B. J., Rogan, R. C., Richardson Jr., J. W., Ma, B. and Balachandran, U., *Stability of the cubic perovskite* $SrFe_{0.8}Co_{0.2}O_{3-\delta}$. Solid State Ionics, **146**: 311–321 (2002).
79. Wachsman, E. D. and Clites, T. L., *Stable mixed-conducting bilayer membranes for direct conversion of methane to syngas.* Journal of the Electrochemical Society, **149**: A242–A246 (2002).
80. Wang, H., Cong, Y. and Yang, W., *Investigation on the partial oxidation of methane to syngas in a tubular* $Ba_{0.5}Sr_{0.5}Co_{0.8}Fe_{0.2}O_{3-[delta]}$ *membrane reactor.* Catalysis Today, **82** (1–4): 157–166 (2003).
81. Tong, J., Yang, W., Cai, R., Zhu, B., Xiong, G. and Lin, L., *Investigation on the structure stability and oxygen permeability of titanium-doped perovskite-type oxides of* $Ba_{0.8}Ti_{0.2}Co_xFe_{1-x}O_{3-\delta}$ *(x = 0.2–0.6)*, Separation and Purification Technology, **32**: 289–299 (2003).
82. Chen, C. S., Kruidhof, H., Bouwmeestera, H. J. M., Verweij, H. and Burggraaf, A. J., *Oxygen permeation through oxygen ion oxide-noble metal dualphase composites.* Solid State Ionics, **86–88**: 569–572 (1996).
83. Chen, C. S., Kruidhof, H., Bouwmeestera, H. J. M., Verweij, H. and Burggraaf, A. J., *Thickness dependence of oxygen permeation through stabilized bismuth oxide-silver composites.* Solid State Ionics, **99**: 215–219 (1997).
84. Chen, C. S. and Burggraaf, A. J., *Stabilized bismuth oxide ± noble metal mixed conducting composites as high temperature oxygen separation membranes.* Journal of Applied Electrochemistry, **29**: 355–360 (1999).
85. Lee, K. S., Lee, S., Kim, J. W. and Woo, S. K., *Enhancement of oxygen permeation by* $La_{0.6}Sr_{0.4}CoO_{3-delta}$ *coating in* $La_{0.7}Sr_{0.3}Ga_{0.6}Fe_{0.4}O_{3-delta}$ *membrane.* Desalination, **147** (1–3): 439–444 (2002).
86. Hendriksen, P. V., Larsen, P. H., Mogensen, M., Poulsen, F. W. and Wiik, K., *Prospects and problems of dense oxygen permeable membranes.* Catalysis Today, **56** (1–3): 283–295 (2000).
87. Tan, X. and Li, K., *Oxidative coupling of methane in a perovskite hollow-fiber membrane reactor.* Industrial and Engineering Chemistry Research, **45** (1): 142–149 (2006).
88. Bouwmeester, H. J. M., *Dense ceramic membranes for methane conversion.* Catalysis Today, **82** (1–4): 141–150 (2003).
89. Dyer, P. N., Richards, R. E., Russek, S. L. and Taylor, D. M., *Ion transport membrane technology for oxygen separation and syngas production.* Solid State Ionics, **134**: 21–33 (2000).
90. Balachandran, U., Dusek, J. T., Maiya, P. S., Ma, B., Mieville, R. L., Kleefisch, M. S. and Udovich, C. A., *Ceramic membrane reactor for converting methane to syngas.* Catalysis Today, **36** (3): 265–272 (1997).
91. Thursfield, A. and Metcalfe, I. S., *The use of dense mixed ionic and electronic conducting membranes for chemical production.* Journal of Material Chemistry, **14**: 2475–2485 (2004).
92. Ikeguchi, M., Mimura, T., Sekine, Y., Kikuchi, E. and Matsukata, M., *Reaction and oxygen permeation studies in* $Sm_{0.4}Ba_{0.6}Fe_{0.8}Co_{0.2}O_{3-[delta]}$ *membrane reactor for partial oxidation of methane to syngas.* Applied Catalysis A: General, **290** (1–2): 212–220 (2005).
93. Martin, M. and Schmackpfeffer, R., *Demixing of doped oxides: influence of defect interactions.* Solid State Ionics, **72** (2): 67–71 (1994).
94. Pei, S., Kleefisch, M. S., Kobylinski, T. P., Faber, J., Udovich, C. A., Zhang-McCoy, V., Dabrowski, B. D., Balachandran, U., Mieville, R. L. and Poeppel, R. B., *Failure mechanisms of ceramic membrane reactors in partial oxidation of methane to synthesis gas.* Catalysis Letters, **30**: 201–212 (1995).

95. Maiya, P. S., Anderson, T. J., Mieville, R. L., Dusek, J. T., Picciolo, J. J. and Balachandran, U., *Maximizing H_2 production by combined partial oxidation of CH_4 and water gas shift reaction.* Applied Catalysis A: General, **196** (1): 65–72 (2000).
96. Jin, W., Gu, X., Li, S., Huang, P., Xu, N. and Shi, J., *Experimental and simulation study on a catalyst packed tubular dense membrane reactor for partial oxidation of methane to syngas.* Chemistry Engineering Science, **55**: 2617–2625 (2000).
97. Jin, W., Li, S., Huang, P., Xu, N., Shi, J. and Lin, Y. S., *Tubular lanthanum cobaltite perovskite-type membrane reactors for partial oxidation of methane to syngas.* Journal of Membrane Science, **166**: 13–22 (2000).
98. Ritchie, J. T., Richardson, J. T. and Luss, D., *Ceramic membrane reactor for synthesis gas production.* American Institute of Chemical Engineers Journal, **47** (9): 2092–2101 (2001).
99. Shao, Z., Dong, H., Xiong, G., Cong, Y. and Yang, W., *Performance of a mixed-conducting ceramic membrane reactor with high hydrogen permeability for methane conversion.* Journal of Membrane Science, **183**: 181–192 (2001).
100. Shao, Z., Xiong, G., Tong, J., Dong, H. and Yang, W., *Ba effect in doping $Sr(Co_{0.8}Fe_{0.2})O_{3-\delta}$ the phase structure and oxygen permeation properties of the dense ceramic membranes.* Separation and Purification Technology, **25**: 419–429 (2001).
101. Tong, J. H., Yang, W. S., Cai, R., Zhu, B. C. and Lin, L. W., *Novel and ideal zirconium-based dense membrane reactors for partial oxidation of methane to syngas.* Catalysis Letters, **78** (1–4): 129–137 (2002).
102. Sammells, A. F., *Catalytic Membrane Reactors for Spontaneous Synthesis Gas Production*, in *2nd Annual UNESCO Training School and Poster Symposium.* (1999) Rosebank, Johannesburg, South Africa.
103. Lin, Y. S. and Zeng, Y., *Catalytic Properties of Oxygen Semipermeable Perovskite-Type Ceramic Membrane Materials for Oxidative Coupling of Methane.* Journal of Catalysis, **164** (1): 220–231 (1996).
104. Zeng, Y. and Lin, Y. S., *Catalytic properties of yttria doped bismuth oxide ceramics for oxidative coupling of methane.* Applied Catalysis A: General, **159** (1–2): 101–117 (1997).
105. Wang, W. and Lin, Y. S., *Analysis of oxidative coupling of methane in dense oxide membrane reactors.* Journal of Membrane Science, **103** (3): 219–233 (1995).
106. Hazbun, E. A., *Ceramic membrane for hydrocarbon conversion, US4791079.* (1988).
107. Hazbun, E. A., *Ceramic membrane and use thereof for hydrocarbon conversion, US4827071.* (1989).
108. Omata, K., Yamazaki, O., Tomita, K. and Fujimoto, K., *Oxidative coupling of methane on an ABO(3) type oxide with mixed conductivity.* Journal of the Chemical Society – Chemical Communications, (14): 1647–1648 (1994).
109. Nomura, K., Hayakawa, T., Takehira, K. and Ujihira, Y., *Oxidative coupling of methane on perovskite oxides, $(Ba,Ca)(Co,Fe)O_{3-\delta}$.* Applied Catalysis A: General, **101**: 63–72 (1993).
110. ten Elshof, J. E., Bouwmeester, H. J. M. and Verweij, H., *Oxidative coupling of methane in a mixed-conducting perovskite membrane reactor.* Applied Catalysis A: General, **130**: 195–212 (1995).
111. Zeng, Y., Lin, Y. S. and Swartz, S. L., *Perovskite-type ceramic membrane: synthesis, oxygen permeation and membrane reactor performance for oxidative coupling of methane.* Journal of Membrane Science, **150** (1): 87–98 (1998).
112. Xu, S. J. and Thomson, W. J., *Perovskite-type oxide membranes for the oxidative coupling of methane.* American Institute of Chemical Engineers Journal, **43** (S11): 2731–2740 (1997).
113. Lu, Y., Dixon, A. G., Moser, W. R., Ma, Y. and Balachandran, U., *Oxygen-permeable dense membrane reactor for the oxidative coupling of methane.* Journal of Membrane Science, **170**: 27–34 (2000).
114. Akin, F. T. and Lin, Y. S., *Oxidative coupling of methane in dense ceramic membrane reactor with high yields.* American Institute of Chemical Engineers Journal, **48** (10): 2298–2306 (2002).

115. Akin, F. T. and Lin, Y. S., *Selective oxidation of ethane to ethylene in a dense tubular membrane reactor.* Journal of Membrane Science, **209** (2): 457–467 (2002).
116. Tan, X., Liu, Y. and Li, K., *Preparation of LSCF ceramic hollow fibre membrane for oxygen production by a phase-inversion/sintering technique.* Industrial and Engineering Chemistry Research, **44**: 61–66 (2005).
117. Tellez, C., Menendez, M. and Santamaria, J., *Simulation of an inert membrane reactor for the oxidative dehydrogenation of butane.* Chemical Engineering Science, **54** (13–14): 2917–2925 (1999).
118. Ioannides, T. and Gavalas, G. R., *Catalytic isobutane dehydrogenation in a dense silica membrane reactor.* Journal of Membrane Science, **77** (2–3): 207–220 (1993).
119. Itoh, N. and Haraya, K., *A carbon membrane reactor.* Catalysis Today, **56** (1–3): 103–111 (2000).
120. Terry, P. A. anderson, M. and Tejedor, I., *Catalytic dehydrogenation of cyclohexane using coated silica oxide ceramic membranes.* Journal of Porous Materials, **6** (4): 267–274 (1999).
121. Chiang, P. H., Eng, D. and Stoukides, M., *Electrocatalytic methane dimerization with a Yb-doped* $SrCeO_3$ *electrolytes.* Journal Electrochemical Society, **138**: L11 (1991).
122. Chiang, P. H., Eng, D. and Stoukides, M., *Electrocatalytic nonoxidative dimerization of methane over Ag electrodes.* Solid State Ionics, **61**: 99–103 (1993).
123. Hamakawa, S., Hibino, T. and Iwahara, H., *Electrochemical methane coupling using protonic conductors.* Journal of the Electrochemical Society, **140** (2): 459–462 (1993).
124. Woldmann, L. S. and Sokolovskii, V. D., *Electrocatalytic methane coupling in the absence of oxygen on a high-temperature proton-conducting electrolyte.* Catalysis Letters, **8**: 61 (1991).
125. White, J. H., Schwartz, M. and Sammells, A. F., *Solid state proton and electron mediating membrane and use in catalytic membrane reactors, US5821185.* 1998.
126. Diakov, V. and Varma, A., *Reactant distribution by inert membrane enhances packed-bed reactor stability.* Chemical Engineering Science, **57** (7): 1099–1105 (2002).
127. Diakov, V., Lafarga, D. and Varma, A., *Methanol oxidative dehydrogenation in a catalytic packed-bed membrane reactor.* Catalysis Today, **67** (1–3): 159–167 (2001).
128. Brinkmann, T., Perera, S. P. and Thomas, W. J., *An experimental and theoretical investigation of a catalytic membrane reactor for the oxidative dehydrogenation of methanol.* Chemical Engineering Science, **56** (6): 2047–2061 (2001).
129. Alfonso, M. J., Menendez, M. and Santamaria, J., *Vanadium-based catalytic membrane reactors for the oxidative dehydrogenation of propane.* Catalysis Today, **56** (1–3): 247–252 (2000).
130. Julbe, A., Farrusseng, D., Jalibert, J. C., Mirodatos, C. and Guizard, C., *Characteristics and performance in the oxidative dehydrogenation of propane of MFI and V-MFI zeolite membranes.* Catalysis Today, **56** (1–3): 199–209 (2000).
131. Kolsch, P., Smejkal, Q., Noack, M., Schafer, R. and Caro, J., *Partial oxidation of propane to acrolein in a membrane reactor – experimental data and computer simulation.* Catalysis Communications, **3** (10): 465–470 (2002).
132. Lu, Y., Dixon, A. G., Moser, W. R. and Hua M. Y., *Oxidative coupling of methane in a modified [gamma]-alumina membrane reactor.* Chemical Engineering Science, **55** (21): 4901–4912 (2000).
133. Ramachandra, A. M., Lu, Y., Ma, Y. H., Moser, W. R. and Dixon, A. G., *Oxidative coupling of methane in porous Vycor membrane reactors.* Journal of Membrane Science, **116** (2): 253–264 (1996).
134. Lafarga, D. and Varma, A., *Ethylene epoxidation in a catalytic packed-bed membrane reactor: effects of reactor configuration and 1,2-dichloroethane addition.* Chemical Engineering Science, **55** (4): 749–758 (2000).
135. Jin, W., Li, S., Huang, P., Xu, N., Shi, J. and Lin, Y. S., *Tubular lanthanum cobaltite perovskite-type membrane reactors for partial oxidation of methane to syngas.* Journal of Membrane Science, **166** (1): 13–22 (2000).

136. Ishihara, T. and Takita, Y., *Partial oxidation of methane into syngas with oxygen permeating ceramic membrane reactors.* Catalysis Surveys from Japan, **4** (2): 125–133 (2000).
137. Shao, Z., Xiong, G., Dong, H., Yang, W. and Lin, Y. S., *Synthesis, oxygen permeation study and membrane performance of a* $Ba_{0.5}Sr_{0.5}Co_{0.8}Fe_{0.2}O_{3-\delta}$ *oxygen-permeable dense ceramic reactor for partial oxidation of methane to syngas.* Separation and Purification Technology, **25**: 97–116 (2001).
138. Dong, H., Shao, Z. P., Xiong, G. X., Tong, J. H., Sheng, S. S. and Yang, W. S., *Investigation on POM reaction in a new perovskite membrane reactor.* Catalysis Today, **67** (1–3): 3–13 (2001).
139. Shao, Z., Xiong, G., Dong, H., Yang, W. and Lin, L., *Synthesis, oxygen permeation study and membrane performance of a* $Ba_{0.5}Sr_{0.5}Co_{0.8}Fe_{0.2}O_{3-[delta]}$ *oxygen-permeable dense ceramic reactor for partial oxidation of methane to syngas.* Separation and Purification Technology, **25** (1–3): 97–116 (2001).
140. Feng, S. J., Ran, S., Zhu, D. C., Liu, W. and Chen, C. S., *Synthesis gas production from methane with* $SrFeCo_{0.5}O_y$ *membrane reactor.* Energy and Fuels, **18** (2): 385–389 (2004).
141. Gu, X. H., Jin, W. Q., Chen, C. L., Xu, N. P., Shi, J. and Ma, Y. H., *YSZ-*$SrCo_{0.4}Fe_{0.6}O_{3-delta}$ *membranes for the partial oxidation of methane to syngas.* American Institute of Chemical Engineers Journal, **48** (9): 2051–2060 (2002).
142. Hamakawa, S., Hayakawa, T. and Mizukami, F., *Research on a ceramics membrane reactor for natural gas conversion.* Catalysis Surveys from Asia, **9** (2): 95–101 (2005).
143. Hamakawa, S., Hayakawa, T., Suzuki, K., Murata, K., Takehira, K., Yoshino, S., Nakamura, J. and Uchijima, T., *Methane conversion into synthesis gas using an electrochemical membrane reactor.* Solid State Ionics, **136–137**: 761–766 (2000).
144. Hu, J., Xing, T., Jia, Q., Hao, H., Yang, D., Guo, Y. and Hu, X., *Methane partial oxidation to syngas in* $YBa_2Cu_3O_{7-x}$ *membrane reactor.* Applied Catalysis A: General, **306**: 29–33 (2006).
145. Kharton, V. V., Yaremchenko, A. A., Valente, A. A., Sobyanin, V. A., Belyaev, V. D., Semin, G. L., Veniaminov, S. A., Tsipis, E. V., Shaula, A. L., Frade, J. R. and Rocha, J., *Methane oxidation over Fe-, Co-, Ni- and V-containing mixed conductors.* Solid State Ionics, **176** (7–8): 781–791 (2005).
146. Lu, H., Tong, J. H., Cong, Y. and Yang, W. S., *Partial oxidation of methane in* $Ba_{0.5}Sr_{0.5}Co_{0.8}Fe_{0.2}$ $O_{3-delta}$ *membrane reactor at high pressures.* Catalysis Today, **104** (2–4): 154–159 (2005).
147. Tsai, C. Y., Dixon, A. G., Moser, W. R. and Ma, Y. H., *Dense perovskite membrane reactors for partial oxidation of methane to syngas.* American Institute of Chemical Engineers Journal, **43** (11): 2741–2750 (1997).
148. Zhang, P., Chang, X. F., Wu, Z. T., Jin, W. Q. and Xu, N. P., *Effect of the packing amount of catalysts on the partial oxidation of methane reaction in a dense oxygen-permeable membrane reactor.* Industrial and Engineering Chemistry Research, **44** (6): 1954–1959 (2005).
149. Wang, H., Cong, Y. and Yang, W., *Oxidative coupling of methane in* $Ba_{0.5}Sr_{0.5}Co_{0.8}Fe_{0.2}O_{3-[delta]}$ *tubular membrane reactors.* Catalysis Today, **104** (2–4): 160–167 (2005).
150. Akin, F. T. and Lin, Y. S., *Oxidative coupling of methane in dense ceramic membrane reactor with high yields.* American Institute of Chemical Engineers Journal, **48** (10): 2298–2306 (2002).
151. Akin, F. T. and Lin, Y. S., *Controlled oxidative coupling of methane by ionic conducting ceramic membrane.* Catalysis Letters, **78** (1–4): 239–242 (2002).
152. Zeng, Y., Akin, F. T. and Lin, Y. S., *Oxidative coupling of methane on fluorite-structured samarium-yttrium-bismuth oxide.* Applied Catalysis A: General, **213** (1): 33–45 (2001).
153. Andersen, A. G., Hayakawa, T., Suzuki, K., Shimizu, M. and Takehira, K., *Electrochemical methane conversion over* $SrFeO_{3-delta}$ *perovskite on an yttrium stabilized zirconia membrane.* Catalysis Letters, **27** (1–2): 221–233 (1994).
154. Hibino, T., Sato, T., Ushiki, K. and Kuwahara, Y., *Membrane reactor for oxidative coupling of* CH_4 *with an oxide ion-electron hole mixed conductor.* Journal of the Chemical Society – Faraday Transactions, **91** (24): 4419–4422 (1995).

155. Wang, H. H., Cong, Y., Zhu, X. F. and Yang, W. S., *Oxidative dehydrogenation of propane in a dense tubular membrane reactor*. Reaction Kinetics and Catalysis Letters, **79** (2): 351–356 (2003).
156. Rebeilleau-Dassonneville, M., Rosini, S., Veen, A. C. van, Farrusseng, D. and Mirodatos, C., *Oxidative activation of ethane on catalytic modified dense ionic oxygen conducting membranes*. Catalysis Today, **104** (2–4): 131–137 (2005).
157. Takehira, K., Sakai, N., Shimomura, J., Kajioka, H., Hamakawa, S., Shishido, T., Kawabata, T. and Takaki, K., *Oxidation of C2-C4 alkanes over MoO3-V2O5 supported on a YSZ-aided membrane reactor*. Applied Catalysis A: General, **277** (1–2): 209–217 (2004).
158. Lofberg, A., Boujmiai, S., Capoen, E., Steil, M. C., Pirovano, C., Vannier, R. N., Mairesse, G. and Bordes-Richard, E., *Oxygen permeation versus catalytic properties of bismuth-based oxide ion conductors used for propene oxidation in a catalytic dense membrane reactor*. Catalysis Today, **91–92**: 79–83 (2004).
159. Wang, H., Cong, Y. and Yang, W., *Partial oxidation of ethane to syngas in an oxygen-permeable membrane reactor*. Journal of Membrane Science, **209** (1): 143–152 (2002).
160. Wang, H. H., Cong, Y. and Yang, W. S., *High selectivity of oxidative dehydrogenation of ethane to ethylene in an oxygen permeable membrane reactor*. Chemical Communications, (14): 1468–1469 (2002).
161. Zhu, W., Han, W., Xiong, G. and Yang, W., *Mixed reforming of heptane to syngas in the* $Ba_{0.5}Sr_{0.5}Co_{0.8}Fe_{0.2}O_3$ *membrane reactor*. Catalysis Today, **104** (2–4): 149–153 (2005).

Index

Note: page numbers in **bold** refer to tables and those in *italic* to figures.